Interdisciplinary Collaboration for Water Diplomacy

This book introduces the concept of Water Diplomacy as a principled and pragmatic approach to problem-driven interdisciplinary collaboration, which has been developed as a response to pressing contemporary water challenges arising from the coupling of natural and human systems.

The findings of the book are the result of a decade-long interdisciplinary experiment in conceiving, developing, and implementing an interdisciplinary graduate program on Water Diplomacy at Tufts University, USA. This has led to the development of the Water Diplomacy Framework, a shared framework for understanding, diagnosing, and communicating about complex water issues across disciplinary boundaries. This framework clarifies important distinctions between water systems – simple, complicated, or complex – and the attributes that these distinctions imply for how these problems can be addressed. In this book, the focus is on complex water issues and how they require a problem-driven rather than a theory-driven approach to interdisciplinary collaboration. Moreover, it is argued that conception of interdisciplinarity needs to go beyond collaboration among experts, because complex water problems demand inclusive stakeholder engagement, such as in fact–value deliberation, joint fact-finding, collective decision making, and adaptive management. Water professionals working in such environments need to operate with both principles and pragmatism in order to achieve actionable, sustainable, and equitable outcomes. This book explores these ideas in more detail and demonstrates their efficacy through a diverse range of case studies. Reflections on the program are also included, from conceptualization through implementation and evaluation.

This book offers critical lessons and case studies for researchers and practitioners working on complex water issues as well as important lessons for those looking to initiate, implement, or evaluate interdisciplinary programs to address other complex problems in any setting.

Shafiqul Islam is Professor of Civil and Environmental Engineering at the Fletcher School of Law and Diplomacy, and the Director of the Water Diplomacy Program at Tufts University, USA.

Kevin M. Smith is a PhD candidate in environmental and water resources engineering and a member of the third cohort of Water Diplomacy graduate students at Tufts University, USA.

Earthscan Studies in Water Resource Management

For more information and to view forthcoming titles in this series, please visit the Routledge website: www.routledge.com/books/series/ECWRM/

Interdisciplinary Collaboration for Water Diplomacy

A Principled and Pragmatic Approach

Edited by Shafiqul Islam and Kevin M. Smith

Routledge
Taylor & Francis Group

LONDON AND NEW YORK

from Routledge

First published 2020
by Routledge
2 Park Square, Milton Park, Abingdon, Oxon OX14 4RN

and by Routledge
52 Vanderbilt Avenue, New York, NY 10017

Routledge is an imprint of the Taylor & Francis Group, an informa business

First issued in paperback 2021

British Library Cataloguing-in-Publication Data
A catalogue record for this book is available from the British Library

Library of Congress Cataloging-in-Publication Data
Names: Islam, Shafiqul, 1960- editor. | Smith, Kevin M., 1990- editor.
Title: Interdisciplinary collaboration for water diplomacy : a principled and
    pragmatic approach / edited by Shafiqul Islam and Kevin M Smith.
Description: Abingdon, Oxon ; New York, NY : Routledge, 2020. |
    Includes bibliographical references and index.
Identifiers: LCCN 2019034206 (print) | LCCN 2019034207 (ebook) |
    ISBN 9781138369283 (hardback) | ISBN 9780429428760 (ebook)
Subjects: LCSH: Water-supply—Management. | Diplomacy. |
    Water-supply—Political aspects. | Water-supply—Government policy. |
    Water-supply—Environmental aspects. | Water resources development.
Classification: LCC TD345 .I539 2020 (print) | LCC TD345
    (ebook) | DDC 333.91—dc23
LC record available at https://lccn.loc.gov/2019034206
LC ebook record available at https://lccn.loc.gov/2019034207

ISBN: 978-1-138-36928-3 (hbk)
ISBN: 978-1-03-208489-3 (pbk)
ISBN: 978-0-429-42876-0 (ebk)

Typeset in Bembo
by Apex CoVantage, LLC

Contents

PART II
Problem-driven interdisciplinary collaboration in action: case studies from the Tufts Water Diplomacy program

Foreword

Anthony P. Monaco

In our interconnected world, water is an integral part of any discussion on food, energy, public health, environment and the future. While innovations in science and technology can expand the available quantity and quality of water, they cannot solve water problems that intersect with the value-laden needs of individuals and institutions. To address these complex water problems, what is needed is an approach that goes beyond the application of science and technology or the implementation of management policies.

Water Diplomacy – rooted in ideas of complexity science and negotiation theory – represents a significant step forward in that direction. The Water Diplomacy Framework proposes a principled, pragmatic approach to negotiation – one that is based on credible scientific knowledge, mediated through equity and sustainability as guiding principles for policy action.

The program in Water Diplomacy at Tufts started more than a decade ago as a University Seminar, bringing together four schools across the university. And in 2009, the importance of this interdisciplinary field and the excellence of Tufts' work in it were recognized by the National Science Foundation with a multimillion-dollar Integrative Graduate Education and Research Traineeship (IGERT) grant. Since its establishment, Tufts' Water Diplomacy program has educated 27 doctoral students representing four of the university's schools. Students, working alongside faculty in the program, so far have published four books (including this volume), 46 refereed publications, made more than 100 national and international conference presentations, and travelled to 24 countries for their research, presentations, and internships.

The success of the Water Diplomacy program is an interdisciplinary accomplishment. It demonstrates what is possible when experts from across disciplines collaborate to find solutions to the complex problems. Thanks to our outstanding Schools of Engineering and Arts and Sciences, as well as the Fletcher School of Law and Diplomacy, Tufts is uniquely positioned to synthesize the science, engineering, policy, and politics of water through Water Diplomacy. To my knowledge, no other graduate program fills this particular interdisciplinary need and niche, allowing students to synthesize knowledge of science, policy, and politics to become Water Diplomats.

The contributions of the Water Diplomacy program at Tufts reflect the fundamental importance of interdisciplinary approaches to the most complex and challenging problems facing our increasingly globalized world. They also reflect the unique value that rigorous academic research and insightful graduate training can make to educating the next generation of leaders we need.

This program reflects Tufts' values and spirit as a university: Our interdisciplinary culture encourages such active involvement of students and faculty from multiple schools. Our focus on active citizenship catalyzes projects with real-world impact. And the global experiences and knowledge of our students and faculty facilitate valuable comparative insights.

I am grateful to Dr. Islam and all of the program's faculty, staff, students, and advisory board members for their dedication and passion for making this program a success. At the same time, as they would be the first to acknowledge, the program's global collaborators have played indispensable parts in its agenda of research and teaching. No single institution could possibly provide the breadth of expertise needed to address these issues. And lasting, truly interactive partnerships with policy makers and citizens alike, with both public authorities and NGOs, are critical to ensuring the real-world grounding of scholarship and the effective dissemination of its outcomes.

Looking ahead, the issues addressed by Water Diplomacy will only grow in importance. Not just as president of Tufts University but as a citizen concerned with sustainability and the equitable and peaceful distribution of resources, I am grateful for the continuing efforts of those associated with our Water Diplomacy program.

Anthony P. Monaco
President, Tufts University, USA
Professor of Biology
Professor of Neuroscience

Preface

Kevin M. Smith and Shafiqul Islam

There is an alluring simplicity to the water cycle as it is shown in textbook figures. A cloud raining on pristine highlands grows into a river meandering to the sea. Over this landscape, we draw arrows depicting water's movement and transformation. A version taught to kindergarteners might go like this: clouds produce rain, which runs over the landscape, eventually evaporating from the land surface or the ocean, becoming a cloud again.

With age, we add a few more processes to our description: evapotranspiration from plants, sublimation from ice, infiltration, and groundwater flow. By the time we've reached a first-year college course in hydrology, there are so many arrows we can hardly see the landscape. Even with all this detail, however, there is an elegance to the figure that makes this description seem definitive and comprehensive. Put a water molecule down anywhere in this landscape, and you can easily trace its fate.

Yet this is a vision of water as a purely physical object, governed only by natural forces. The true journey of a water molecule often includes as many societal transformations as it does natural. Where in the above depiction can we see vapor become a sacred river? When do rising floodwaters become an agricultural boon or a residential hazard? Is the groundwater in the picture on its way to being a human right or a tradable commodity?

These are not just sentimental considerations. Whether water is a human right or a private property has a bearing on who decides, where, how much, and with what quality it will flow. There is, for instance, nothing in the water cycle that shows us that water can move against gravity when pulled by money; no depiction of water pumped miles below the earth's surface to flush out hydrocarbons from fractured shale. We cannot use the water cycle to tell the full story of why the Aral Sea is fading into the desert or why so much water sits sequestered behind concrete dams.

To tell these stories well, we must acknowledge that the forces that govern water are often as much societal as they are natural. We need to recognize the difference between water as a physical object (hydrogen and oxygen make water) and a resource (when limited, who gets more water: fish, farmers, or urban development?). What's more, we need to abandon the idea that it is sensible to talk about the natural and societal aspects of these forces as if they are separable. On the contrary, there is a growing body of literature to suggest that the natural and societal dynamics of these phenomena are intrinsically coupled in complex ways.

With the growing awareness of coupling between many natural and human systems, researchers and practitioners working on environmental and social issues find themselves facing previously unrecognized challenges. For many, these challenges quickly exceed the boundaries of their knowledge or practice. Unsurprisingly, calls for broad interdisciplinary collaboration have become the standard response to the boundary-crossing problematics of coupled natural and human systems.

However, the widespread promotion of interdisciplinary collaboration has not come with clarity of intention or with an instruction manual. Are collaborators expected to synthesize their disciplinary knowledge into a general theory that spans their disciplines? Or are they simply coming together to solve a problem that they couldn't otherwise address in isolation? This ambiguity creates epistemological and operational pitfalls for collaborators engaged in interdisciplinary work.

We certainly did not escape these challenges and ambiguities in our own interdisciplinary efforts. Indeed, they appeared regularly over the course of the decade-long evolution of the Water Diplomacy program at Tufts University. This program brought together faculty and students from across the university to develop and apply new ways of thinking about complex water challenges in an interdisciplinary setting. At the heart of the Water Diplomacy program was an interdisciplinary doctoral pathway that brought in new cohorts of students each year. Unsurprisingly, with each new cohort, we found ourselves with new questions regarding our motivations and intentions for promoting interdisciplinary work.

We arrived at our current thinking on interdisciplinarity and Water Diplomacy through this evolutionary process of adaptive learning by doing. We fully expect (indeed, we hope) that our understanding of these issues will continue to evolve and transform our learning by doing for years to come. Nevertheless, we believe we have come to a point at which it may be useful to share what we have learned so far with the greater community of interdisciplinary explorers, both to help them in their pursuits and to facilitate our own learning through feedback from interdisciplinary thinkers and practitioners.

To understand what we hope to achieve with this book, it is useful to contrast it with the findings of a much earlier interdisciplinary research program on water. In the early 1950s, several professors at Harvard University began to consider how they could create a research and training program that would empower a generation of water management professionals. Over the next decade, these early conversations grew into an influential and conceptually transformative initiative known as the Harvard Water Program.

This enterprise brought together professionals from several federal agencies and hosted them at the Harvard Graduate School of Public administration alongside academics from engineering, economics, government, and public administration. Their goal was to create planning and design methodologies for maximizing the extractable benefits from "multiunit, multipurpose water resource systems." In 1962, members of the Harvard Water Program shared their findings in *Design of Water-Resource Systems: New Techniques for Relating Economic Objectives, Engineering Analysis, and Government Planning.*

Design of Water-Resource Systems changed the way water management was taught and practiced in the United States. Over the course of more than 600 pages, it offered a series of operational recipes for the design and management of large-scale hydraulic infrastructure projects, many of which evolved into standard "best practices" within the engineering community. While there is no doubt that the book was an ambitious, important, and well-executed project, it is emblematic of a bygone era.

In the 1960s, large infrastructure projects were the answer to almost every water-related challenge. The question was how to execute these projects economically and efficiently. Planners and managers did not concern themselves much with the problematic nature of pluralistic societal values, the competing demands of industry and ecosystems, or political jockeying over transboundary water. Tellingly, although the Harvard Water Program was directed by a professor of government and was designed for economists and administrators in addition to engineers, the non-engineering aspects of water resources management were only briefly discussed in the closing chapter "System Design and the Political Process: A General Statement."

Our book, on the other hand, starts from a different premise. We take the societal and political dynamics to be intrinsic and omnipresent factors in the planning and management of water as an object as well as a natural resource. In the classical wave-particle duality in physics, light may appear to be a particle if we ask a particle-like question and a wave if we ask a wave-like question. A similar duality exists in our either–or framing of water as an object or a resource. Many water problems are rooted in both numbers and narratives. Yet we tend to obscure one to highlight the other.

This narrowness of thinking diminishes the effectiveness of either approach. We need a representation (vocabulary and framework) in which these are not either–or options but compatible aspects of a problem. Numbers may show trends and patterns. Narratives capture local culture and values as well as highlight ambiguity and confusion. With our emerging understanding of the coupled natural and human aspects of water challenges, there is an increased recognition that the objective and subjective facets of the problem can't be easily decoupled. That is to say, the objectivity of the scientific method can't be separated from subjectivity in the choice of data, models, and metrics for assessment. In such situations, the scientific method loses its primacy of objectivity and autonomy and becomes interpretive and pluralistic.

When scientific methods fail to provide reproducible findings and the consensus of how to act is absent, a negotiated interpretation of the diagnosis and prescription – a convergence of numbers and narratives – is required to address the problem. Such a convergence will require attention to both principles and pragmatism. Science alone is not sufficient. Nor is policy making that doesn't take science into account. Candidate solutions evolve from rigorous scientific study and a pragmatic awareness of the context, capacity, and constraints of the chosen problem context. In this process, stakeholders are engaged to bring in a plurality of values to bear on the proposed solutions with the hopeful outcome that an effective and socially acceptable and politically feasible intervention can be negotiated.

As such, in this book you will find no ready-made and cookbook-type "best practices" that can be deployed for repeatable success. Indeed, the last five decades have shown that such recipes seldom achieve their intended consequences in the messy context of the real world. Instead, this book offers a different perspective of what it means to be an interdisciplinary water professional in a world in which many natural and human systems are understood to be coupled in complex ways. Rather than providing definitive answers, we hope to provide insight about the factors that we've found to be critical to the success and failure of interdisciplinary collaboration to address socioenvironmental issues. In making this case, we draw upon existing literature, our take on complexity science and principled pragmatism, examples of student projects, and the reflections and evaluations of faculty engaged in the process.

We argue that the dominant implicit mode of interdisciplinary collaboration is theory driven: it seeks to create some form of generalizable theory that will bridge disciplines. While this is a powerful mode of interdisciplinary practice effective for problems arising in simple and complicated systems, we argue that there is little hope for finding generalizable theory when it comes to problems arising from complexity. Instead, we argue that complexity demands a problem-driven interdisciplinary approach that integrates knowledge – with an explicit recognition of the nature, capacity, and constraints inherent to the chosen problem – for actionable, measurable, and trackable outcomes. In lieu of seeking a *theory* to bridge disparate disciplines, we argue that teams should seek shared *frameworks* that facilitate agreement on the nature of the problem context (e.g., simple, complicated, or complex) as well as provide a common language for describing and diagnosing problems (e.g., the Water Diplomacy Framework). We see such frameworks as an effective means of catalyzing interdisciplinary research and practice on complex water issues.

We hope you'll find this book to be a valuable resource, whether you are reading it as a new way of thinking about problem-driven interdisciplinary collaboration to resolve complex problems; as a brief introduction to coupled natural and human systems; as a survey of interdisciplinary responses to address complex water challenges; or as a companion guide to planning, evaluating, or enhancing an interdisciplinary research program of your own. While the chapters of the book are arranged to build on each other thematically, we have structured the book in a way that each chapter can be approached independently to suit our readers' particular interests.

Part I of the book provides both a contextual backstory to the Tufts interdisciplinary Water Diplomacy graduate program and summarizes our latest thinking on how to respond to the complexity of coupled natural and human systems (CNH). Our current understandings are a product of nearly a decade of organizational learning and the input of both students and faculty engaged in the program. In Chapter 1, we tell the story of our effort to educate a new generation of interdisciplinary water professionals, who we call Water Diplomats. Given the inherent complexity of CNH systems, our Water Diplomats would need to be able to work effectively in interdisciplinary teams as well as with a broad range of stakeholders in the process of inclusive fact–value deliberation, joint fact-finding, and consensus-oriented decision making. Our approach, therefore, combined a depth of disciplinary expertise with a breadth of common understanding with a principled and pragmatic approach to pursue problem-driven research and practice. In the subsequent chapters in Part I, we explore the motivation and rationale to conceptualize and operationalize this approach for a general class of complex problems using Water Diplomacy as an example.

In Chapter 2, we argue that many of our most pressing socioenvironmental issues arise from the intrinsic complexity of CNH systems. To highlight the importance of recognizing this complexity, we explore how the identification of a system as simple, complicated, or complex puts important constraints and considerations on the way these problems need to be addressed. We use the example of an ongoing sanitation crisis in Lowndes County, Alabama, to help make these distinctions clear.

In Chapter 3, we outline why interdisciplinarity is a reasonable response to address the complexity of CNH systems. More importantly, we argue that certain forms of interdisciplinary collaboration are more likely to have success when dealing with CNH systems than others. In particular, we make the case for what we call a problem-driven approach, in which the collaboration is motivated by a particular problem rather than the pursuit of bridging theories or finding generalizable solutions. We further argue that developing shared frameworks instead of general theories can be an effective means of catalyzing interdisciplinary problem-driven research for timely and actionable outcomes. We find this is a struggle many interdisciplinary teams will continue to face. Our own interdisciplinary work and experience have been greatly facilitated and enriched by the co-development of the Water Diplomacy Framework as a shared language for describing and understanding complex water problems rather than a unifying theory for solving them.

While interdisciplinary problem-driven collaboration among experts is critical, actionable outcomes on complex water issues also require inclusive fact–value deliberation, joint fact-finding, and collective decision-making processes that include broad ranges of stakeholders. In Chapter 4, we explore how Water Diplomats need to go beyond problem-driven collaboration among experts into the messy world of finding negotiated resolutions to complex water issues. To effectively engage in such situations, we argue that Water Diplomats will need to be principled and pragmatic in the synthesis and application of four domains of knowledge: the formal knowledge of *episteme*; the practical wisdom of *phronesis*; the practiced technique of *techne*; and the thoughtful practice of *praxis*. We provide three archetypes from recent literature on professional practice – the Honest Broker, the Humble Analyst, and the Democratic Professional – as motivating examples of the concrete ways in which Water Diplomats can effectively engage in stakeholder-centered deliberative processes.

In Part II of the book, we turn to concrete examples of research conducted as part of the Tufts Water Diplomacy Program. We think these examples exemplify complex challenges faced by water professionals working on a range of socioenvironmental issues. Moreover, these chapters demonstrate that it is possible to engage effectively in problem-driven interdisciplinary work without establishing or finding general theories. Instead, students were able to harness the Water Diplomacy Framework as a shared language of description that allowed them to converse fluently with their peers across disciplines. Finally, we see these chapters as evidence that problem-driven collaboration can be approached with humility and attention to both principles and pragmatism. A more comprehensive introduction to these case studies is provided in Chapter 5.

In Part III of the book, we turn to evaluations and reflections of our interdisciplinary graduate program on Water Diplomacy, as well as invite perspectives on the practice of Water Diplomacy in the future. In these chapters, we share lessons learned throughout the conceptualization, implementation, and evaluation of our interdisciplinary graduate research program on Water Diplomacy. While there is no way to guarantee specific outcomes – irrespective of contexts – from an interdisciplinary collaboration, there are critical factors that we have identified that we believe impact the likelihood of success across contexts. We hope that these reflections will offer practical insights for readers looking to establish or improve existing interdisciplinary research programs.

We didn't write this book because we have all the answers. Even after a decade of learning by doing, we are not free of every doubt about the why, how, or what of interdisciplinary research and practice, nor do we feel we have arrived at a definitive conceptualization of complex water resource challenges. What we are sure of is this: many of our most pressing socioenvironmental challenges emerge from an underlying complexity that cannot be addressed through science alone. We are also convinced that interdisciplinary collaboration will be required to meet these challenges, even as we continue to be open to debate about what forms of interdisciplinary collaboration are most appropriate. What we promote is a problem-driven mode of interdisciplinary collaboration with a goal to make our ideas clearer and provide actionable options with trackable outcomes in a timely fashion.

Finally, we want to be up front about the fact that there is nothing easy about interdisciplinary work. Nevertheless, that is not to say it is not personally rewarding to those who practice it. In our own experience, we have seen interdisciplinary collaboration offer its practitioners unique opportunities for personal and professional development, as well as lasting friendships. We are grateful for the enduring bonds we have forged with our Water Diplomacy family at Tufts and beyond, and we look forward to working together for years to come.

Contributors

Agustín Botteron is the Deputy Chief Resilience Officer in the government of the City of Santa Fe, Argentina.

Jessica Rozek Cañizares is a PhD candidate in biology at Tufts University, USA, researching wetland loss and conservation of migratory birds.

Enamul Choudhury, PhD, is Professor of Urban Affairs and Geography at Wright State University, USA.

Laura Corlin, PhD, is Assistant Professor of Epidemiology and Biostatistics at Tufts University, USA.

Gabriela Marie Garcia is a PhD candidate in biology at Tufts University, USA, researching the socioecological resilience of coffee agroecosystems.

Timothy S. Griffin, PhD, is Associate Professor of Nutrition, Agriculture, and Sustainable Food Systems and Division Chair of Agriculture Food and Environment at the Friedman School of Nutrition Science and Policy, Tufts University, USA.

Shafiqul Islam, ScD, is Professor of Civil and Environmental Engineering at the Fletcher School of Law and Diplomacy and the Director of the Water Diplomacy Program at Tufts University, USA.

Greg Koch is a globally recognized leader in sustainable development, adaptation and resiliency, water resource management and corporate water stewardship, and related policy and finance solutions, as well as a technical director with the consultancy Environmental Resources Management.

Ashley C. McCarthy is a PhD candidate in the Agriculture, Food and Environment program at Tufts University, USA.

Anthony P. Monaco, PhD, became president of Tufts University, USA, in 2011 after two decades as a faculty member and senior administrative leader at the University of Oxford, UK. Dr. Monaco is a distinguished scientist in the area of genetics and neurodevelopmental disorders.

William Moomaw, PhD, is Professor Emeritus of International Environmental Policy at the Fletcher School of Law and Diplomacy at Tufts University, USA.

Glenn G. Page is President/CEO of SustainaMetrix and led the external evaluation of the Tufts University Water Diplomacy Program, USA.

Wahid Palash, PhD, is a consulting water resources specialist who earned his doctorate in hydrometeorology, environmental, and water resources engineering from Tufts University, USA.

Kent E. Portney, PhD, is the Bob Bullock Chair of Public Policy and Finance and Director, Institute for Science, Technology and Public Policy at the George H. W. Bush School of Government and Public Service at Texas A&M University, USA.

Laura Read, PhD, is a water resources engineer at the National Center for Atmospheric Research whose current research focuses on hydrologic modeling for the purposes of flood forecasting.

J. Michael Reed, PhD, is Professor of Biology at Tufts University, USA, where he works on the effects of human alteration of landscapes on extinction risk of birds.

Amanda C. Repella was the Water Diplomacy Network Coordinator at Tufts University, USA, from 2012–2016 and is currently Data Program Lead and Workforce Economist for Larimer County, Colorado, Economic and Workforce Development.

Michael Ritter, PhD, is Assistant Professor of International Development at Houghton College, USA.

Michal Russo is a PhD candidate in the School of Environment and Sustainability at the University of Michigan, USA.

Gregory N. Sixt, PhD, is Research Manager at the Abdul Latif Jameel Water & Food Systems Lab at the Massachusetts Institute of Technology, USA.

Kevin M. Smith is a PhD candidate in environmental and water resources engineering at Tufts University, USA.

Lawrence Susskind, PhD, is Ford Professor of Urban and Environmental Planning at MIT and Vice Chair of the Program on Negotiation at Harvard Law School, USA.

Charles B. van Rees, PhD, is a Fulbright Postdoctoral Scholar at the Doñana Biological Station in Seville, Spain.

Part I

Problem–driven interdisciplinary collaboration

A principled pragmatic approach to addressing complex problems using Water Diplomacy as an example

1 Origins

Conceptualization, implementation, and evolution of an interdisciplinary graduate program on Water Diplomacy

Shafiqul Islam, Kent E. Portney,
J. Michael Reed, Timothy S. Griffin,
and William Moomaw

Prelude

Interdisciplinarity has been hailed as one of the most promising and inspiring contemporary academic pursuits (Klein 1990; National Academy of Sciences 2005). In its ideal form, interdisciplinarity is the answer to achieving integrative understanding and action in the midst of disciplinary fragmentation. However, despite its widespread adoption as an idea, whether this promise is borne out in reality remains highly contested (e.g., Leahey et al. 2017). One of the central issues in measuring the success of interdisciplinary efforts is disagreement on what constitutes interdisciplinarity in both theory and practice. Is interdisciplinarity simply an integration of disciplinary depth across a breadth of methods, assumptions, and tools? Is it really integration that is at stake, or is it the ability to converse freely across previously guarded borders that broadens the possibility of exploring and executing actionable ideas? What kind of conversations and actions count as truly interdisciplinary?

Many books, articles, and reports have been written about interdisciplinarity. We didn't write this book because we have developed a new theory of interdisciplinary research. Nor have we found the perfect recipe for effective interdisciplinary collaboration. We wrote this book, particularly this opening chapter, because we found – whether one is a champion or a critic of interdisciplinarity or whether one agrees on the why, what, and how of interdisciplinarity – that there is a widespread consensus that interdisciplinarity broadens the solution space by remaining open to a diversity of values, interests, and tools. The idea that diversity fosters creativity has deep roots. Cognitive diversity – differences in perspectives, methods, and models from different domains and disciplines – improves performance at problem solving and reduces bias toward the received wisdom of "conventional" approaches. However, introducing greater diversity into the problem–solving process is almost bound to increase the variance in successful outcomes. This finding is consistent with the mixed results recorded by many interdisciplinary pursuits. This leads to a central question in interdisciplinary studies: what factors improve the success rate of interdisciplinary programs without jeopardizing their requisite diversity? Through the stories we share in this book, we hope to share a few lessons about the factors we've found to be important.

Why do we believe we need to improve interdisciplinarity in the water space? By way of comparison, consider the 1962 book *Design of Water-Resource Systems: New Techniques for Relating Economic Objectives, Engineering Analysis, and Governmental Planning* that summarized the evolution of the Harvard Water Program (Maass et al. 1962). In the 1960s, we addressed our water problems with reservoirs and treatment facilities. At that time, a water professional's objectives could all be written down and quantified in objective functions. Today, we are tasked with satisfying both quantitative and qualitative objectives, such as value-laden water claims made by and on behalf of individuals, industries, and ecosystems; a prudent balance between the utilization and conservation of our natural resources; the facilitation of industrial and economic growth; and ensuring a sustainable future for our next generations. Satisfying these increasingly diverse objectives demands the integration of scientific learning with the complex political reality of real-world problem solving. Water professionals cannot easily translate solutions born out of scientific findings into the messy context of the real world. We need to bridge this divide between theory and practice and resolve complex water management problems – where natural, societal, and political elements cross multiple boundaries and interact in unpredictable ways.

It is often difficult to distinguish between multidisciplinarity, interdisciplinarity, and transdisciplinarity (Klein 2010). In fact, a single project can go through various types and stages of disciplinary integration. We neither think interdisciplinary knowledge is per se better than disciplinary nor do we think numbers are better or worse than narratives. Different kinds of problems and questions require different kinds of integrative understanding, thinking, and acting. The nature of complex water problems, however, demands a problem-driven, principled, and pragmatic approach to interdisciplinary research and practice. This interdisciplinarity needs to go beyond collaboration among experts, as professionals will need to participate in inclusive stakeholder-centered processes (e.g., fact–value deliberation, joint fact-finding, collective decision making, and adaptive management) in order to achieve sustainable and equitable outcomes. In this chapter, we explore the story of how we conceived of and eventually implemented an integrative graduate program aimed at training professionals who could be effective in such situations. Our program was funded in large part through a grant from the National Science Foundation (NSF) through their (now-retired) Integrative Graduate Education and Research Traineeship (IGERT) program.

This is neither a systematic literature review nor a formal methodology to develop a theory of interdisciplinarity. This is the telling of stories of the development of our program and what we have learned from those experiences.

These stories cost a huge monetary investment from NSF and tremendous time investment from the faculty and participants; it represents a decade from our academic life and about five or more years from each of the students who came along for the ride in a shared experience. We feel this shared experience – gained through our decade-long trials, tribulations, and excitement – can be best communicated and understood by weaving together stories from the conceptualization, implementation, and evolution of the program. While details of our experience may be contextual to Tufts, we hope our choice of storytelling lens will evoke multiple meanings and provide varied perspectives of interdisciplinarity for actionable outcomes across different contexts and boundaries.

Conceptualization of an interdisciplinary program

Acknowledging boundaries

Water doesn't care about disciplinary or jurisdictional boundaries. Whether it takes the form of a storm surge flooding roadways, a river exceeding its banks, or contested flows across arbitrary human–drawn lines on a map, this fact is the same. Water's indifferent disregard for the barriers we build, the lines we draw, and the human and ecological needs to which we allocate specific quantities are often at the core of the most pressing challenges and intractable conflicts involving water.

Several years before the Water Diplomacy IGERT was funded and the first cohort was recruited (see Figure 1.1 for a timeline of program development and evolution), we were considering how we could take water – a catalyst for cooperation and a compounding factor in conflicts – and transform it to a vehicle for fruitful outcomes to contentious water problems. Specifically, we were considering how this could be achieved through conversation and action facilitated by engaged, scholarly professionals who could think and act across boundaries. This was envisioned not as a technical interdisciplinary collaboration that merges multiple sciences but as something broader: a merging of science and engineering with diplomacy and negotiation – what current intelligentsia refer to as "convergence" (National Research Council 2014). Much in the way water will rise to flood a field, the next generation of scholars and practitioners working within the context of contested waters would need to fill the organizations involved in addressing these challenges with new ways of conceptualizing and seeking actionable resolution to seemingly intractable and increasingly complex water problems.

Minding the gaps

We set out an ambitious goal to break disciplinary boundaries interfacing water science, policy, and politics. We sought to achieve high levels of convergence between sciences and engineering on one hand and diplomacy and policy on

WATER DIPLOMACY PROGRAM TIMELINE

OUR FIRST NSF PROPOSAL FOR A WATER AND DIPLOMACY FOCUSED IGERT

2006

The team of 8 national and 5 international partner organizations sparked what would become Water Diplomacy. Faculty from Tufts bridged across 7 departments. This version of the grant was not funded, but provided a starting point for future collaboration.

2008

WATER C3B SEMINAR: WATER CONSTRAINTS, CONFLICTS AND COOPERATION AT BOUNDARIES

This graduate level Civil Engineering seminar "recognized that triple constraints (quantity, quality, and ecological integrity) on water lead to conflicts."

WATER AND DIPLOMACY UNIVERSITY SEMINAR - WATER AND DIPLOMACY: INTEGRATION OF SCIENCE, ENGINEERING, AND NEGOTIATIONS

Shafik Islam and Bill Moomaw present and Facilitate a University Seminar for Graduate and upper level Undergraduate students.

IGERT GRANT AWARDED

We were awarded a grant by the NSF to implement our IGERT proposal.

2010

IGERT PHD PROGRAM BEGINS

The initial cohort had three students enrolled in Civil and Environmental Engineering and two at the Fletcher School of Law and Diplomacy.

FIRST WATER DIPLOMACY WORKSHOP

28 water professionals from 20 countries participated in the inaugural water diplomacy workshop.

2012

FIRST WATER DIPLOMACY BOOK PUBLISHED

Water Diplomacy: A Negotiated Approach to Managing Complex Water Networks focused on the Water Diplomacy Framework, what it was, and how it could be used to address complex water problems.

BOSTON WATER GROUP ESTABLISHED

Tufts Water Diplomacy members, along with faculty at Harvard, Boston University and Northeastern University initiate, the Boston Water Group, a professional group for practitioners to network, share, and reflect on critical water issues.

2014

SPECIAL WATER DIPLOMACY ISSUE OF JOURNAL OF CONTEMPORARY WATER RESOURCES AND EDUCATION

Contributors to this issue were primarily Water Diplomacy program affiliates.

FINAL IGERT COHORT ADMITTED

The final IGERT added six students to the program: two biologists, three environmental engineers, and one student from the Friedman School of Nutrition Science and Policy.

2016

SECOND WATER DIPLOMACY BOOK PUBLISHED

Water Diplomacy In Action: Contingent Approaches to Managing Complex Water Problems.

WATER DIPLOMACY ROUNDTABLE

As a capstone to the Water Diplomacy IGERT, we held a two day roundtable that brought current and former students, faculty, external evaluators and invited guest speakers together to discuss the future of interdisciplinary scholarship and practice on Water Diplomacy.

2018

THIRD WATER DIPLOMACY BOOK PUBLISHED

Complexity of Transboundary Water Conflicts Enabling Conditions for Negotiating Contingent Resolutions

PRESENT VOLUME PUBLISHED

Figure 1.1 A timeline detailing some of the major milestones in the development and evolution of the Water Diplomacy Program at Tufts University.

the other. We recognized that scientists and engineers are trained to avoid making value–based judgments, while real–world water problems have value–laden aspects that cannot be disregarded or disaggregated. Indeed, when the way forward is presented as a competition between facts and values, we are faced with a false choice. Achieving actionable outcomes in the messiness of the real world does not entail decision between facts *or* values; rather, it requires an integration of facts *and* values. As a result, pure science-based approaches to complex water resource management that don't engage in fact–value deliberation will seldom work in practice. Unfortunately, the training pipeline for top scientists and expert negotiators has little overlap between disciplines or even between subdisciplines within the same broader field, leading to multiple sets of highly trained professionals that cannot even speak the same technical language.

This lack of overlap is perhaps the most critical gap to address boundary-crossing water conflicts with competing needs. Here, existing laws, economic instruments, or best management practices may not be sufficient. We argued that these problems can only be adequately addressed through effective negotiation processes that include multiple perspectives and preferences of stakeholders and a broad range of technical professionals to frame, understand, and ultimately resolve their competing and often conflicting needs.

Different facets of science related to water and its multiple uses traditionally encouraged curricula with strong disciplinary emphases. Historically, scientifically trained water managers aspired to rise above politics, because in the political realm, science is construed to be free of value judgments. In this view, science is politically neutral, data are objective, and every problem has a true, unique, and well-defined solution. In reality, however, this has become increasingly problematic with the proliferation of institutional boundaries and organized special-interest groups. Opposing interest groups may interpret the same data to arrive at conflicting conclusions or dismiss science that does not support their preferred outcome (e.g., Fort et al. 2003; Biswas 2004; Lankford and Cour 2005; Pahl-Wostl et al. 2007).

At least two important gaps continue to persist in the design and implementation of water programs that aspire to interdisciplinarity (e.g., Groves and Moody 1992; Pahl-Wostl et al. 2007). **First**, many of these programs simply draw members from a mixture of disciplines without any systematic attempt to bridge their distinct vocabularies. Consequently, these programs offer limited *inter-* or *cross*-disciplinary opportunities and may amount to little more than a contest of ideas originating from individual disciplines. Since conflicts of vocabulary are not addressed, it is hard to assess whether there is even joint agreement on the nature of the problem under consideration. Current literature on human learning suggests that to achieve interdisciplinary solutions to boundary-crossing problems, we need to jointly define a problem with an explicit recognition of interdependencies. **Second**, while science-based management of water issues has received much attention in designing water programs, other elements – such as negotiation in the context of competing needs and different governance systems – have not.

We envisioned that Water Diplomats would be a new type of water professional who would be competent in multiple water-related academic and applied fields. A key challenge was how to develop these competencies as an interdisciplinary skill set for students entering the program with diverse undergraduate preparation in natural sciences, social sciences, and engineering. While we saw this as a particularly needed skill set, it was additionally imperative that students exiting the program had an array of appropriate career opportunities open to them, as "Water Diplomat" was not a recognizable career option. Therefore, each student had to be an expert in his or her discipline as well as having the skills to work effectively across disciplines.

Prior to securing funding for our program, special university funding supported a Tufts-wide graduate seminar run by Islam and Moomaw, two of the founding members of the IGERT program. The seminar was entitled "Water Constraints, Conflicts and Cooperation at Boundaries." In it, students explored the idea that the boundaries and constraints in biophysical, social, and political processes are woven into water resource management challenges. This laid the groundwork for further exploration of the integration of science, engineering, and negotiation in water-related problem solving.

Special NSF funding was obtained to catalyze what would become the inaugural Water Diplomacy Workshop, an intensive five-day course designed and taught by Islam and Susskind. In this workshop, we shepherded mid-career water practitioners and scholars from more than 40 countries through a process designed to teach them how to apply ideas from complexity science and key theories of mutual-gains negotiation to the water problems they interacted with in their work. A book (Islam and Susskind 2013) was drafted and published to assemble these ideas together for the first time. And the refinement of ideas through these experiments and activities helped us to hone the initial design and proposal for a grand experiment in graduate educational excellence. Our IGERT proposal was funded in 2009, and we thought we were prepared to begin breaking down boundaries.

What were the thematic basis and unifying aspects?

We began from a recognition that science alone will not solve major water problems nor will policy operating in a vacuum without inputs from science. Moreover, we recognized the importance of a mutual-gains approach to negotiating sustainable and equitable outcomes for water problems with competing stakeholder interests and boundary-crossing aspects. We needed professionals who could think across boundaries, integrate knowledge, link it to action from multiple perspectives, and develop creative options for increasing the distribution of benefits among partners through mutual-gains negotiations.

To start this, we needed a framework. The initial framing was that natural systems, (constrained by quantity, quality, and the ecological function of water) and societal systems (with feedback and interdependencies among values and norms, economic factors, and governance) are intrinsically coupled. Attempts to address these coupled water problems would need to account for these constraints, linkages, and feedbacks using a systems view to derive synthesized water knowledge to support negotiated solutions (see Figure 1.2). The program focused on two challenges: **(i)** mechanisms to integrate knowledge from various science and policy domains to **formulate** and **frame** the questions and **(ii)** methods to produce **actionable knowledge** by involving

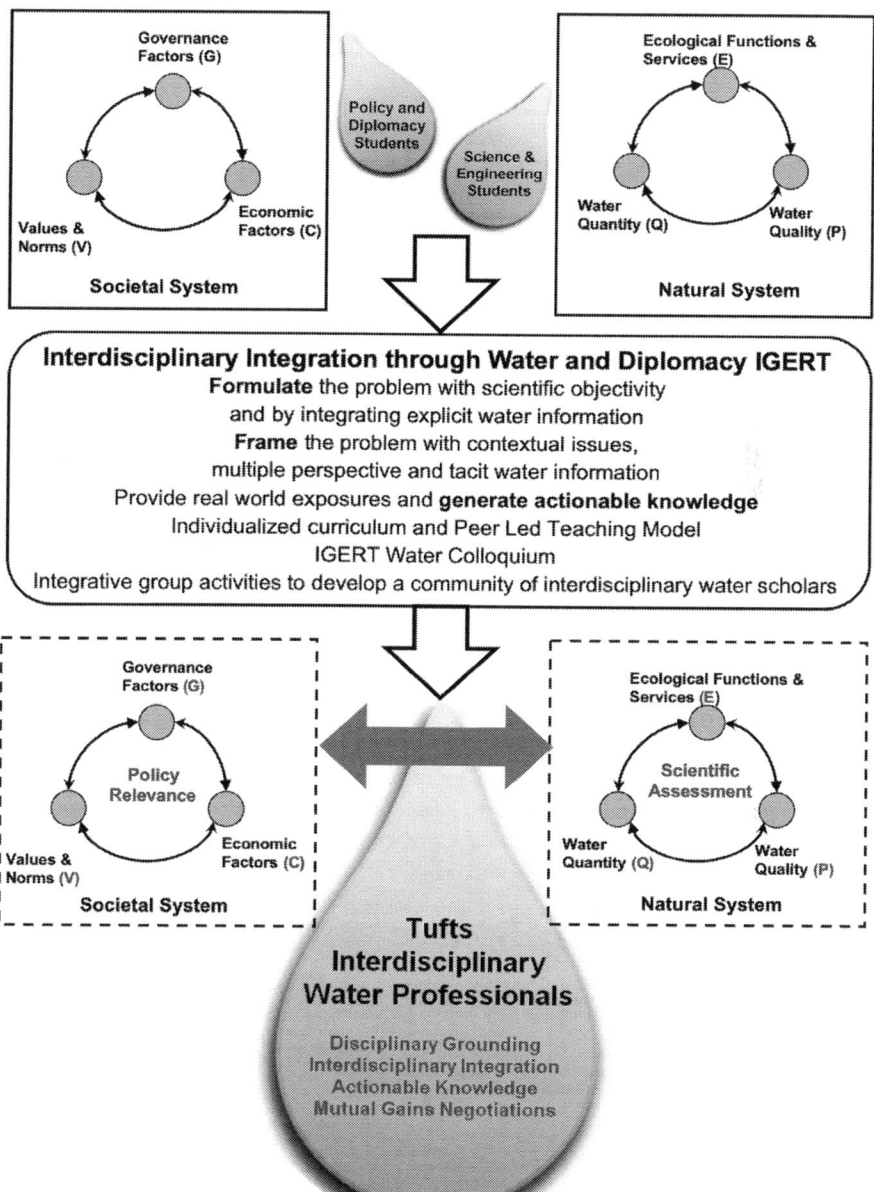

Figure 1.2 The original program schematic diagram for the Water Diplomacy IGERT. This figure from the Water Diplomacy IGERT proposal illustrates our lofty goals for creating a previously unknown type of water professional via an interdisciplinary PhD.

relevant users and knowledge producers in the development of negotiated resolutions to water problems with competing needs.

What did we want to create?

We recognized that simply connecting experts, creating more scientific knowledge, developing models, and sharing data was not enough to effectively manage 21st-century water problems. These problems require water professionals and decision makers to find intervention points at which framing water issues and envisioning a different water future could be planned and executed. We recognized that understanding contemporary and emerging water problems required knowledge and thinking skills that transcended traditional disciplines. Such understanding demanded that we acknowledge the multifaceted and coupled nature of natural and societal systems (NSSs). Traditionally, students pursuing water careers developed expertise in the societal or natural domain (Figure 1.2: top panel) with rigid disciplinary boundaries. Our integrative efforts attempted to create porous boundaries between natural and societal domains (Figure 1.2: bottom panel with dotted boundary and explicit interactions between NSSs) and encouraged sharing of explicit and tacit knowledge between domains and disciplines.

What was our vision for future water professionals?

The Water Diplomacy Framework (WDF) was created to help resolve water-related conflicts (see Figure 1.3). It acknowledges that traditional problem-solving frames are adequate to address simple problems when reasonable

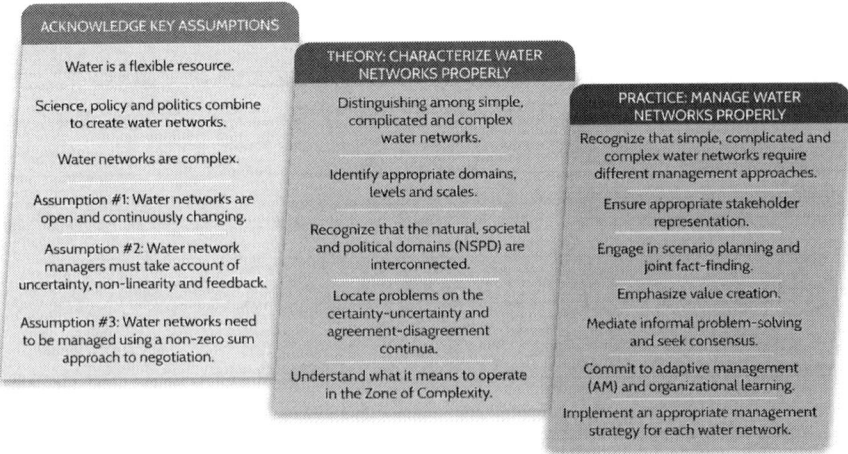

Figure 1.3 The Water Diplomacy Framework as originally conceived in Islam and Susskind (2013).

scientific certainty and consensus about effective interventions exists. When water challenges stem from complex – interconnected, uncertain, unpredictable, and boundary-crossing – system dynamics with feedback, traditional frames for problem solving can be limiting or counterproductive. A recurrent factor for such limitations is that traditional problem-solving frames often separate the observation-based technical (what is) from the value-based sociopolitical (what ought to be) dimensions of the problem. The WDF emphasizes that when dealing with complex problems, these dimensions cannot be decoupled. The WDF acknowledges both the limits to knowledge – objectivity of observations and subjectivity of interpretation – and the contingent nature of our action. The WDF proposes a principled pragmatic approach to negotiation that acknowledges and addresses interconnectedness among natural, societal, and political domains.

Faculty reflections on the implementation and evolution of the program

What were some of the general and special challenges?

The Water Diplomacy Program was designed to confront and address a number of challenges to interdisciplinary doctoral training. Achieving convergence between science and diplomacy involving number of related disciplines is no small task. We discovered, for example, that many of the entering students whose academic and professional experience was rooted in engineering disciplines expected to be able to immediately and directly apply their knowledge of systems theory to societal systems. Although there are examples where such mergers have been accomplished with some success, such as incorporating economic systems models of water demand, extending this approach to process-oriented issues such as policy making and conflict negotiation has been far less successful. Our curriculum was designed to address these issues particularly in the water economics and policy course by incorporating readings about applications of systems models to water demand, as well as readings about theories and analyses of water policy making and policy implementation. Since much of the policy making, implementation, and negotiation literature is rooted in behavioral science, students received a healthy dose of exposure to the analysis and methodologies of such research. After all, what tends to make water issues complex is the coupling of human or societal elements with natural variables.

Many students came into the program with strong statistical analytic tools, but few had ever learned how these tools are typically or potentially applied to research on the behavioral side of policy and diplomacy. Readings that served as exemplary cases of empirical analysis of social issues of water were highlighted and discussed. Inevitably, this highlighted significant similarities among different disciplinary approaches to similar problems and often provided students with a common vocabulary to refer to concepts that are known to each discipline by different names. If our goal was to produce well-trained engineers who could engage in policy and negotiation research or well-trained social scientists who could engage in hydrologic or water engineering research, we probably could never succeed. Such a goal would have probably been far too ambitious. Our goal, however, was to move toward deeper understandings across disciplines, and we undoubtedly achieved some modicum of success in this regard. Preparing students to effectively use research from natural and social science disciplines in applied settings was accomplished as outlined below.

Implementing the Water Diplomacy Program faced a number of specific and sometimes idiosyncratic institutional challenges not necessarily different from those that might be encountered in any interdisciplinary, multidisciplinary, or transdisciplinary curriculum. A number of the disciplines whose expertise was needed were reticent to commit resources to a curriculum outside of their regular offerings. In some cases, there were no faculty who had a particular set of expertise. For example, there was no faculty member in the economics department whose expertise was ecological economics. Fortunately, such expertise was available elsewhere at the university or, in the case of expertise in negotiation, through collaboration with a partner at a nearby university. Another challenge arose when efforts were made to admit highly qualified students through specific doctoral programs. Although the core doctoral program in civil and environmental engineering was highly receptive to admitting such students, others were far less so. And because a number of disciplines relevant to Water Diplomacy did not have existing doctoral programs at all, alternative admission pathways had to be defined. At Tufts, for example, the political science, sociology, anthropology, and urban planning departments did not have doctoral programs. So students who otherwise were highly qualified to enter the Water Diplomacy Program had to do so through other departments or programs.

How did we get where we are now?

It was a combination of thinking before acting, attracting excellent students from a variety of disciplines, creating an atmosphere of interaction and deliberate feedback, and being adaptive to feedback and changing circumstances helped us navigate through this long journey. The original group of faculty had a vision of the program, and we actually sat down and actively exchanged ideas and listened to each other. This might sound banal or trivial, but it is neither – it was central to our success. Although we had faculty turnover for a variety of reasons, we were able to maintain a high level of communication. We had the same extensive communication with the graduate students in our program. None of this happened by accident. We had weekly colloquia involving faculty and students at which ideas from students and faculty were shared and critiqued, annual retreats designed specifically to foster feedback and planning for the future, and team-taught classes where all of the faculty members in a course were present at every lecture regardless of who was presenting. Graduate students from previous cohorts were involved in recruiting and vetting applicants for future cohorts, and specific feedback was elicited regarding revising courses to improve the learning experience.

Another key to success is that even though there was a "Gang of Five" faculty co-running the program and doing most of the teaching, there was a single leader (Islam) throughout the tenure of the program who served as the de facto leader. We also had a full-time staff person to coordinate details and do paperwork (e.g., organize a place and food for retreats, do the massive amounts of paperwork required to fund travel for many student internships). The head of the program led the constant battle with university administration to ensure that our students had joint space for their offices – which was key for many interactions, securing university support and resources to match NSF requirements for funding, acting as the interface with our various outside evaluation committees, and was the focus for the many administrative duties required in university persistence. Having a single person as a constant thread was critical to the program's success. And having a staff assistant who was thoughtful and competent was critical to the program leader's success.

From a practical standpoint, what do others going down a parallel road of interdisciplinary education need to know about how to implement this based on our case study?

Some of the answers to this question are simple and predictable: you need money, space for graduate students, dedicated faculty (including a point person to lead and coordinate), carefully selected graduate students, a clear vision that all of the faculty involved strongly support (not just casually agree to), and institutional support. This last requirement can be a tough one because it is largely out of faculty control. To have successful interdisciplinary programs that are supported across departments rather than being housed within a single department is fraught with pitfalls. For example, space is an issue at most universities – who has it, who wants it, and who decides. One of the most difficult problems that we have seen in creating and maintaining an interdisciplinary program is having a place for members to interact – regularly, daily, constantly. This is how ideas get generated and problems get solved – through interactions. To do this, you need space – space that you can count on year after year, that the administration will protect from departments and other interdisciplinary programs. This means that your program has to have the ear (and support) of a top university person who is in a position to make this happen. Every year, defending space was a struggle for us, and before our program even ended, another cross-university group was encroaching on our graduate student and administrative space. The moment our grant ran out, the space was taken away from us. This makes it almost impossible to continue the program.

Also be warned that agreements made by one administration might not be valued or backed by subsequent administrations. It sometimes does not matter what you have written on paper. That's just reality and a constant potential threat. (Note that it can also be an opportunity if your current administration is not sufficiently interested.)

One aspect of administrative support that is more subtle and potentially difficult to deal with is the influence of departments that are the official homes of the participating faculty and graduate students in the program. Promotion, faculty raises, course offerings, and in some cases graduate student support are decided at the department level. If a department chair does not value the interdisciplinary program that is drawing resources (faculty time at a minimum) away from the department, the program will struggle or even fail. Having the upper administration tell departments to value a program is insufficient and might even be counterproductive (depending on the relationship between the administration and the department). Consequently, no matter how appropriate an assistant professor might be to a program – leave that person alone! Tenure is given within departments, and at research universities, it is given for research within that discipline. So unless a faculty member's hiring is specifically to be part of an interdisciplinary program and promotion is driven by it, leave untenured faculty alone.

The one solution to most of these problems is to have lots of program money, and keep bringing in lots of money from grants and contracts. Not an easy task, but as far as we can tell, money is an essential ingredient.

Were there any points during the progress of the program at which it changed direction or where there was an appreciable difference in the velocity of its evolution?

We don't recall any monumental changes in our program – but there were many small changes throughout the entire duration of the program. For example, we taught three classes that were the core of the WD Program, and after the first year, we reversed the order of two of the classes. We did this because of feedback from the students, who felt that the learning experience worked better that way. We changed a basic requirement of our internship program so that it no longer required internships to be overseas. Students came to us with excellent within-country internship opportunities that either built beautifully into their PhD research programs or provided networking opportunities for career development. With the benefit of their feedback, we realized that one did not need to travel abroad to gain the types of experiences we wanted students to have.

Over the years, we played with the format of our weekly colloquium, from discussing water case studies in detail to student presentations of their research to covering issues in the field to using it as planning time and changing up how it was chaired (rotating among students, run by faculty). We finally returned to our original plan of student presentations with time for feedback and discussion. This was viewed by the graduate students as the most valuable format for them, both as presenters and for learning across disciplines. However, as student feedback in this chapter shows, our several adaptations to the colloquium format did not settle all concerns about how to make the best use of our weekly meeting. In order to keep students and faculty engaged, we had to strike a careful balance between being adaptive to feedback and creating a space for discussion that had a predictable format that matched expectations.

As part of our IGERT program, we hired an outside consultant to do regular reviews of our students and our program. This external evaluator (Glenn Page of SustainaMetrix) provided regular feedback that we used to change our program in different ways. Page was deeply involved in the program – coming to all of our annual retreats, meeting individually with all of the graduate students and faculty, facilitating dialogue, and making presentations. Page also met with us at other times during the year so that we could provide updates on the program, develop plans for the future, and make decisions about how to modify the Water Diplomacy Program going forward. Since there is a chapter later in this book written by Page (Chapter 13), we won't discuss the details here – but it suffices to say that there was regular (formal and informal) time spent planning, reflecting on, and revising our program.

How did the program elements change from what was proposed?

Perhaps surprisingly, the Water Diplomacy Program changed very little in its fundamental program elements. This might be because we went through several proposal revisions as we made our way through funding cycles and because during this time two of the founding faculty "test drove" a Water Diplomacy course a couple of times. So the basics of the program elements were interdisciplinary learning in the classroom, cross-disciplinary advising, common office space for the students, professional internships, and lots of structured time for exchanging ideas and professional development (colloquia, retreats). These fundamental program elements did not change, though the details within each did.

What constitutes our successes? What were our failures?

This probably can be summed up fairly simply. Our greatest successes have been our students that finished the program. As a group, they have a staggering number of professional publications, they are conversant in the concepts of Water Diplomacy, and those who have finished have started successful careers (university faculty, postdocs, jobs in their field, a Fulbright scholar, etc.). These successes were brought about by good screening for student admissions, a well-developed plan and sufficient support for internships, supervised teaching experience, presenting papers at national and international meetings, research publications, and the hard work of the students. We demonstrated that an interdisciplinary graduate program on Water Diplomacy could work and that the professional development and training could succeed. Another success is that the core faculty involved in the program developed a great deal professionally;

the program altered research programs and teaching portfolio for several faculty involved in the program.

We had two types of failures. First, there were four students who entered the program but did not flourish, and they eventually dropped out of the program (and typically out of their departmental PhD programs as well). There were a variety of reasons for each failure; given the high investment of time, money, and opportunity costs in each student – by faculty as well as by the students themselves – each one was a tough loss. The second failure was that we were unable to secure outside support to permanently fund a broader Water Diplomacy Program within the university. Further large grants were not forthcoming, and the university's foundations division was not able to find donor support for the program, despite how well it was performing.

Were there situations in which we were naive? How so? And most importantly, what did we learn?

Most of the changes we made during the program were the result of systematic feedback and adaptive implementation. So one thing we learned was that framing of our program through a complexity lens was very helpful. It allowed us to be reflective of our action and adapt to changing circumstances. One place at which we were naive was thinking that two years of graduate support would be a strong enticement to attract the best graduate students and that finding funding for the duration of the graduate students' time at the university would not be a problem. It turned out that some excellent students turned us down to go to programs that offered more years of support and that the two years we secured was not enough for some students. Some students or their advisors found other means of support, or their home department supported the students on teaching assistantships, but this type of support was not available in all departments.

Another place we were naive was in thinking that all of the graduate students would engage in the Water Diplomacy Program to the same degree the faculty did – in particular, to the degree that several faculty did. What we found was a real mix. Some students became completely engaged, and they became one backbone to the successes of our program. They were completely engaged and continued their engagement throughout their PhDs. Other students ended most of their interactions with the Water Diplomacy Program as soon as their funding ended. That was not entirely surprising, but certainly disappointing. And there was a small fraction of students who were never sincerely engaged in the program even while being funded by it. In a small program (we admitted around five students per year), it is important that all students be engaged – and the longer they stay engaged, the better for the program.

The other point of naivete (at least for some of the faculty) was the lack of interest by some of the administration to help our program excel or even to protect our space from year to year, and the difficulties of managing a grant between different colleges within the same university (engineering vs. arts and sciences). These, however, are infrastructure issues rather than program issues.

Student reflections on the value of key program elements

We expected students from our IGERT Water Diplomacy Program to develop a broader understanding of key ideas, assumptions, and principles of the Water Diplomacy Framework through individualized curricula, the weekly colloquium, a PhD dissertation that fulfilled the discipline-specific requirement but also contained a Water Diplomacy issue, joint advising on the dissertation by both a natural and societal domain faculty, and an internship experience with relevance to Water Diplomacy. Our students, who are now out in the wild as practicing Water Diplomats across a variety of roles in academia, professional practice and public services, interacted with these elements through unique experiences along highly individualized pathways.

While we may have had preconceptions of what these programmatic elements would impart, our students' reflections from the other side of their Water Diplomacy graduate experience tell us about how they value aspects of the program. Some of these are quite surprising. Our weekly colloquium supported both topical explorations with guests and presentation opportunities for students for feedback. Students were invited to bring rough presentations and half-baked ideas to clarify, refine, and improve with their peers and professors. Some students found the colloquium experiences to be uniquely beneficial:

> "*I think one unique strength I bring to my current position is the capacity to call different departments directors to the table and convey meetings around complex local-government issues, aiming at developing and delivering solution projects. I strongly feel this is a skill I have learned from my participation in WD Colloquium.*"
>
> "*I think having a colloquium group who were mostly fluent in Water Diplomacy concepts was important in accelerating the exchange of ideas.*"

Others were critical of the format, wishing that it had been more efficient or effective toward other end goals:

> "*I think the colloquium was a wonderful opportunity to learn about fellow students' projects and research activities, but these [colloquia] could have been better organized by including opportunities for external experts and partners working with WD students could be more involved.*"
>
> "*I think some of the 'research update' colloquia could have been replaced with trainings, guest speakers and other activities teaching us new skills.*"
>
> "*I have mixed feelings about colloquium as an effective use of time. I think the student presentations were a useful forum for feedback and presentation practice. On one hand I think these conversations did help develop student skills in discussing*

complex topics with scholars from a wide range of backgrounds. On the other hand, it was a significant weekly time commitment that could have been made more efficient. Better facilitation and planning I think would have allowed the group to accrue the benefits of Colloquium with less frequent perhaps but still regular meetings taking some of the time burden off the students."

The internship component was also cited as a valuable aspect, particularly for the benefit it imparted to their CVs and employability.

"The internship experience, especially, has been seen as valuable by many professional employers, who looked very favorably on that aspect of my training."

"My Water Diplomacy research internship ... ultimately led to my current position."

We asked Water Diplomacy graduates which of their experiences they felt would be most universally transferrable to other interdisciplinary PhD programs. The aspects of the program that supported informal cross-disciplinary interactions and developed a sense of community were particularly valuable.

"One of the strongest parts of my experience in the Water Diplomacy IGERT was that from day one my academic peer group spanned multiple disciplines. Shared office space and shared meals created room for the free exchange of ideas without the artificial boundaries of what would have passed for reasonable conversation in a less academically diverse group. I think every PhD student could benefit from being part of a multidisciplinary 'cohort,' even if they do not have the benefit of shared office space or a common interdisciplinary purpose (such as Water Diplomacy). PhD programs could convene these groups at graduate orientation and provide incentives to engage them in continued dialogue (e.g., reimbursement for shared lunches, coffee)."

"Both as a graduate student and as a post-doc, I shared office space with others from different disciplines and I had opportunities to regularly discuss research ideas in formal settings (e.g., colloquia). Not only have I gained new insights, tools, and perspectives, I've been able to work on many rewarding interdisciplinary research projects as a result of these informal and formal connections."

"Being a part of an interdisciplinary cohort of PhD students was central to my experience in the Water Diplomacy Program. While I was also part of a cohort of CEE graduate students, being a part of the IGERT program expanded my network both professionally and socially. The formal activities (such as shared classes and the annual retreat) helped to build a sense of community. However, the shared office space and the frequent but informal interactions it facilitated were even more influential."

A challenge in such a broad interdisciplinary program is that each student was a member of a group that had remarkably different career goals in a broad variety of fields. It was not surprising to learn that several of our students found that the need to develop both disciplinary and interdisciplinary communication skills enhanced their broader PhD experience and was a unique value of the Water Diplomacy Program:

> *"Coming up with common definitions to help communication across disciplines was a basic but very important exercise. I've found many times after Tufts, these seemingly simple tasks are those that are overlooked, but cause the most confusion. Communication and wanting to effectively explain across disciplines or sectors is much of what Water Diplomacy is to me."*

The surprising aspect of this is not that this experience provided broader exposure to communication demands for different audiences but that multimedia techniques (e.g., maps, data visualizations, and storytelling) were found to be critical aspects of successful interdisciplinary communication:

> *"Data visualization and geospatial mapping provided a lingua franca for our group of WD students, helping bridge semantic gaps between our 'depth' disciplines. Visual communication and storytelling were critical for connecting to audiences outside of our group too. When preparing summaries of my dissertation work for job applications, I found that my data graphics grabbed attention and propelled engaging discussions. Today I use data visualization and visual storytelling extensively in my work. The skills honed in the Water Diplomacy Program help me to communicate my technical work to wide audiences including farmers, environmental policy advocates, and hedge fund managers."*

One outcome that matches exactly what we hoped for is that our Water Diplomats are able to intercede in the framing of a problem that allows for opportunities to create value. This can occur through technical processes:

> *"Every discipline has its form of tunnel vision. My experience has been that my engineering collaborators often consider their role in the context of a 'linear' model of science and policy. A common starting assumption is that the world is filled with economically rational agents who believe in being governed by policies that rank best under cost-benefit analysis ... It is assumed that the last mile of translating this theory into action is beyond the engineer's purview. After all, in a rational world, what more is to be done after one has proven the dominance of a strategy in quantitative terms? What other arguments could be offered? The Water Diplomacy Program emphasized that many water challenges will require collaborative and adaptive management strategies. These collaborative approaches necessitate creating space for both 'numbers and narratives.' I have been able to guide my collaborators to diagnose why the gap between 'the best available technologies' for storm water engineering and 'what is ordinarily done in practice' is so large. This allowed us to avoid the line of thinking 'how can we improve the best available technologies' and focus instead on 'what can we do to improve the collaborative processes in which municipalities balance capital expenditures against*

perceived risk?' This question is still very much within the realm of engineering practice, but it avoids assuming that the bottleneck for achieving actionable outcomes is that the 'best available technologies are not good enough.' Instead it emphasizes looking for bottlenecks in the last mile of translating available technologies into options that can work in the world we live in."

It also gave our graduates practical skills for working in private firms in which client relationship management is paramount to continued business success.

"Perhaps the most readily applicable notion from Water Diplomacy has been to look beyond tradeoffs and seek opportunities to create value. From a business perspective, delivering clients value is how you ensure an ongoing business relationship. To deliver what is explicitly asked for is not enough. If the clients don't feel like we have created value for them, they are unlikely to return. It is my understanding that these ideas are common in business school courses, but are not part of the regular curricula of any non–business-oriented academic programs. In this sense, I would have been unlikely to be exposed to these ideas outside of Water Diplomacy, at least until much later on in my career."

Concluding remarks

In this chapter, we have shared the story of the conceptualization and implementation of an interdisciplinary graduate program that sought to educate a new generation of Water Diplomats who could bring both principles and pragmatism to bear on their work in complex water management challenges. We have shared the reflections of key faculty involved in the program, as well as shared findings from former students on what elements of the program they felt were most valuable. There is nothing easy about interdisciplinary work, but it is not without its rewards. More importantly, however, we see principled, pragmatic, problem-driven interdisciplinary collaboration as an essential mode of working when confronted with problems that arise from a coupling of natural and human systems, and we will use the remainder of Part I to explore this idea.

In Chapter 2, we argue that coupled natural and human systems are inherently complex – as opposed to being simple or complicated – which has significant implications for how problems arising in these systems need to be approached by researchers and practitioners. In Chapter 3, we argue that theory-driven modes of collaboration are seldom able to reach actionable outcomes on problems arising in coupled natural and human systems but that problem–driven approaches that utilize shared frameworks (e.g., the WDF) can be effective. In Chapter 4, we examine how professionals can utilize principles and pragmatism to go beyond interdisciplinary collaboration among experts and embrace inclusive fact–value deliberation, joint fact-finding, collective decision making, and adaptive management to achieve actionable, sustainable, and equitable outcomes.

Acknowledgments

In closing, we fondly note that our Water Diplomacy journey has benefitted from discussions with and comments and critiques from many, including L. Abriola, A. Akanda, L. Bacow, J. Bandyopadhyay, A. Biswas, J. Barucha, Y. Bar-Yam, E. Choudhury, A. Jutla, J. Kenny, P. Mollinga, A. Najam, A. Repella, B. Roach, P. Rogers, D. Small, J. Shimshack, L. Susskind, R. Vogel, and A. Wolf, to whom we are grateful. More importantly, our heartiest gratitude and congratulations go to our Water Diplomats, who have helped us learn and grow together to make an impact in a growing field that we coined as Water Diplomacy many years ago. We are grateful to the US National Science Foundation for their generous support through two grants (NSF-IGERT 0966093 and NSF-RCN 1140163) to make this a reality.

References

Biswas, A.K., 2004. Integrated water resources management: A reassessment. *Water International*, 29 (2), 248–256.

Fort, D., Lawford, R.G., Hartman, H.C., and Eden, S., 2003. *Water: Science, Policy, and Management*. Washington, DC: American Geophysical Union.

Groves, J.R., and Moody, D.W., 1992. A survey of hydrology course content in North American universities. *Journal of the American Water Resources Association*, 28 (3), 615–621.

Islam, S., and Susskind, L.E., 2013. *Water Diplomacy: A Negotiated Approach to Managing Complex Water Networks*. New York: RFF Press.

Klein, J.T., 1990. *Interdisciplinarity: History, Theory, & Practice*. Detroit, MI: Wayne State University Press.

Klein, J.T., 2010. A taxonomy of interdisciplinarity. *In*: J.T. Klein and C. Mitcham, eds. *The Oxford Handbook of Interdisciplinarity*. Oxford: Oxford University Press.

Lankford, B., and Cour, J., 2005. *From Integrated to Adaptive: A New Framework for Water Resources Management of River Basins*. The Proceedings of the East Africa River Basin Management Conference.

Leahey, E., Beckman, C.M., and Stanko, T.L., 2017. Prominent but less productive: The impact of interdisciplinarity on scientists' research. *Administrative Science Quarterly*, 62 (1), 105–139.

Maass, A., Hufschmidt, M.M., Dorfman, R., Thomas, H.A., Jr., Marglin, S.A., and Fair, G.M., 1962. *Design of Water-Resource Systems: New Techniques for Relating Economic Objectives, Engineering Analysis, and Governmental Planning*. Cambridge, MA: Harvard University Press.

National Academy of Sciences, 2005. *Facilitating Interdisciplinary Research*. Washington, DC: National Academies Press.

National Research Council, 2014. *Convergence: Facilitating Transdisciplinary Integration of Life Sciences, Physical Sciences, Engineering, and Beyond*. Washington, DC: The National Academies Press.

Pahl-Wostl, C., Craps, M., Dewulf, A., Mostert, E., Tabara, D., and Taillieu, T., 2007. Social learning and water resources management. *Ecology and Society*, 12 (2).

2 Making distinctions

The importance of recognizing complexity in coupled natural and human systems

Kevin M. Smith and Shafiqul Islam

Natural and human systems

Historically, it has been commonplace to draw conceptual boundaries between "natural" systems and "human" systems. Roughly speaking, this distinction has been made between systems governed by invariant physical laws (natural systems) and those involving norms and behaviors (human systems). Natural systems are often taken to be describable in context independent ways, while descriptions of human systems are highly context dependent. While rarely stated explicitly, these sorts of distinctions are often enforced implicitly in academia as role-defining "boundary-work" that separates practitioners in the natural and social sciences (Gieryn 1983).

While we think it is important that our readers are aware that such distinctions remain prevalent in professional practice today, we will not examine the problematics of the disciplinary "border zone" in detail (Brown 1963), as the precision of the distinction between natural and human systems is not critical to our argument. Nor do we require that readers accept the designations of *natural* and *human* as the appropriate system signifiers. Rather, we adopt these terms as a colloquial shorthand that allows us to distinguish between the physical laws that suggest that water will flow downhill (natural) and the social norms that suggest it may also flow uphill when drawn toward money and power (human).

The traditional approach has been to treat such phenomena as coexistent but largely independent. However, a growing body of literature on the coupling of natural and human systems suggests this assumption of independence is misplaced (Liu et al. 2007; Alberti et al. 2011; Chen and Liu 2014). With apparently increasing frequency, researchers are identifying phenomena that defy the dichotomous description of natural or human. These phenomena also resist description by parts: they cannot merely be "built up" from physical or social components. This evidence points to an irreducible and intrinsic condition that exists in coupled natural and human systems. This condition of emergence, which evades precise definition, is widely considered a defining attribute of *complex* systems (Bechtel and Richardson 1992; Holland 2002; Fromm 2005).

That is not to say that the association of emergence and complexity is without controversy. Indeed, one researcher in the journal *Complexity* has referred to the idea of emergence of a "venerable concept in search of a theory" (Corning 2002). While these are important debates, we find that quite often these sorts of arguments rest on differing interpretations of what precisely is meant by "complexity" and "emergence" (Islam and Choudhury 2018). Our use of complexity refers to a particular condition of a system that gives rise to: ambiguous and non-prospective cause–effect relationships; broad and often-irreducible predictive uncertainty; and emergent "surprises" that dominate system behavior and response.

Some non–complex systems (which we will soon distinguish as "complicated") may also demonstrate fuzzy cause–effect relationships, predictive uncertainty, and "novel" properties at particular scales. However, what distinguishes complex systems from those of complicated systems is that the "fuzziness" of the cause–effect relationships cannot be easily clarified through traditional deterministic or probabilistic calculus; the predictive uncertainty cannot be statistically characterized; and the emergent behaviors cannot be prespecified or forced to recur.

Therefore, in describing and identifying coupled natural and human systems as *complex*, we are making a critical distinction that has important implications for how they should be approached by researchers and practitioners. In subsequent chapters, we will develop our ideas about the specific types of responses we think are warranted (i.e., problem-driven interdisciplinarity, principled pragmatism). For the remainder of the chapter, however, we will attempt to get around the semantic messiness of the word "complexity" by offering our readers a concrete example of a coupled natural and human system. By progressively attempting to see this system as *simple*, then as *complicated*, and finally as *complex*, we hope to demonstrate why such distinctions are necessary and important in diagnosing and responding to many contemporary water problems.

Vignette: Lowndes County, Alabama

On June 19, 2018, the United States of America announced its withdrawal from the United Nations Human Rights Council (UNHRC). While announcing the departure, Nikki Haley (the then US ambassador to the United Nations) figuratively referred to the UNHRC as a "cesspool" (Morello 2018). Two days later, Philip Alston, the United Nations Special Rapporteur on Extreme Poverty and Human Rights, reciprocated by using Haley's own words to call attention to what he saw as ongoing human rights shortfalls in the United States. In an address to the UNHRC on June 21, Alston condemned the open cesspools that he observed on a recent trip throughout Alabama, noting that "raw sewage poured into the gardens of people who could never afford to pay $30,000 for their own septic systems" (Alston 2018a).

Lawyers studying the poor sanitation conditions in rural Alabama have referred to it as "America's Dirty Secret" (Winkler and Flowers 2017). Lowndes County, Alabama, has emerged as the focal point for the broader crisis after being called out in multiple UNHRC reports (Albuquerque 2011; Alston 2018b) and receiving coverage from several domestic (Smith 2017; Flowers 2018) and foreign news agencies (Cleek 2015; Pilkington 2017). Only about 20% of Lowndes County residents are served by publicly funded municipal sewerage (Albuquerque 2011, p. 6; Tavernise 2016). The remaining 80% of residents are required by law to install privately funded on-site wastewater systems (Alabama State Code § 22–26–1 2014; Alabama State Code § 22–26–2 2016). However, the subtropical climate and clay-rich soils in Lowndes County are not amenable to traditional septic systems, meaning legal compliance often requires the installation of higher-tech "engineered" septic systems at an unbearable cost (He et al. 2011; Meza 2018).

With a median household income of less than $30,000 (U.S. Census Bureau 2018), the $9,000-to-$30,000 price tag of a compliant engineered septic system remains out of reach for most Lowndes County residents (Izenberg et al. 2013). At the higher end, the cost of such a system is on par with the median value of owner-occupied mobile homes, which make up some 35% of the county's housing stock (Meza 2018, p. 7). For those who can pay, the investment does not guarantee performance or compliance. Some residents report being taken advantage of by fly-by-night contractors (Carrera 2014, p. 114), and many others lament that their systems have never worked as expected (Smith 2017).

Even well-designed systems are under constant threat of displacement and damage due to the extreme contraction and expansion cycles typical of local soils (Meza 2018, p. 6). Overall, the Alabama Department of Public Health estimates that between 40% and 90% of households in Lowndes County have either an inadequate septic system or no such system at all (Albuquerque 2011, p. 6). For many residents, disposing of wastewater in open ditches or piping it to the edge of a property line is considered an unfortunate but inescapable reality. In neighboring Wilcox County, one University of Alabama researcher estimates that half a million gallons of untreated sewage enter the county's waterways every day (Jones 2017).

What is the impact of all of this untreated sewage on Alabama's ecosystems? No systematic research on the topic has been published to date. However, news coverage of the situation often shows visuals of surface waters covered with thick algal blooms. These blooms are likely caused by an excess of nutrients, a condition known as eutrophication. As these blooms die, the sinking biomass will be consumed by bacteria in a process that can drop dissolved oxygen to critical levels. Low dissolved oxygen levels (a condition known as hypoxia) can lead to fish kills and other die-offs in aquatic life (Mallin and McIver 2012). Failing septic systems have also been linked to local elevations in dissolved nitrate concentrations (Katz et al. 2011), a contaminant that is "probably carcinogenic to humans" when ingested (International Agency for Research on Cancer 2010). Research has also pointed to a link between fecal contamination and the potential for increased antibiotic resistance in aquatic microorganisms, the long-term effects of which are uncertain but concerning (Graves et al. 2002). In aggregate, ecosystems subjected to a sustained exposure to leachate from failed septic systems can suffer a loss of biodiversity and a reduction of critical ecosystem services (Withers et al. 2014).

In 1999, with concerns mounting, officials from the Alabama Department of Public Health began criminally prosecuting dozens of noncompliant households. This campaign ended in public outcry, but not before the arrest of several residents who could not afford to comply (Albuquerque 2011, p. 7; Carrera 2014, chap. 4; Tavernise 2016). In addition to the risk of prosecution, residents with inadequate on-site sanitation face an increased risk of exposure to parasites and pathogens. A recent study of Lowndes County households practicing open sewerage showed that one in three individuals tested positive for hookworm infection (McKenna et al. 2017), a disease long considered to be eradicated in the United States (Pilkington 2017). Even some residents connected to municipal sewerage face exposure, as publicly owned wastewater lagoons and spray fields routinely back up and overflow (Cleek 2015). Other communities that remain unconnected from the municipal system are nevertheless close enough to the treatment works to be directly impacted by overflows (NewsHour 2018). Some worry that a changing climate will exacerbate the situation by bringing more frequent high-intensity rain events to the area (Meza 2018, p. 6).

All of this is unfolding against a backdrop of racial disparity inseparable from the legacies of slavery, lynching, disenfranchisement, and wrongful convictions. The same clay-rich soils that stymie septic systems today encouraged the development of cotton plantations in the 1800s (Pilkington 2017). By 1860, 70% of the population in Lowndes County were enslaved persons (Hergesheimer 1861). Today, a similar percentage of residents identify as Black or African American (U.S. Census Bureau 2018). The demographics in Lowndes County are part of a broader pattern of communities founded on slavery that cuts across the so-called "Black Belt" of the Southern United States. These regions consistently exhibit deeply entrenched poverty and higher mortality rates, especially among Black Americans (Barry-Jester 2017).

Simple, complicated, or complex?

While it is noncontroversial to say that there are problems that need to be fixed in Lowndes County, the exact nature of those problems, their viable solutions, and who the responsible parties are remains embroiled in debate. This debate is an important one to get right, because proposals that mistakenly identify the underlying nature of the problem are destined to fail, potentially in harmful ways. What do we mean by the underlying nature of the problem? At the highest level, we make a fundamental distinction between problems that arise from simple, complicated, and complex systems. These labels are not intended to describe the difficulty of the problem to be solved but rather to characterize the intrinsic properties of the system in which it is embedded.

Simple systems, for instance, are characterized by observable and reliable cause–effect relationships. In general, it is possible to link an effect to a specific cause. In such systems, epistemic uncertainty – the form of uncertainty that comes from a limited understanding of the system – can be reduced to infinitesimal levels through rigorous research and experimentation. Furthermore, a simple system's aleatory uncertainty – the form of uncertainty that comes from random processes in the system – can be set down in definitive mathematical quantities. Since the uncertainty in the system is well characterized, high-fidelity prediction is possible.

Viable solutions to problems in simple systems can often be developed almost entirely "in the lab" with minimal adjustment before deployment "in the field." Moreover, these solutions are generally transferable to similar systems exhibiting similar problem characteristics, leading to effective "best practices." Problems stemming from simple systems are not necessarily "easy" to solve. Indeed, they may be economically expensive, time consuming, and technically challenging. Nevertheless, when sufficient resources are applied, we can have high confidence of success.

Complicated systems, on the other hand, exhibit cause–effect relationships that are more troublesome. In general, these systems do not allow attribution of a particular effect to a particular cause. Rather, they tend to suffer from equifinality, a condition in which there are many possible causal chains that could have led to the same outcome. Our understanding of the system may be improved through careful study, but the level of detailed measurements and experiments required to reach generalizable conclusions may not be feasible. Randomness in the system may also be quantified, but these conclusions are often drawn from and contingent on a partial understanding of the system, so the system may occasionally exhibit surprising and inexplicable behavior.

On the whole, prediction may be sufficiently reliable to describe the most common complicated system behaviors, but it is not to be expected to reliably predict low-probability and potentially high-consequence outcomes. Unlike simple systems, effective solutions for problems in complicated systems are not directly transferable between apparently similar systems. Instead, they often require informed opinion of experts with local and contextual knowledge. Moreover, value judgments are often required to determine the level of redundancy and safety factors that should be applied to compensate for the uncertain risk of low-probability events.

For complex systems, on the other hand, the conventional notion of cause–effect relationships may not be an appropriate framework, since the relationships are often too ambiguous to be useful. Not only is equifinality rampant, but in general it is not practically possible to prespecify a particular system behavior or force it to occur. This is because in complex systems, "surprises" are the rule rather than the exception. Evolution – in both space and time – in complex systems is often described as emergent, a designation that is meant to imply that it cannot be predicted by independent study of its constituent parts.

Typical modes for reducing our epistemic uncertainty about the nature of the systems often fail when they are applied to complex systems. For simple and complicated systems, divide-and-conquer research strategies can be effective. Using this technique, teams study individual system processes in detail and later assemble their findings into a unified model that reproduces system behavior with reasonable accuracy. However, for complex systems, the model that is assembled will generally fail to reproduce system behaviors in any appreciable sense.

It is probably a mistake to aspire to find predictable *solutions* to problems that arise from complex systems. Instead, *adaptive management* is what is reasonably attainable for complex systems. Problems arising from complex systems cannot be resolved with any sort of finality, but they can be mitigated through continuous intervention. Moreover, since the behaviors exhibited by complex systems evolve over space and time, interventions to manage the impacts of these behaviors must be continually revised. To remain effective, management policies must be structured with contingencies that allow them to be periodically revisited and reassessed.

The importance of system identification

Proceeding with the design of an intervention must be undertaken only after sufficient consideration has been given to the identification and characterization of the underlying system (simple, complicated, or complex). The misidentification of a system as simple when it is complicated or complex will lead to an intervention that doesn't fit the actual demands of the system. Far from being just ineffective, such errors in judgment can be harmful. In addition, a misidentification of a simple system as complicated or complex can be just as costly. Approaching a simple system through the lens of complexity will leave many simple problems unresolved, since much time and energy will be unnecessarily lost to overanalysis.

So what type of system is producing the problems we see in Lowndes County, Alabama? We will eventually argue that it is a complex system. However, to make our argument, we will proceed by first assuming the system is simple and show what sort of conclusions and interventions such an assumption might lead to. We will then highlight where these assumptions fall short, repeating the exercise again on the assumption that the system is complicated. In turn, we will show how this assumption of a complicated system may also fall short and why we characterize the Lowndes County system to be indeed complex. We will conclude with a brief discussion of what this characterization of complexity means and how such complex problems should be approached, a topic we will give more treatment to in Chapter 3 (i.e., utilizing problem-driven interdisciplinary approaches) and Chapter 4 (i.e., applying principled pragmatism).

Seeing sanitation in Lowndes County as a simple system

At the outset, we would like to emphasize again that to describe a system as *simple* is not to imply anything about the *difficulty* of solving the problems that arise from it. A problem arising from a simple system can be very hard to solve in practice, even when reliable interventions are available. According to global estimates assembled by the World Health Organization, only two out of every five people across the globe have access to "safely managed" sanitation services (WHO 2017, p. 4). The problems that leave 60% of the world population without access to high-quality sanitation services are manifold in nature, as are the systems that they are embedded in. These problems are persistent, even though not all of them arise from complicated or complex systems. Some of these failing sanitation systems only suffer from a lack of capital investment and could be remedied with transferable interventions that are well understood. Nevertheless, investment bottlenecks mean that problems with straightforward solutions can still be difficult to address.

Let's assume that the failing sanitation system in Lowndes County, Alabama, is an example of a simple system. For a simple system, it should be possible to provide a clear "root cause" diagnosis of the problem, articulate scalable and repeatable interventions, and predict the outcomes of interventions with high accuracy.

One such description might be as follows: only 20% of Lowndes County residents are served by municipal sewerage, and the remainder are responsible by law for acquiring and operating an on-site sewage treatment and disposal system. Nearly all of these systems are suspected to be absent, inadequate, or failing. With an estimated 17% of the world's population using septic systems (WHO 2017, p. 16), the design of functional systems, even in challenging environments, is well understood. In the case of Lowndes County, the local clay-rich soils do not permit sufficient percolation of septic system leachate. However, since the 1970s, designs for "mound" septic systems have been available that utilize imported fill (usually sand) to overcome the limitations of native soils (Bouma et al. 1975). All of this excavation and hauling of material comes at a high cost, which points to why so few systems have been built to the adequate specification. The problem in Lowndes County is therefore not a lack of know-how, but insufficient investment.

On this view, the sanitation crisis in Lowndes County is widespread but not systemic. It can be addressed through isolated and repeatable investments in adequately designed septic systems. The state has already tried to enforce that homeowners make such investments through the threat of arrest (Albuquerque 2011, p. 7; Carrera 2014, chap. 4; Tavernise 2016), but in a county where the up-front capital costs of a "mound" system often exceed a household's annual income, there is little that punitive measures can achieve except amplifying distress and hardship. Instead, the government could better meet its compliance goals by developing a mechanism for amortizing and subsidizing the high capital costs of these systems to make them more affordable. Accounting for the environmental and public health impacts of inadequate sanitation lends additional credibility to the argument for public assistance in the provision of these systems.

How much capital would need to be raised? Recall our assumption is that the system is simple and that our interventions (i.e., installing "mound" septic systems) are isolated and repeatable. On this basis, the total capital required is simply the per-household cost of a "mound" septic system multiplied by the number of systems installed. The most recent available census data suggests there are approximately 5,200 housing units in Lowndes County (U.S. Census Bureau 2018). An estimated 80% of those housing units are currently not tied into municipal sewerage, leaving about 4,200 housing units in need of some sort of on-site sewage treatment and disposal. Since there is a high degree of uncertainty as to how many of these systems are failing, one can derive a worst-case estimate by assuming that all of them are in need of replacement. Moreover, because there is little information available regarding site variability, we will also assume that each site will require the high-end estimate of $30,000 per septic system. Under these assumptions, the only thing standing in the way of achieving universal access to adequate sanitation in Lowndes County is a capital investment of $126 million.

Is $126 million a feasible figure when it is nearly 28 times the county's $4.5 million annual general budget (Association of County Commissions of Alabama 2016, p. 54)? With an expected lifetime of 20 years (U.S. Environmental Protection Agency 1999), a septic system represents a long-term infrastructure investment. Municipalities in the United States routinely raise funds for such investments by issuing debt instruments known as general obligation bonds. In exchange for up-front capital, municipalities agree to issue amortized principal and interest payments to bondholders over the lifetime of the bond. These are backed by the authority of the municipality to raise funds through taxation if it can't meet its obligations by other means.

Lowndes County has issued such bonds in the past. Moody's Investors Service most recent bond rating for the county was in 2013, when its 2011 series bonds were downgraded from Aa3 to A2 (Moody's Investors Service 2013). Assuming this A2 rating still stands, we only need to choose a bond length to estimate the effective interest rates the county will have to yield to investors in order to be attractive in today's market. Let us assume a bond length of 20 years so that the fulfillment of the county's debt obligations will coincide with the expected lifetime of the septic systems and the need for another major round of investment in sanitation infrastructure. As of June 2019, the 20-year municipal bond yield for A2 ratings was about 2.5%. Neglecting the overhead expenses associated with issuing bonds, this suggests that the county should be able to raise capital at the cost of 2.5% interest, compounded annually over 20 years.

If the county were successful in raising the full $126 million, it would need to find $8 million a year to service this debt (assuming the bond pays out in equal semi-annual installments over 20 years). This debt service would be nearly double the county's current general operating budget of $4.5 million. However, if a household sanitation fee was assessed to all 5,200 households in Lowndes County, the $8 million per year could be raised at a household cost of less than $5 per day. While this is more than double the $2 per day that a household in Jacksonville, Mississippi, would pay to cover 400 gallons worth of daily sewer fees, it is less than the $7 per day a comparable household would pay in Atlanta, Georgia (Circle of Blue 2010).

When we start from the premise that the system is simple, arriving at a conclusion like the one above is straightforward and rational. Indeed, the argument reads like a long-form answer to a homework problem in an engineering textbook. While there is room to quibble over overlooked alternatives (e.g., composting toilets) or whether the financing strategy is feasible (e.g., can a county of 11,000 raise $126[1] million in debt financing?), if you accept the premise that a lack of investment in existing technology is the root of the problem, it is not hard to see how in the matter of a few paragraphs, an approximate solution can be worked out.

Of course, if the premise is wrong – if the problem isn't merely due to a lack of investment in existing technology – carrying out the prescription would be devastating for the county. The goalpost of achieving sanitation goals would end up further out of reach than ever before. Residents would find themselves with the burden of financing the county's debt while forgoing the promised quality-of-life improvements. Moreover, any previously earned trust and willingness to participate in public sanitation efforts would likely be decimated.

Seeing sanitation in Lowndes County as a complicated system

We've already foreshadowed that we will argue against the premise of the previous section. We will attempt to do so without implying that there is inherent virtue in what may appear to be a more sophisticated and nuanced description of the problem. Effective problem solving is about correctly identifying the nature (e.g., simple, complicated, or complex) of the underlying system and devising appropriate responses in light of it. Some water systems are in fact simple, and to categorize them as complicated or complex can prevent otherwise actionable interventions from moving forward.

In the case of Lowndes County, however, we believe it is a mistake to see the county's sanitation system as a simple one. In this section, we will entertain the idea that the system might instead be complicated as well as explore the differences in approach this classification entails.

To see the sanitation system in Lowndes County as a complicated system is first and foremost to embrace the idea that there is no singular "root cause" for the sanitation crisis. Underinvestment is a factor, but it is one among many. Indeed, one homeowner reported that "he could afford any system, yet he could not find one that actually worked" (Subcommittee on Water Resources and Environment 2019). Stories of fly-by-night contractors (Carrera 2014,

p. 114) and systems that never functioned as promised (Smith 2017) show the risks faced by homeowners who do manage to raise sufficient funds for a new system. Accusations of "willful negligence" have been lodged against a local engineering firm for failing to deliver results on a $4.8 million public sanitation contract in nearby Perry County (Cleek 2015). Such stories cast doubt on whether there is sufficient oversight capacity at the county or state level to ensure that a systematic septic system replacement effort would not fall victim to deceptive practices.

Indeed, even the publicly owned infrastructure in Lowndes is suffering. Consider the presumably fortunate 20% of Lowndes County residents who are connected to municipal sewerage. One researcher estimated that nearly a third of these sewer-connected households have regular backups that expose them to risks comparable to those faced by residents with absent or failing septic systems. The researcher concluded: "this finding is meaningful, as it shows the complicated nature of raw sewage exposure and wastewater treatment in Lowndes County" (Meza 2018, p. ii). While it is doubtful the author is using "complicated" with the specific meaning we are here, the general sentiment holds: insufficient investment in on-site wastewater treatment and disposal is not the "root cause" of the sanitation crisis in Lowndes.

Just as there is no single cause of the sanitation crisis, there will be no single remedy. As the *Montgomery Advertiser* notes, "a one-size-fits-all solution won't work" (Brown 2019). Instead, adopting a complicated vision of the system makes space for local experimentation. In the words of local activist Catherine Coleman Flowers, "we want to test out different ideas" rather than "put something down and produce a lot of them, and if it doesn't work, we've lost the money" (Young 2016). This call to action has been heeded by several major universities, including the University of Alabama, the University of Southern Alabama, Michigan State University, Georgia Tech University, and Duke University.

Among the early insights developed through these research partnerships was that, while rural, the housing density in Lowndes County is far from uniformly dispersed. Indeed, housing tends to be clustered, with several segments of an extended family living on a single parcel. Groups of three or more manufactured homes on a single lot are common (Meza 2018, p. 7). This clustering introduces challenges as well as opportunities. On the one hand, it exacerbates the demand that will be placed on any on-site wastewater treatment options; on the other, it offers an opportunity to pool investment resources in a way that makes previously unconsidered treatment options feasible. In particular, this clustered scale of treatment may benefit from technologies being developed as part of military and hydrofracking operations (Brown 2019).

Much of this research remains speculative, since as one University of Alabama engineering professor put it, "there are a lot of anecdotal stories but a lack of definition about how big this problem is" (Brown 2019). The Alabama Department of Public Health's low-end estimate is that 40% of households lack a functional on-site wastewater treatment system, but their high-end estimate of 90% is more than double that figure (Albuquerque 2011, p. 6). However, this uncertainty in scale is merely emblematic of the deeper and more challenging lack of local knowledge. Recall that in complicated systems, problems do not clearly arise from a single cause. Merely observing a problem (e.g., a failing septic system) is not a sufficient basis for prescribing an intervention. Instead, an understanding of locally relevant factors is required to ensure the intervention is appropriate to the specific case.

Research groups from several different universities have conducted independent house-to-house surveys to gather data on demographics and the condition of on-site treatment systems in Lowndes and the surrounding areas (Elliott et al. 2017; Meza 2018). However, since "site surveys are expensive and time-consuming," universities engaged in research on the problem in Lowndes County are trying to develop localized insights by exploiting publicly available datasets (Elliott 2019). Researchers at the University of Alabama are hopeful that by leveraging "engineering and geology methods," they can create maps and a geospatial database that will allow them to "start prepping reasonable solutions" (Brown 2019).

For problems arising from simple systems, established solutions are highly transferable. For complicated systems, problems generally require solutions that are tailored to fit local conditions using expert knowledge. If this knowledge is absent, it will take time and effort to develop, so it is sensible that as of yet, there is no actionable guidance for Lowndes County or its residents. Since there can be a variety of causal elements leading to the same apparent outcome, detailed site-level knowledge can lead to better outcomes. However, an approximate understanding of local conditions, such as those derived from geospatial datasets, may be sufficient to produce adequate solutions.

To compensate for this approximate knowledge, such interventions incorporate safety factors, such as oversizing and redundancy. Since this "overdesign" comes at a cost, it is often infeasible to attempt to eliminate all risk of failure. Instead, designers seek to quantify the risk of failure and develop solutions that provide a certain *level of service*. A level of service is a quantitative measure that can take many forms. In the water systems literature, levels of service have been expressed in terms of maximum allowable "downtime" of a public water supply (Asefa et al. 2015), the maximum volume of contaminated water allowed to enter a public water supply prior to detection (Kessler et al. 1998), and the allowable level of departure from optimal levels of chlorine residuals in a water distribution system (Li et al. 2015). Demonstrating that a problem in a complicated system has been "solved" is often reducible to demonstrating that the desired level of service has been met.

Of course, establishing the desired level of service requires value judgments. These judgments are often codified in government regulations and standards set by intergovernmental bodies. This codification gives the subjective notion of "adequate" an apparently objective and measurable basis. As an example close to home for the authors, the City of Cambridge, Massachusetts, is not currently required to eliminate all raw sewage discharge into the Charles River from its Mount Auburn outfall. Its obligations under the Clean Water Act are considered to be met so long as it discharges fewer than four times a year and with a total volume of less than 0.84 million gallons (U.S. Environmental Protection Agency 2009). On one hand, the city's discharge permit establishes a legally binding requirement for a minimum level of service. On the other hand, as long as the level of service is met, the city enjoys a certain degree of impunity to external objections that suggest the problem of raw sewage discharge is "unsolved."

The problematics of establishing an acceptable level of service also appears in the setting of international standards by intergovernmental bodies. For example, the World Health Organization's highest attainable service level for water and sanitation is referred to as "safely managed." To achieve a "safely managed" level of service requires the "use of improved facilities that are not shared with other households and where excreta are safely disposed of in situ or transported and treated offsite" (WHO 2017, p. 8). Included in this list of qualifying "improved facilities" are "composting toilets" and "pit latrines with slabs" (WHO 2017, p. 8). While it is unlikely researchers involved in the Lowndes County sanitation crisis would recommend pit latrines as a viable option, the idea of composting toilets has already been raised. Indeed, the candidacy of composting toilets was only redacted after a local activist explained to researchers that this option was "not far from the outhouses of her youth" (Weinthal et al. 2018).

That activist, Catherine Coleman Flowers, grew up in Lowndes County and is one of the leading voices calling for creative collaborations on the sanitation crisis. As the face of the Alabama Center for Rural Enterprise, Flowers has become the de facto representative of county residents in collaborations with "historically white and Southern academic institution[s]" (Weinthal et al. 2018). While Flowers's efforts are admirable, her emergence as a figurehead and apparent sole interface between the impacted community and the researchers from far-flung universities highlights the absence of a robust forum for stakeholder engagement on the issue. Even if only from a practical standpoint, the direct engagement in stakeholders on sanitation issues is critical to achieving sustainable and equitable outcomes. Global accounting of sanitation projects suggests that between 30% and 70% fail within the first several years, with insufficient stakeholder participation held as a key contributing factor (Morales et al. 2014, p. 2817). As we will detail in the subsequent section, the coupled natural and social dynamics of the Lowndes County sanitation crisis give rise to a complexity that is likely to be answered with inclusive and sustained engagement of stakeholders (e.g., in fact–value deliberation, joint fact-finding, collective decision making, and adaptive management).

Seeing sanitation in Lowndes County as a complex system

Seeing sanitation in Lowndes county as a simple system may lead one to the tempting but misguided conclusion that the crisis can be settled for $5 per household per day. Seeing the system as complicated leads one to be optimistic about the application of locally tailored technologies designed by universities according to international standards. On the other hand, to identify and characterize sanitation in Lowndes County as a complex system is to recognize not only the presence of critical social dynamics but their inseparability from the natural dynamics that those researchers are primarily trying to address. This inseparability implies that no matter how advanced, bespoke, or affordable these technologies are, they are unlikely to succeed if not developed in direct and ongoing collaboration with the impacted communities.

Empirical evidence for this claim is found widely in the international planning and development literature (Seppälä 2002; Lüthi et al. 2010; van Vliet et al. 2011; Hendriksen et al. 2012; Lopes et al. 2012). Despite what seems to be a widespread understanding, many projects continue to move forward without stakeholder engagement or try to bring stakeholders in only at the last moment. When these "technically sound" projects fail, the blame is commonly passed on to the communities that are on the receiving end of the intervention. This is often done subtly, and perhaps unconsciously, by generically blaming the failures on "contexts" and "social factors." In the spirit of the proverb that "there's no accounting for tastes," post-hoc evaluations often point to "something strange and exotic in the community's values or institutions which can account for their lack of cooperation" (Cairncross and Feachem 2005, p. 153).

Of course, with this sleight of hand, those responsible for the project have obscured (even, perhaps, from themselves) the fact that in designing their intervention, they have brought their own perspectives, interests, and values to bear on the problem, as objective as they purport to be (Cairncross and Feachem 2005, p. 153). What is exotic to the circumstances of the problem, then, is not the values of the communities receiving the intervention but the values of the project planners and administrators. Recall that of the universities engaged in active research on Lowndes – the University of Alabama, the University of Southern Alabama, Michigan State University, Georgia Tech University, and Duke University – the majority are outside of Alabama, and those that are in Alabama are at least a two-hour drive away from Lowndes County.

We do not intend to malign these universities or their researchers in any way. Rather, we wish to express concern that without inclusive, early, and persistent stakeholder engagement, the implicit values, interests, and perspectives of the researchers may lead them, like so many projects before, to invest in "solutions" that are not viable. Moreover, we do not wish to suggest that viability of projects is a simple matter of aligning the values of one community with another. Stakeholder engagement is also a matter of dignity, agency, and power redistribution. Consider, for a moment, the power asymmetries established and reaffirmed when financial support flows to the universities researching problems in distant communities (Brown 2019; Elliott 2019) rather than going to the communities themselves.

None of this should suggest that researchers at these universities or anywhere else are acting with anything but the best of intentions. However, without ongoing stakeholder consultation and a co-production of knowledge, it is easy to be misled about the actual requirements of an intervention. For example, researchers in economics have often suggested that flat-rate schemes (e.g., in water supply and sanitation) should be avoided, in part because they disproportionately burden the poor. However, it may be that those same communities that the economists were trying to save from the burden of "flat rates" will decide to embrace them, since "a commitment to paying the same as everybody else may be a statement of political equality" (Curtis 1995, pp. 117–118). In the same sense, while sanitation researchers universally acknowledge the importance of human dignity in their work, dignity is often portrayed as "a question of technology and the aesthetics of its individual experience" (Redfield and Robins 2016, p. 153). While there is no doubt providing a "dignified" experience is important, researchers often overlook the fact that for those they are trying to help, "dignity remains a matter of political relations" (Redfield and Robins 2016, p. 153). Rather than looking for the deliverance of a "technology" they may, more importantly, be looking for a deliverance of "justice" (Redfield and Robins 2016, p. 153).

What should be drawn from these lessons is this: even the most well-intentioned received wisdom (e.g., that the poor should be sheltered from flat taxes; that dignity is about user experience) is an inadequate substitute for knowledge co-produced in partnership with the community. After all, to "accept" sanitation technology that has not been co-produced is to effectively yield one's political leverage in the matter (Redfield and Robins 2016, p. 157). If the nature of the technology is significantly different than what would be considered mainstream, those accepting the technology risk losing one more "form of belonging" in society (Anand 2011, p. 543). In this sense "the problem of sanitation simultaneously connects and divides human populations. It unites them at a species level only to distinguish them at a social one" (Redfield and Robins 2016, p. 146).

Ideally, inclusive approaches that engage stakeholders in joint fact-finding and in long-term collaborative adaptive management can create space for the open consideration of these issues in a way that does not reduce stakeholders' dignity, agency, or political leverage. In such a space, there is room for scientific understanding, local knowledge, and collective decision making on issues such as acceptable risk. There is a great deal of literature exploring both the theory and practice of such models, including several "handbooks" [e.g., Susskind et al. (1999), Gastil and Levine (2005)]. Rather than attempt to provide a definitive summary of this substantive body of literature, we will use the space of the next two chapters to elevate two key aspects of this type of collaboration that we feel have been underdeveloped: problem-driven interdisciplinarity (Chapter 3) and principled pragmatism (Chapter 4).

Closing remarks

We have used the ongoing sanitation crisis in Lowndes County, Alabama, as an example of a problem that has arisen out of a coupling of natural and human systems. We've argued that this coupling creates intrinsic complexity that cannot be adequately approached by methods and techniques developed for complicated or simple systems. We've attempted to avoid describing the attributes of complex systems purely in the abstract (e.g., as systems that possess nonlinear boundary-crossing feedbacks at multiple scales, etc.). Too often these abstract descriptions lead to semantic debates about what is and isn't complex. While valuable in some academic contexts, within the scope of the present work, these debates serve as a distraction from the more important question: why do such distinctions matter?

In this chapter, we have endeavored to explore this question. Our key takeaway is this: the way that we characterize a system shapes our response to its problems. If we see the system as simple, we expect that we can rely on experts (or the marketplace) to develop repeatable solutions that treat root causes and adopt widely available and tested "best practices." If we identify the system as complicated, we seek diverse expertise to create locally tailored solutions that meet a required level of service. If we identify the system as complex, we recognize that expertise is no substitute for the knowledge and understandings derived from inclusive, early, and ongoing stakeholder engagement with an explicit recognition of the capacity and constraints imposed by the context. Misidentification of a system in either direction (i.e., simple as complex or complex as simple) can lead not only to the misallocation of time, treasure, and talent but also to outcomes that are harmful to the communities we are trying to serve.

While we focused primarily on the importance of stakeholder engagement as a response to complexity in this chapter, it is only one aspect of the principled and pragmatic approach to problem-driven interdisciplinary collaboration that we advocate for going forward. Our aim in this chapter was to introduce readers to the idea that natural and human systems can become coupled in complex ways and to highlight the stark differences in the ways researchers and practitioners might respond when a system is identified as simple, complicated, or complex. Having done so, Chapter 3 will transition into a discussion about the importance of problem-driven interdisciplinarity as a response to the complexity of coupled natural and human systems.

Note

1 All currencies mentioned in this book denote US$.

References

Alabama State Code § 22-26-1, 2014.
Alabama State Code § 22-26-2, 2016.
Alberti, M., Asbjornsen, H., Baker, L.A., Brozovic, N., Drinkwater, L.E., Drzyzga, S.A., Jantz, C.A., Fragoso, J., Holland, D.S., Kohler, T.A., Liu, J., McConnell, W.J., Maschner, H.D.G., Millington, J.D.A., Monticino, M., Podestá, G., Pontius, R.G., Jr., Redman, C.L., Reo, N.J., Sailor, D., and Urquhart, G., 2011. Research on Coupled

Human and Natural Systems (CHANS): Approach, challenges, and strategies. *Bulletin of the Ecological Society of America*, 92 (2), 218–228.

Albuquerque, de C., 2011. *Report of the Special Rapporteur on the Human Right to Safe Drinking Water and Sanitation*. United Nations, No. A/HRC/18/33/Add.4.

Alston, P., 2018a. *Oral Statement by Mr. Philip Alston Special Rapporteur on Extreme Poverty and Human Rights*. United Nations. Available from: https://www.ohchr.org/EN/News Events/Pages/DisplayNews.aspx?NewsID=23243

Alston, P., 2018b. *Report of the Special Rapporteur on Extreme Poverty and Human Rights on His Mission to the United States of America*. United Nations, No. A/HRC/38/33/Add.1.

Anand, N., 2011. Pressure: The politechnics of water supply in Mumbai. *Cultural anthropology: Journal of the Society for Cultural Anthropology*, 26 (4), 542–564.

Asefa, T., Adams, A., and Wanakule, N., 2015. A Level-of-service concept for planning future water supply projects under probabilistic demand and supply framework. *Journal of the American Water Resources Association*, 51 (5), 1272–1285.

Association of County Commissions of Alabama, 2016. *Comparative Data on Alabama Counties*. 9th Edition.

Barry-Jester, A.M., 2017. Patterns of death in the south still show the outlines of slavery. *FiveThirtyEight*. Available from: https://fivethirtyeight.com/features/mortality-black-belt/

Bechtel, W., and Richardson, R.C., 1992. Emergent phenomena and complex systems. *In*: Ansgar Beckermann Hans Flohr, ed. *Emergence or Reduction? Essays on the Prospects of Nonreductive Physicalism*. Berlin: Walter de Gruyter Verlag, 257–288.

Bouma, J., Converse, J.C., and Otis, R.J., 1975. A mound system for onsite disposal of septic tank effluent in slowly permeable soils with seasonally perched water tables. *Journal of Environmental Quality*, 4 (3), 382–388.

Brown, M., 2019. With new grant, UA team mapping wastewater woes in Black Belt. *Montgomery Advertiser*, 26 Apr.

Brown, R., 1963. *Explanation in Social Science*. Chicago: Aldine Publishing Company.

Cairncross, S., and Feachem, R.G., 2005. *Environmental Health Engineering in the Tropics: An Introductory Text*. 2nd ed. Chichester: Wiley.

Carrera, J., 2014. *Sanitation and Social Power in the United States*. Urbana and Champaign: University of Illinois.

Chen, J., and Liu, Y., 2014. Coupled natural and human systems: A landscape ecology perspective. *Landscape Ecology*, 29 (10), 1641–1644.

Circle of Blue, 2010. Wastewater treatment costs in major U.S. cities. *Water News*.

Cleek, A., 2015. Filthy water and shoddy sewers plague poor Black Belt counties. *Al Jazeera America*.

Corning, P.A., 2002. The re-emergence of 'emergence': A venerable concept in search of a theory. *Complexity*, 7 (6), 18–30.

Curtis, D., 1995. Power to the people: Rethinking community development. *In*: N. Nelson and S. Wright, eds. *Power and Participatory Development*. London: Intermediate Technology Publications, 115–124.

Elliott, M., 2019. *EPA Grant Assists in Understanding Wastewater Issues in Rural Alabama*. University of Alabama News Center. Available from: www.ua.edu/news/2019/04/epa-grant-assists-in-understanding-wastewater-issues-in-rural-alabama/

Elliott, M., Das, P., Blackwell, A., Aytekin, E., Hu, Y., White, K., Jones, R., and Lu, Y., 2017. *Surface Discharge of Raw Wastewater Among Unsewered Homes in Central Alabama*. Presented at the AGU Fall Meeting.

Flowers, C., 2018. A county where the sewer is your lawn. *The New York Times*, 22 May.

Fromm, J., 2005. Types and forms of emergence. *arXiv [nlin.AO]*.

Gastil, J., and Levine, P., eds., 2005. *The Deliberative Democracy Handbook: Strategies for Effective Civic Engagement in the Twenty-First Century.* San Francisco: Jossey-Bass.

Gieryn, T.F., 1983. Boundary-work and the demarcation of science from non-science: Strains and interests in professional ideologies of scientists. *American Sociological Review,* 24 (6), 781–795.

Graves, A.K., Hagedorn, C., Teetor, A., Mahal, M., Booth, A.M., and Reneau, R.B., Jr, 2002. Antibiotic resistance profiles to determine sources of fecal contamination in a rural Virginia watershed. *Journal of Environmental Quality,* 31 (4), 1300–1308.

He, J., Dougherty, M., Zellmer, R., and Martin, G., 2011. Assessing the status of onsite wastewater treatment systems in the Alabama Black Belt soil area. *Environmental Engineering Science,* 28 (10), 693–699.

Hendriksen, A., Tukahirwa, J., Oosterveer, P.J.M., and Mol, A.P.J., 2012. Participatory decision making for sanitation improvements in unplanned urban settlements in East Africa. *Journal of Environment & Development,* 21 (1), 98–119.

Hergesheimer, E., 1861. *Map Showing the Distribution of the Slave Population of the Southern States of the United States.* Compiled from the Census of 1860.

Holland, J.H., 2002. Complex adaptive systems and spontaneous emergence. *In:* A.Q. Curzio and M. Fortis, eds. *Complexity and Industrial Clusters.* Heidelberg: Physica-Verlag HD, 25–34.

International Agency for Research on Cancer, 2010. *Ingested Nitrate and Nitrite and Cyanobacterial Peptide Toxins.* IARC Monographs on the Evaluation of Carcinogenic Risks to Humans, 94.

Islam, S., and Choudhury, E., 2018. Complexity and contingency: Understanding transboundary water issues. *In:* E. Choudhury and S. Islam, eds. *Complexity of Transboundary Water Conflicts: Enabling Conditions for Negotiating Contingent Resolutions.* New York: Anthem.

Izenberg, M., Johns-Yost, O., Johnson, P.D., and Brown, J., 2013. Nocturnal convenience: The problem of securing universal sanitation access in Alabama's Black Belt. *Environmental Justice,* 6 (6), 200–205.

Jones, A., 2017. *A Raw Deal.* University of Alabama News. Available from: www.ua.edu/news/2017/12/a-raw-deal/

Katz, B.G., Eberts, S.M., and Kauffman, L.J., 2011. Using Cl/Br ratios and other indicators to assess potential impacts on groundwater quality from septic systems: A review and examples from principal aquifers in the United States. *Journal of Hydrology,* 397 (3), 151–166.

Kessler, A., Ostfeld, A., and Sinai, G., 1998. Detecting accidental contaminations in municipal water networks. *Journal of Water Resources Planning and Management,* 124 (4), 192–198.

Li, C., Yu, J.Z., Zhang, T.Q., Mao, X.W., and Hu, Y.J., 2015. Multiobjective optimization of water quality and rechlorination cost in water distribution systems. *Urban Water Journal,* 12 (8), 646–652.

Liu, J., Dietz, T., Carpenter, S.R., Alberti, M., Folke, C., Moran, E., Pell, A.N., Deadman, P., Kratz, T., Lubchenco, J., Ostrom, E., Ouyang, Z., Provencher, W., Redman, C.L., Schneider, S.H., and Taylor, W.W., 2007. Complexity of coupled human and natural systems. *Science,* 317 (5844), 1513–1516.

Lopes, A.M., Fam, D., and Williams, J., 2012. Designing sustainable sanitation: Involving design in innovative, transdisciplinary research. *Design Studies,* 33 (3), 298–317.

Lüthi, C., McConville, J., and Kvarnström, E., 2010. Community-based approaches for addressing the urban sanitation challenges. *International Journal of Urban Sustainable Development,* 1 (1–2), 49–63.

Mallin, M.A., and McIver, M.R., 2012. Pollutant impacts to Cape Hatteras National Seashore from urban runoff and septic leachate. *Marine Pollution Bulletin*, 64 (7), 1356–1366.

McKenna, M.L., McAtee, S., Bryan, P.E., Jeun, R., Ward, T., Kraus, J., Bottazzi, M.E., Hotez, P.J., Flowers, C.C., and Mejia, R., 2017. Human intestinal parasite burden and poor sanitation in rural Alabama. *The American Journal of Tropical Medicine and Hygiene*, 97 (5), 1623–1628.

Meza, E., 2018. *Examining Wastewater Treatment Struggles in Lowndes County, AL*. Master of Environmental Management, Duke University.

Moody's Investors Service, 2013. *Rating of Lowndes County, Alabama General Obligation Warrants, Series 2011*. Available from: https://www.moodys.com/

Morales, M., del, C., Harris, L., and Öberg, G., 2014. Citizenshit: The right to flush and the urban sanitation imaginary. *Environment & Planning A*, 46 (12), 2816–2833.

Morello, C., 2018. U.S. withdraws from U.N. Human Rights Council over perceived bias against Israel. *The Washington Post*, 19 June.

NewsHour, P.B.S., 2018. The story of American poverty, as told by one Alabama county. *PBS NewsHour*. Available from: www.pbs.org/newshour/show/the-story-of-american-poverty-as-told-by-one-alabama-county

Pilkington, E., 2017. Hookworm, a disease of extreme poverty, is thriving in the US South. Why? *The Guardian*, 5 Sept.

Redfield, P., and Robins, S., 2016. An index of waste: Humanitarian design, 'dignified living' and the politics of infrastructure in Cape Town. *Anthropology Southern Africa*, 39 (2), 145–162.

Seppälä, O.T., 2002. Effective water and sanitation policy reform implementation: Need for systemic approach and stakeholder participation. *Water Policy*, 4 (4), 367–388.

Smith, C., 2017. Meet the Americans who live with open sewers in their yard. *Huffington Post*, 15 Dec.

Subcommittee on Water Resources and Environment, 2019. *The Clean Water State Revolving Fund: How Federal Infrastructure Investment Can Help Communities Modernize Water Infrastructure and Address Affordability Challenges*. Washington, D.C.: U.S. Government Publishing Office.

Susskind, L.E., McKearnen, S., and Jennifer, T.L., 1999. *The Consensus Building Handbook: A Comprehensive Guide to Reaching Agreement*. Thousand Oaks: SAGE Publications.

Tavernise, S., 2016. A toilet, but no proper plumbing: A reality in 500,000 U.S. homes. *The New York Times*, 26 Sept.

U.S. Census Bureau, 2018. *QuickFacts: Lowndes County, Alabama*. Available from: www.census.gov/quickfacts/fact/table/lowndescountyalabama,US/PST045218

U.S. Environmental Protection Agency, 1999. *Decentralized Systems Technology Fact Sheet: Septic Tank – Soil Absorption Systems*. No. EPA 932-F-99-075.

U.S. Environmental Protection Agency, 2009. *Authorization to Discharge Under the National Pollutant Discharge Elimination System Copy*. No. MA0101974.

van Vliet, B.J.M., Spaargaren, G., and Oosterveer, P., 2011. Sanitation under challenge: Contributions from the social sciences. *Water Policy*, 13 (6), 797–809.

Weinthal, E., Albright, E.A., Flowers, C.C., and Stewart, E., 2018. *Solution-centered Collaborative Research in Rural Alabama*. Social Science Research Council. Available from: https://items.ssrc.org/insights/solution-centered-collaborative-research-in-rural-alabama/

WHO, 2017. *Progress on Drinking Water, Sanitation and Hygiene: 2017 Update and SDG Baselines*. World Health Organization.

Winkler, I.T., and Flowers, C.C., 2017. America's dirty secret: The human right to sanitation in Alabama's Black Belt. *Columbia Human Rights Law Review*, 49, 181.

Withers, P.J.A., Jordan, P., May, L., Jarvie, H.P., and Deal, N.E., 2014. Do septic tank systems pose a hidden threat to water quality? *Frontiers in Ecology and the Environment*, 12 (2), 123–130.

Young, M., 2016. Experts discuss solutions for Lowndes County sewage issues. *WSFA12 News*, 6 Nov.

3 Working together

An argument for problem–driven interdisciplinary collaboration

Kevin M. Smith and Shafiqul Islam

Problems that cut across boundaries

The lines that demarcate *environmental problems* from *societal problems* are getting thinner. Processes such as industrialization, urbanization, and agricultural intensification muddle our ability to partition phenomena into neat categories such as *natural* or *human-made*. Indeed, a growing body of literature suggests that many natural and human systems have become coupled in ways that are intrinsically *complex*. In the previous chapter, we explored the significance of this complexity and why such a distinction matters in terms of how researchers and practitioners approach their work. Briefly: systems that are complex present unique challenges, such as irreducibility, unpredictability, and noncontrollability. These properties arise from scale-dependent interactions of coupled processes whose variety easily exceeds the reach of a single discipline.

We subscribe to the contemporary view that these are not the sort of challenges that can be answered by disciplines working in isolation. We also recognize and highlight that such issues "cut right across the boundaries of subject matter or discipline" (Popper 2002, p. 88). A key question, as we see it, is not whether these challenges demand an interdisciplinary approach but rather which modes of interdisciplinary collaboration will achieve actionable outcomes that matter and are effective?

Today the dominant mode of interdisciplinary collaboration is what we would call primarily theory driven. It starts from the premise that the problem under study is systemic in nature. In turn, it is taken for granted that solving the problem requires a theory that can predict or at the very least explain the behavior of the system. On this view, the impetus for interdisciplinary collaboration is the recognition that "a full predictive or even descriptive understanding requires the use of many disciplines" (Institute of Medicine et al. 2005, p. 30). The hopeful outcome is that the theories from various disciplines can be synthesized into a theory of the whole. Interdisciplinary collaboration of this sort has had a key role in the rapid growth of technological innovation in the last half century (see e.g., Chapter 3 of Institute of Medicine et al. 2005). However, the vast majority of these successes were from the domain of physical and biological systems where prediction and control are possible because system behavior is governed by fundamental laws of nature. We would categorize most of these systems as simple or complicated, but not complex (for details of these distinctions, please see Chapter 2).

If a reliance on underlying laws of nature is prerequisite for theory-driven collaboration, one must question whether such a premise can serve us well in addressing our most pressing social, political, and economic challenges where natural and human systems are coupled. For millennia, humans have made water – in addition to a physical object – a cultural, religious, political, and economic object. As a result, many water issues are inherently complex and undermine reliable prediction and control. There are no generalizable laws that can govern water's access, allocation, use, and transformations. Attribution of causality in such systems is fraught with practical and philosophical difficulties. We challenge the dominant premise that any amount of synthesis and tweaking of disparate subject matter is likely to improve this condition. Rather, in the case of complex water issues, we heed Popper's advice to be "not students of some subject matter, but students of problems" (Popper 2002, p. 88).

In this chapter, we will develop the case for problem-driven collaboration on problems arising from coupled natural and human systems, using water as a motivating example. We will not dwell too long on the value of *interdisciplinarity* as a mode of problem solving itself, as we believe there is widespread and growing agreement on this point. We focus, instead, on what is lacking: an explicit consensus on the modes of interdisciplinarity that will be useful or lead to actionable outcomes in a given circumstance. We argue that the dominant implicit mode of interdisciplinary collaboration is theory driven: it seeks to create some form of generalizable theory that will bridge disciplines. While this is a powerful mode of interdisciplinary practice effective for problems arising in simple and complicated systems, we argue that there is little hope for finding generalizable theory when it comes to problems arising from complexity. Instead, we argue that complexity demands a problem-driven interdisciplinary approach that integrates knowledge – with an explicit recognition of the nature, capacity, and constraints inherent to the chosen problem – for actionable, measurable, and trackable outcomes. In lieu of seeking a shared *theory* to bridge disparate disciplines, we argue that teams should seek shared *frameworks* that facilitate agreement on the nature of the problem context (e.g., simple, complicated, or complex) as well as provide a common language for describing, diagnosing, and resolving problems (e.g., the Water Diplomacy Framework).

Our use of "framework," "theory," and "model" is aligned with the distinctions made by Elinor Ostrom. Briefly, frameworks "organize diagnostic and prescriptive inquiry," "provide a metatheoretical language that can be used to compare theories," and "help analysts generate the questions that need to be addressed when they first conduct an analysis" (Ostrom 1999, p. 25). Theories, on the other hand, "make specific assumptions that are necessary for an analyst to diagnose a phenomenon, explain its processes, and predict outcomes" (Ostrom 1999, p. 25). Models are even more specific, "making precise assumptions about a limited set of parameters and variables" (Ostrom 1999, p. 26). This nesting of specificity means that "several theories are usually compatible with any framework" and "multiple models are compatible with most theories" (Ostrom 1999, pp. 25–26). Ultimately, this nested structure allows experts from across different disciplines to engage in problem-driven collaboration by adopting a common framework. By finding a framework compatible with their disparate theories, collaborators can focus on the problem at hand rather than focus on bringing these theories into agreement. Likewise, the ability of multiple models to be compatible with any given theory indicates that there is space for negotiation and creative thinking in the production of actionable solutions. In this way, we see shared conceptual frameworks (such as the Water Diplomacy Framework) as an effective means of catalyzing interdisciplinary research on complex water issues.

Interdisciplinarity – a warmly persuasive word

In 1976, Raymond Williams published a pithy and well-received history of a few dozen words that hold powerful but often ambiguous cultural significance in the English language. If a similar effort were conducted today, one might reasonably expect to see an entry for *interdisciplinarity*. It appears to have achieved the status of what Williams referred to as a "warmly persuasive word" – one whose particular meaning may be open to debate but whose general virtue is taken for granted in public discourse (Williams 1985, p. 76). In our view, interdisciplinarity is often warranted on pragmatic grounds and requires no strict theoretical support.[1] Garry Brewer put it succinctly: "the world has problems, but universities have departments" (Brewer 1999, p. 327).

However, while we hold that interdisciplinarity is *often necessary*, we also regard it as *rarely sufficient* to resolve contemporary challenges in science and society. In this sense, we emphasize Georges Gusdorf's warning not to be lulled into the view of interdisciplinarity as an "epistemological panacea" (Gusdorf 1977, p. 580) as cited in Graff (2016). Rather, we suggest that interdisciplinarity needs to be regarded as a high-level strategy for innovation; its success or failure depends largely on the details of how it is pursued relative to the context in which it is applied. When it comes to interdisciplinarity, one size does not fit all.

So what sort of interdisciplinarity will help us be effective in addressing problems in the space of coupled natural and human systems? What sort of requirements would it have? First, we note that social and environmental problems have immediate and compounding impacts that demand a rapid response. To this end, any collaboration must yield timely and actionable outcomes. Second, as we saw in the previous chapter, complex systems are open and dynamic. For this reason, we require a form of collaboration that acknowledges the nonfinality and fallibility of its interventions by intentionally incorporating flexibility and capacity for contingent adaptive management. Finally, we must also recognize that the problems we seek to address often have dramatic effects on livelihoods that are not our own as academics or professional practitioners. In this sense, our chosen approach to collaboration must have explicit provision for inclusive stakeholder engagement and be sensitive to differences in culture and values.

All of these requirements, in our view, point to a strategy we call principled and pragmatic problem–driven collaboration. In the remainder of this chapter, we focus on the value of the problem–driven aspect of this strategy. In the next chapter, we will more closely examine the importance of principles and pragmatism in carrying out this work.

Why dichotomize interdisciplinarity?

In order to outline the advantages of a problem–driven approach to collaboration, we will distinguish it from what we call a theory–driven approach. In Chapter 1, we warned against the dangers of dichotomous thinking. Too often we are presented with the false choices between facts or values, numbers or narratives, objectivity or subjectivity. We do not wish to repeat the same mistake here. Indeed, we would be doing a grave injustice to our readers and the interdisciplinary community at large to suggest that these are the only two ways of conceiving of interdisciplinarity or that a choice must be made between one *and* the other. In reality, projects of any significant size or duration may engage in multiple modes of interdisciplinary work (Stokols et al. 2008). Our aim in making a distinction between "theory–driven" and "problem–driven" approaches is to draw attention to what we see as an often-neglected concern: how to structure interdisciplinary work in a way that it respects the constraints imposed by complex systems using water problems as a motivating example.

We do not have the hubris or the ambition to suggest we could provide a thorough treatment of the diverse classifications of interdisciplinary approaches that have been proposed over the last several decades. Instead, we refer readers to Klein (2017) for rigorous but concise analytical comparisons, Chettiparamb (2007) for a comprehensive literature review including key excerpts from widely-cited papers, and Hoffmann–Riem et al. (2008) and Hearn (2003) for nuanced historical accounts of the development of interdisciplinary ideas. These chapters and articles point to the diversity of perspectives that are active in common interdisciplinary practice. However, as many observers note, the particular words used to describe these practices are applied haphazardly, suggesting their significance often goes unspoken or assumed as part of an implicit hierarchy (Graff 2016).

This suggests that many interdisciplinary collaborations proceed without much critical attention to whether the structure, function, and objectives of the team are aligned to the context in which they will be working. For collaborators engaged in problems arising from coupled natural and human systems, we propose a strategy of explicitly aligning the team's objectives around solving particular problems rather than deriving general solutions. On the basis of this alignment, we split the great diversity of interdisciplinary approaches into their "problem-driven" and "theory-driven" variants.

The taxonomic tree of interdisciplinarity bears rich foliage. Our intention in using this dichotomy is not to diminish this diversity. Rather, we are suggesting that a variety of conceptualizations of interdisciplinarity can flourish supported by two distinct trunks, which fork sharply in their intentions. In our assessment, these contrasting objectives – "problem-driven" versus "theory-driven" – have significant implications for interdisciplinary collaborations on important social and environmental issues. Others have noted similar schisms in interdisciplinary intentions, but to our knowledge, they have not drawn explicit attention to the importance of matching a team's objectives to the context of their work. For example, the distinction we have made between "problem-driven" and "theory-driven" interdisciplinarity maps closely to the "instrumental" and "conceptual" divide identified by Liora Slater and Alison Hearn originally in Salter and Hearn (1996) and revisited in Hearn (2003).

Theory-driven collaboration

What we describe as theory-driven collaboration has ontological roots in general notions of the unity of science. The objective of such theory-driven work is "epistemological bridge building" (Chettiparamb 2007) across multiple disciplines to arrive at a more general theory (e.g., molecular biology). For some, it is a matter of repairing a unity that should never have been broken. In the words of one researcher: "the domains into which we have carved up reality can be united under one umbrella" (Gabora and Aerts 2005, p. 81). Central to the unity of science thesis is that phenomena at all levels of organization (particle to universe; individual to population; agent to institution) are governed by common and fundamental laws. While coming to a theory of everything is widely considered intractable,[2] the unity-of-science thesis suggests all specialized theories are susceptible to some form of synthesis into a more general theory.

Theory-driven approaches can work to solve simple and complicated problems

Theory-driven collaboration to bridge disparate fields of knowledge has a history of notable successes, stretching back centuries. Indeed, authors have argued that Kepler's first law of planetary motion was an artifact of such an effort: an innovation requiring synthesis of recent astronomical theories (and data) with theorems from geometry that had been available for thousands of years, but that had gone unrecognized as relevant (Nissani 1995, p. 123). Likewise, Gregor Mendel's work with peas in the late 1800s inspired his contemporaries to work across disciplines to develop generalizable and bridging theories. Together their efforts led to powerful and unified theories that gave rise to the new field of genetics (Nissani 1995, p. 123). A similar story can be seen in the history of molecular biology, a field that was established through collaboration between physicists and biologists who shared dreams of a unified theory (Pohl et al. 2008, p. 413).

The feeling engendered by these successes, past and present, is that "we are on the threshold of forging a new unity in science and engineering, a process that will drive our progress more than ever before" (Colwell and Eisenstein 2001, p. 59). Despite widespread lamenting of siloed thinking, scientists and engineers working on cutting-edge science and technology problems readily acknowledge the need to work outside their narrow fields to arrive at higher levels of prediction and control. Indeed, those who study management and innovation in detail point to theory-driven interdisciplinary work as everyday practice in successful science and engineering organizations. For example, the National Aerospace Plane (NASP) program was designed as an interdisciplinary effort to develop the theory required to create aircraft capable of flight beyond Mach 5. This speed barrier required a paradigm shift between theories and technologies that could be developed independently and those that needed to be unified and synthesized, since the theoretical results "valid at the threshold of Mach 5 are no longer valid beyond Mach 5" as "certain physio-chemical laws are reversed" (Foray and Gibbons 1996, p. 269).

Proponents of theory-driven interdisciplinarity sometimes give the sense that we are at the dawn of an intellectual gold rush. Stories like the one about Kepler stumbling upon millennia-old geometric theories suggest that the answers are out there; we are just unable to see them because of our disciplinary blinders. Indeed, there seems to be "promethean potential" in "undiscovered public knowledge" (Fuller 2017, p. 1). To harness this power, all we need to do is to "stray beyond path-dependencies of established paradigms" (Fuller 2017, p. 1) and embrace attempts to unify our knowledge in natural sciences and apply it to all problems.

Of course, it is important to note that successes of unification of knowledge from the natural sciences are primarily applicable to simple and complicated systems (see Chapter 2 for a thorough distinction between simple, complicated, and complex systems). Nevertheless, given the evident success and the excitement around the potential of theory-driven interdisciplinary collaboration for simple and complicated systems, it is no wonder that attempts have been made to extend this mode of thinking and working into the complex domain of coupled natural and human systems. The hope is that – in the same way we have synthesized general theories of planetary motion, genetics, and molecular biology – we might be able to synthesize general theories for complex systems like ensuring human rights, sustainable development, or water management.

Theory-driven approaches fall short for complex problems

Most theory-driven work may be done implicitly, as researchers may miss the critical distinction between problems that arise from simple and complicated systems and those that arise from systems that are intrinsically complex. Others, however, are explicit in their attempts to develop unifying theories where the power of prediction and control can be brought to complex socioenvironmental issues. For example, Max-Neef et al. (1992) explicitly call for interdisciplinary collaboration to establish a time-invariant, culture-invariant *theory of human needs*. The authors hold that these needs are not only invariant but are "few, finite, and classifiable." What changes across time and across cultures, the authors argue, are the ways in which these needs are satisfied. The temporal and cultural dynamics of basic human needs are therefore reduced to parameters in a general model that predicts how well a particular society is satisfying them.

We deeply support the notion that there is a humanitarian imperative to find orientations of interdisciplinary work that move beyond solving problems within industry and the academy and embrace "the less appreciated power of interdisciplinary protocols to solve complex social problems" (Meek et al. 2001, p. 124). However, on our understanding of the complexity inherent in these problems, theory-driven approaches are not likely to be effective in obtaining this goal. We maintain that there is simply no general theory that can capture "the dynamic interactions between nature and society, with equal attention to how social change shapes the environment and how environmental change shapes society" (Clark and Dickson 2003, p. 8059). This is partly because the conventional meaning of causality is mismatched with the intrinsic properties of coupled natural and human systems.

For instance, theories and models based on cause and effect in natural systems (e.g., water flowing downhill due to the law of gravity, demonstrating no agency or intentionality) are quite different from those in the societal domain, where the nature of human agency (e.g., water flowing uphill toward power, demonstrating intentionality) is an integral part of the observation and explanation of cause and effect. In addition, the conventional mode of causal explanation has a built-in reductionist logic based on a conception of reality as an aggregation of elements (e.g., it implies there is an equivalence between the aggregation of particles to form the universe and the aggregation of agents to form an institution). On the other hand, if reality is emergent, as in the case of a complex system, then the notion of reductionism becomes logically impossible to hold and cannot function as a theoretical basis for explanation. Thus, the limitation of the conventional meaning of causality opens up logical and theoretical space for considering additional conditions for explanation. The meaning of intentionality in human systems and the absence of such intentionality for natural systems has far-reaching implications in framing policy problems and designing implementable solutions for coupled natural and human systems using frameworks, theories, models, and tools primarily developed for natural systems.

So any approach to address complexity in coupled natural and human systems requires us to anticipate the need for contingent resolutions (Islam and Madani 2016; Islam and Choudhury 2018). However, "theory oriented research aims at getting rid of contingent factors" despite the fact that effectively operating in complex systems requires us "to learn as much as possible about their specificity" (Krohn 2008, p. 370). In attempting to apply theory-driven attempts to create "actionable knowledge for collective action in order to mitigate or resolve sustainability problems," researchers have noted a lack of ability to create generalizable knowledge that would "allow transferring, multiplying, and scaling up the solutions" (Lang et al. 2012, p. 38).

Problem–driven collaboration

Problem-driven collaborations, on the other hand, are undertaken without the objective of coming to such general solutions. Rather, the objective of our proposed problem-driven interdisciplinary collaboration is to bring a diversity of expertise to bear on a particular multifaceted problem in order to achieve timely and actionable outcomes. Such an effort is undertaken without the expectation of ever weaving these disparate strands of knowledge into a cohesive or comprehensive whole.

Indeed, we argue that the best way to leverage diverse expertise in interdisciplinary collaborations is not to attempt to build a unifying theory of the system but to have interdisciplinary experts closely inspect the system and prudently identify the elements and interactions relevant to resolving a particular problem. This should not be confused with advocacy for brash reductionism. Rather, the stance of the problem-driven interdisciplinary collaboration we promote is humble, principled, and pragmatic. Professionals ought to approach complex water systems with an unreserved reverence for the systems' emergent qualities and explicit recognition of their own disciplinary blinders yet remain cognizant of the importance of timely and actionable outcomes.

The resultant findings and interventions, while informed and supported by the best available information, should not be expected to be exact or generalizable. Rather, problem-driven interdisciplinary work on complex water issues leads to a set of actionable options that are contingent, experimental, and subject to negotiation and informed consent of the affected parties. It achieves this by focusing not through the development of a shared theory but a shared understanding of a specific problem. In achieving a shared understanding, we do not mean to suggest that all participants need to agree on the fundamental and generalizable features of a problem (i.e., agree on a theory) but rather that they understand the ways in which their own perspectives may be in conflict with others. Moreover, they need to be able to gauge whether these conflicts in perspective matter to the resolution of the problem or if they are practically inconsequential. Even if irreconcilable conflicts of perspectives exist, a problem-driven approach can still yield contingent pathways to resolution that embrace the uncertainty inherent in these differences in perception. In such situations, a portfolio of mutually agreed-upon actionable interventions that represents the interests of all the parties can be a strong indicator of the feasibility of successful long-term collaborative management strategies.

Consider the case of the sanitation crisis in Lowndes County, Alabama, that we discussed in Chapter 2. By Alabama state law, the provision of adequate on-site household wastewater infrastructure is a private responsibility that carries penalties for noncompliant homeowners (specifically, Alabama State Code § 22–26–1 2014; Alabama State Code § 22–26–2 2016). Local activists, however, have argued that the provision of sanitation infrastructure should not be seen as a privilege or private responsibility but as a human right recognized by the United Nations in both the Sustainable Development Goals and Resolution 64/292 (Anzilotti 2019; United Nations General Assembly 2010). This bifurcation represents a fundamental divergence in stakeholder perceptions about the "responsibility" of provisioning adequate on-site household wastewater treatment infrastructure in rural Alabama.

Is it possible to move forward collaboratively without agreement on who bears the responsibility for rural sanitation infrastructure? One viable path forward may be to set aside the questions of responsibility over *infrastructure* and frame the problem instead as a crisis in *public health*. The Alabama Department of Public Health already has a formal mandate to "promote, protect, and improve Alabama's health" and has made commitments to "be flexible in our approach to solving problems and providing services" as well as to "build and maintain internal and external partnerships to address public health challenges" (Alabama Department of Public Health 2018, p. 4). Collaboration may therefore proceed in light of potentially irreconcilable differences over questions of responsibility of provisioning *infrastructure*, by appealing to the shared values embedded in a preexisting institution that has a stated mission of promoting, protecting, and improving *public health* through innovative community partnerships.[3]

Facilitating problem–driven collaboration with shared frameworks

In order for interdisciplinary teams to work effectively in the problem–driven mode idealized above, professionals need to overcome significant communication hurdles, as their highly specialized vocabularies will rarely overlap – or, worse, will have words that overlap with subtly different meanings. "Not speaking the same language" is a "ubiquitous complaint" in interdisciplinary collaboration (Pohl et al. 2008, p. 415). Indeed, "whether it is in terms of a broader culture and way of thinking, or the down-to-earth use of words, languages are key" (Calvert 2010, p. 8).

Adopting a shared language is not a matter of convenience but rather a prerequisite for success. Interdisciplinary research "succeeds by building joint visions of the issue of concern, by finding a common language, by jointly discussing the trade-offs that result from particular choices, and above all through collaborative learning" (Pohl et al. 2008, p. viii). Progress can't be made until interdisciplinary teams "create [a] joint understanding and definition of the ... problem to be addressed" (Lang et al. 2012, p. 29).

To be clear, language barriers may exist in theory-driven collaboration as they do in problem-driven work. However, in theory-driven work, collaborators can fall back on the newly evolving vocabulary that belongs to their unified theory. In lieu of this unified theory, we advocate that problem–driven collaborations explicitly dedicate time to create, adapt, or adopt a shared framework for understanding, describing, and explaining the problems they will be addressing. One such framework is the Water Diplomacy Framework (WDF). The WDF was originally introduced by Islam and Susskind (2013) and has co-evolved throughout the process of adopting and adapting it in a problem–driven interdisciplinary research setting.

The WDF is far from a general theory of solving complex water issues. Instead, it provides a shared vocabulary of common concepts encountered when confronting problems arising from the complexity of coupled natural and human systems, offers a suite of strategies that require the presence of certain enabling conditions for the effectiveness and resilience of negotiated resolutions (Islam and Choudhury 2018), and identifies a set of principles (i.e., respecting equity and sustainability) that are essential to address complex water problems. The effectiveness of this approach is illustrated by the case studies in Chapters 6 through 12, all of which have utilized (or augmented) the WDF as a framework for understanding and addressing the problems they discuss. In this sense, the WDF has proved to be a highly transferable and adaptable framework for arriving at joint understanding to resolve complex water issues.

Towards transferable frameworks instead of transferable solutions

The appeal of theory-driven collaboration is a set of context-independent generalizable and transferable outcomes. Indeed, this is a realizable goal for many simple and complicated problems. However, responses to complex problems are contingent and highly context dependent and often require negotiated agreement about the nature of the problem and related interventions. Consequently, we can't expect generalizable and transferable "solutions" to complex socioenvironmental issues. However, frameworks for understanding and approaching problems in the complex domain have no such limitation. Indeed, we see the development of frameworks like the WDF as a critical way to reduce the length of the interdisciplinary research cycle and improve its outcomes.

We see this not just in our own experience but in frameworks such as the International Classification of Functioning, Disability and Health (ICF), which was established by the World Health Organization as an "interprofessional language with which to communicate health information across disciplinary and geographic boundaries" (Allan et al. 2006, p. 236). Analogous to the WDF: the ICF provides "a common language, terms and concepts" (World Health Organization 2013, p. 7), a suite of strategies to "facilitate cooperation and integrate different perspectives" (World Health Organization 2013, p. 102) and a set of inviolable principles for professional practice (i.e., respect and confidentiality; World Health Organization 2013, p. 10).

The ICF is not intended to serve as a theory of medical practice but rather as a practical means to facilitate it. It can be seen as "a meta–language to help clarify the relationship between data, information and knowledge, and to build a shared understanding and interpretation of concepts" (World Health Organization 2013, p. 7). As such, it can be utilized in problem-driven interdisciplinary contexts to "facilitate collaboration among key players, while matching the interventions with the needs or the purpose of the collaboration" (World Health Organization 2013, p. 44). Indeed, the ICF has received warm welcome from the international medical community because "the provision of a shared framework and language are fundamental steps in achieving the goal of interprofessional collaboration" (Allan et al. 2006, p. 236).

However, ICF is not only a language for health care professionals. It is also designed for use by "people experiencing disability, providing relevant services, or working with disability data and information" (World Health Organization 2013, p. 7). This respects the dignity of the person experiencing the disability, because it allows them to advocate for themselves and communicate more clearly across the "many professionals and systems, for example health, education and social care" that they may encounter for assessment and treatment (World Health Organization 2013, p. 7). While the WDF has not been formalized to the level of the ICF, we too hope that as the framework matures, it can meet the ICF's goal of being intelligible for all stakeholders.

Such a goal is critical, because broad stakeholder engagement is essential to arriving at actionable outcomes on complex water issues. For brevity and thematic clarity in this chapter, we have considered "interdisciplinarity" primarily in the traditional sense of collaboration among experts from different disciplines. However, to arrive at sustainable and equitable outcomes, such collaborations will ultimately need to be embedded in inclusive processes that engage a broad array of stakeholders in fact–value deliberation, joint fact-finding, and consensus-oriented decision making. This goes beyond the traditional requirements placed on professionals practicing interdisciplinarity. It will require them to apply both principles and pragmatism in the synthesis and application of four domains of knowledge: the formal knowledge of *episteme*, the practical wisdom of *phronesis*, the practiced technique of *techne*, and thoughtful practice of *praxis*. We explore these ideas in detail in the following chapter.

Conclusion

Responding effectively to complex socioenvironmental problems requires timely and actionable outcomes. In this chapter, we explored the shortcomings of theory-driven approaches to interdisciplinary collaboration to this end. We argued that theory-driven collaboration can work well in simple and complicated problem contexts through the application of frameworks, theories, models, and tools developed primarily for natural systems. However, the introduction of coupled societal dynamics into natural water systems creates complexity that resists description by a unifying theory. Problem-driven collaborations focus on leveraging interdisciplinary expertise to identify critical system dynamics that are important to the problem at hand. To do so, a shared understanding of the problem must be established among the stakeholders.

This is facilitated by adopting, adapting, or creating a shared conceptual language (framework) for understanding the problem.

We have shown that while the specific problem-oriented outcomes of these collaborations are not expected to be generalizable, shared frameworks (such as the WDF and ICF) can nevertheless be highly transferable between problems. Overall, the presence of a shared framework can reduce the length of the interdisciplinary research cycle and improve its outcomes. We take the case studies in Chapters 6 through 12 to be evidence of this. Moreover, shared frameworks also allow collaborators to come to common understandings of essential professional values. For the ICF, these values are respect and confidentiality (World Health Organization 2013, p. 10). For the WDF, they are equity and sustainability (Choudhury and Islam 2018). The operationalization of these values in the professional practice of Water Diplomacy requires what we call principled pragmatism, which we will cover in detail in the following chapter.

Notes

1 Some scholars have made narrower arguments acknowledging that while interdisciplinarity is necessitated by complexity, its use should be restricted in other contexts. The concern is that broad and unprincipled promotion of interdisciplinarity will continue to damage its conceptual clarity and credibility (Newell 2001). We believe interdisciplinarity can be applied effectively (albeit in different ways) to simple, complicated, and complex systems, as we show in this chapter.

2 In the oft-quoted words of Kenneth Boulding: "Such a theory would be almost without content, for we always pay for generality by sacrificing content, and all we can say about practically everything is almost nothing" (Boulding 1956).

3 Recent actions by Earthjustice (an environmental group) and the Alabama Center for Rural Enterprise (a nonprofit group focused on economic development in rural Alabama) demonstrate a movement toward reframing the crisis in terms of public health. These groups have filed a federal civil rights complaint with the United States Department of Health and Human Services, arguing the Alabama Department of Public Health and the Lowndes County Health Department have "failed to protect black residents from inadequate sewage systems, which has led to water contamination and an outbreak of infectious disease" (Baptiste 2018). When there are significant power asymmetries, pressure from marginalized groups in the form of adversarial litigation may be required to get collaborative processes off the ground. As Judith Innes has noted, "consensus building is not, in any case, the place for redistributing power"; instead "this can and, in many cases, should be done outside of and before a consensus building process" (Innes 2004, p. 12). Indeed, "often lawsuits and social movements are necessary to assure that these weaker players have something to bring to the table" (Innes 2004, p. 12).

References

Alabama Department of Public Health, 2018. *Strategic Plan 2019–2023*. Available from: www.alabamapublichealth.gov/about/assets/adph-strategic-plan-2019-2023.pdf

Alabama State Code § 22-26-1, 2014.

Alabama State Code § 22-26-2, 2016.

Allan, C.M., Campbell, W.N., Guptill, C.A., Stephenson, F.F., and Campbell, K.E., 2006. A conceptual model for interprofessional education: The international classification of functioning, disability and health (ICF). *Journal of Interprofessional Care*, 20 (3), 235–245.

Anzilotti, E., 2019. Why is sanitation still a privilege, not a right? *Fast Company*. Available from: www.fastcompany.com/90347337/why-is-sanitation-still-a-privilege-not-a-right

Baptiste, N., 2018. This black community has filed a civil rights complaint over unsanitary sewage conditions. *Mother Jones*. Available from: www.motherjones.com/environment/2018/10/this-black-community-has-filed-a-civil-rights-complaint-over-unsanitary-sewage-conditions/

Boulding, K.E., 1956. General systems theory: The skeleton of science. *Management Science*, 2 (3), 197–208.

Brewer, G.D., 1999. The challenges of interdisciplinarity. *Policy Sciences*, 32 (4), 327–337.

Calvert, J., 2010. Systems biology, interdisciplinarity and disciplinary identity. *In*: J.N. Parker, N. Vermeulen, and B. Penders, eds. *Collaboration in the New Life Sciences*. Oxon: Routledge, 201–218.

Chettiparamb, A., 2007. *Interdisciplinarity: A Literature Review*. HEA Interdisciplinary Teaching and Learning Group, Centre for Languages, Linguistics and Area Studies, University of Southampton.

Choudhury, E., and Islam, S., 2018. *Complexity of Transboundary Water Disputes: Enabling Conditions for Negotiating Contingent Resolutions*. New York: Anthem Press.

Clark, W.C., and Dickson, N.M., 2003. Sustainability science: The emerging research program. *Proceedings of the National Academy of Sciences of the United States of America*, 100 (14), 8059–8061.

Colwell, R., and Eisenstein, R., 2001. From microscope to kaleidoscope: Merging fields of vision. *In*: J.T. Klein, R. Häberli, R.W. Scholz, W. Grossenbacher-Mansuy, A. Bill, and M. Welti, eds. *Transdisciplinarity: Joint Problem Solving Among Science, Technology, and Society: An Effective Way for Managing Complexity*. Basel: Birkhäuser Basel, 59–66.

Foray, D., and Gibbons, M., 1996. Discovery in the context of application. *Technological Forecasting and Social Change*, 53 (3), 263–277.

Fuller, S., 2017. The military-industrial route to interdisciplinarity. *In*: R. Frodeman, ed. *The Oxford Handbook of Interdisciplinarity*. 2nd ed. Oxford: Oxford University Press.

Gabora, L., and Aerts, D., 2005. Evolution as context-driven actualisation of potential: Toward an interdisciplinary theory of change of state. *Interdisciplinary Science Reviews: ISR*, 30 (1), 69–88.

Graff, H.J., 2016. The 'problem' of interdisciplinarity in theory, practice, and history. *Social Science History*, 40 (4), 775–803.

Gusdorf, G., 1977. Past, present and future in interdisciplinary research. *International Social Science Journal*, 29 (4), 580–600, 77.

Hearn, A., 2003. Interdisciplinarity/extradisciplinarity: On the university and the active pursuit of community. *History of Intellectual Culture*, 3 (1), 1–15.

Hoffmann-Riem, H., Biber-Klemm, S., Grossenbacher-Mansuy, W., Hadorn, G.H., Joye, D., Pohl, C., Wiesmann, U., and Zemp, E., 2008. Idea of the handbook. *In*: G.H. Hadorn, H. Hoffmann-Riem, S. Biber-Klemm, W. Grossenbacher-Mansuy, D. Joye, C. Pohl, U. Wiesmann, and E. Zemp, eds. *Handbook of Transdisciplinary Research*. Dordrecht, The Netherlands: Springer, 3–17.

Innes, J.E., 2004. Consensus building: Clarifications for the critics. *Planning Theory*, 3 (1), 5–20.

Institute of Medicine, National Academy of Engineering, National Academy of Sciences, Engineering, and Public Policy Committee on Science, and Committee on Facilitating

Interdisciplinary Research, 2005. *Facilitating Interdisciplinary Research*. Washington, D.C.: National Academies Press.

Islam, S., and Choudhury, E., 2018. Complexity and contingency: Understanding transboundary water issues. *In*: E. Choudhury and S. Islam, eds. *Complexity of Transboundary Water Conflicts: Enabling Conditions for Negotiating Contingent Resolutions*. New York: Anthem Press.

Islam, S., and Madani, K., eds., 2016. *Water Diplomacy in Action: Contingent Approaches to Managing Complex Water Problems*. New York: Anthem Press.

Islam, S., and Susskind, L.E., 2013. *Water Diplomacy: A Negotiated Approach to Managing Complex Water Networks*. 1st ed. New York: RFF Press.

Klein, J.T., 2017. Typologies of interdisciplinarity. *In*: R. Frodeman, ed. *The Oxford Handbook of Interdisciplinarity*. 2nd ed. Oxford: Oxford University Press.

Krohn, W., 2008. Learning from case studies. *In*: G.H. Hadorn, H. Hoffmann-Riem, S. Biber-Klemm, W. Grossenbacher-Mansuy, D. Joye, C. Pohl, U. Wiesmann, and E. Zemp, eds. *Handbook of Transdisciplinary Research*. Dordrecht, The Netherlands: Springer, 369–383.

Lang, D.J., Wiek, A., Bergmann, M., Stauffacher, M., Martens, P., Moll, P., Swilling, M., and Thomas, C.J., 2012. Transdisciplinary research in sustainability science: Practice, principles, and challenges. *Sustainability Science*, 7 (1), 25–43.

Max-Neef, M., Elizalde, A., and Hopenhayn, M., 1992. Development and human needs. *Real-Life Economics: Understanding Wealth Creation*, 197–213.

Meek, J., Wentworth, J., and Sebberson, D., 2001. The practice of interdisciplinarity: Complex conditions and the potential of interdisciplinary theory. *Issues in Interdisciplinary Studies*, 19 (1), 123–132.

Newell, W.H., 2001. A theory of interdisciplinary studies. *Issues in Interdisciplinary Studies*, 19 (1), 1–25.

Nissani, M., 1995. Fruits, salads, and smoothies: A working definition of interdisciplinarity. *The Journal of Educational Thought (JET)/Revue de la Pensée Éducative*, 29 (2), 121–128.

Ostrom, E., 1999. Institutional rational choice: An assessment of the institutional analysis and development framework. *In*: P.A. Sabatier, ed. *Theories of the Policy Process*. Boulder: Westview Press, 21–64.

Pohl, C., van Kerkhoff, L., Hadorn, G.H., and Bammer, G., 2008. Integration. *In*: G.H. Hadorn, H. Hoffmann-Riem, S. Biber-Klemm, W. Grossenbacher-Mansuy, D. Joye, C. Pohl, U. Wiesmann, and E. Zemp, eds. *Handbook of Transdisciplinary Research*. Dordrecht, The Netherlands: Springer, 411–424.

Popper, K., 2002. *Conjectures and Refutations: The Growth of Scientific Knowledge*. 2nd ed. New York: Routledge.

Salter, L., and Hearn, A., 1996. *Outside the Lines: Issues in Interdisciplinary Research*. Montreal: McGill-Queen's University Press.

Stokols, D., Hall, K.L., Taylor, B.K., and Moser, R.P., 2008. The science of team science: Overview of the field and introduction to the supplement. *American Journal of Preventive Medicine*, 35 (2 Suppl), S77–S89.

United Nations General Assembly, 2010. *Resolution 64/292: The Human Right to Water and Sanitation*. 64th Session.

Williams, R., 1985. *Keywords: A Vocabulary of Culture and Society*. Revised, Subsequent ed. Oxford: Oxford University Press.

World Health Organization, 2013. *How to Use the ICF: A Practical Manual for Using the International Classification of Functioning, Disability and Health (ICF)*. Geneva: WHO.

4 Principled pragmatism

How Water Diplomats approach complex water issues

Kevin M. Smith and Shafiqul Islam

Water diplomats: the water professionals of tomorrow

Contemporary water problems often reflect a coupling of natural and human (CNH) systems. This coupling creates conditions of "essential complexity" (Clark and Dickson 2003, p. 8059) that is intrinsic and irreducible. As a consequence, many problems arising in CNH systems are manifestations of emergent phenomena that defy traditional reductionist and decompositional problem-solving techniques (see Chapter 2 for more on the distinction between simple, complicated, and complex problem domains). The ability of these socio-environmental problems to resist treatment by traditional expert-driven planning and management has widely been described as "wicked" (Rittel and Webber 1973). Responses to this wickedness "require a broad systemic response, working across boundaries and engaging citizens and stakeholders in co-producing policy-making and implementation" (Ferlie et al. 2011, p. 308) as well as engaging in "messy" fact–value conversations.

What does this mean for the water professionals of tomorrow? What skills do they need to be effective in situations that require *working across boundaries*? What self-understanding do they need in order to conceptualize their role as an expert engaged in *inclusive fact–value conversations* and in the *co-production of knowledge with stakeholders*? In order to think about these requirements and where they are lacking in traditional training programs, we will adopt an often-used taxonomy (which has been adapted significantly and is now somewhat removed from its origins in the work of Aristotle) that identifies four distinct categories of *knowing*: episteme, phronesis, techne, and praxis.[1] Tomorrow's water professionals will need to have effective command of them all.

That is a tall order. The emphasis on specialization that exists in many education and training programs suggests one's professional life ought to be dominated by one form of knowing (Wagner 1998). For scientists, this is often the formal knowledge of *episteme* (Sullivan 1991). Artists and engineers are often judged on their *techne*: the practiced technique required to materialize ideas (Mitcham and Schatzberg 2009). Those charged with important societal decisions (e.g., elected officials) are expected to act with the prudence and practical wisdom of *phronesis* (Morgan 1988; Morgan and Kass 1989). Managers, on the other hand, need to epitomize *praxis*: the thoughtful practice required to guide their teams through the ever-changing perils and decision points common to dynamic organizations (Shrivastava 1986).[2]

In contrast to this notion that equates professionalism with absolute specialization, we argue that the notion of *principled pragmatism* can guide professionals towards an effective integration and application of knowledge and skills from all four domains. Resolutions to complex problems require interdisciplinary collaboration, fact–value deliberation, and joint fact-finding with a broad array of stakeholders, as well as contingent, collaborative, and adaptive operational strategies. For many professionals, embracing the pluralism present in these situations can be uncomfortable and challenging when coming from training and education programs in which their specialized expertise was valued unconditionally. They may identify any departure from their discipline's best practices as foolhardy and unacceptable.

However, we argue that when professionals view themselves as *principled pragmatists*, their value as experts need not be threatened by inclusive fact–value deliberation, joint fact-finding, and decision-making processes in which power is shared and knowledge is co-produced. This framing allows them to understand their role as a mediator between the extremes: the infinite reticence of uncompromising dogma on one hand and the brash and foolhardy application of supposedly value-neutral science and technology on the other. Instead, they embrace the humble middle path required by complex problems – one that attends to both principles and pragmatism. Indeed, they recognize that a strict adherence to principles without pragmatism is often not actionable and that pure pragmatism exercised without guiding principles is not sustainable and is unlikely to be equitable.[3]

We highlight that our practice of Water Diplomacy does not take place in the context of what Schön (1983) refers to as the "high ground" – the neat and orderly place where "practitioners can make effective use of research-based theory and technique" (Schön 1983, p. 42). Rather, Water Diplomacy is practiced in "the swampy lowland where situations are confusing 'messes' incapable of technical solutions" (Schön 1983, p. 42). Professionals working in the high ground are "hungry for technical rigor, devoted to an image of solid professional competence, or fearful of entering a world in which they feel they do not know what they are doing ... [and as such] they choose to confine themselves to a narrowly [defined] technical practice" (Schön 1983, p. 43). In contrast, professionals like Water Diplomats working in the swampy lowlands "deliberately involve themselves in messy but crucially important problems and, when asked to describe their methods of inquiry, they speak of experience, trial and error, intuition, and muddling through" (Schön 1983, p. 43).

Several archetypes of professionalism discussed in recent literature embrace Schön's notion that there is important value in working in the swampy lowland of stakeholder-centered negotiated approaches to complex water problems. These descriptions highlight several of the roles Water Diplomats may be asked to perform in practice.[4] The *Honest Broker* mode of practice leverages professional expertise in the development of creative options but does not advocate for a particular outcome. Instead, the Honest Broker tries to identify as clearly as possible the costs, benefits, and uncertainties associated with each option in order to position stakeholders to make an informed decision (Pielke 2007). The Honest Broker approach maintains the primacy of the expert in the knowledge–production and option–generation process but curtails its primacy in the decision making process. On the other hand, the ideal *Humble Analyst* recognizes that knowledge is socially constructed and must be co-produced through joint fact-finding. The Humble Analyst works as a member of a team of stakeholders throughout all stages of the process: fact–value deliberation, problem definition, data collection, analysis, modeling, option generation, and decision making. Humble Analysts "accept a supporting rather than a starring role for professional scientific inquiry" (Andrews 2002, p. 30). The vision of the *Democratic Professional* is similar in practice to that of the *Humble Analyst*, but its intentions and motivations are different. Where the Humble Analyst seeks to acknowledge the epistemological importance of co-producing knowledge for action, the Democratic Professional seeks to engage in "task sharing" as a means of redistributing power and legitimizing the role of the expert professional in society (Dzur 2008).

There are many distinct ways in which professionals will be called to engage in the search for negotiated resolutions to complex water issues. They may be contracted strictly to produce creative options (e.g., as an Honest Broker), or they may be engaged as an expert stakeholder for the entirety of a consensus-oriented process of fact–value deliberation, joint fact-finding, and decision making (e.g., as a Humble Analyst). As consultants in long–term collaborative and adaptive management processes, they may be asked to share parts of their professional responsibility with members of the public (e.g., as a Democratic Professional) to ensure that newly formed governance institutions do not succumb to technocratic management. Later, we explore how these modes of professional practice are examples of principled pragmatism in action. As such, they can serve as useful archetypes for professionals seeking to understand how they can attend to both principles and pragmatism in their work on complex water management issues and move toward the practice of what we call Water Diplomacy.

Four essential domains of knowledge

Professionals working on complex water issues will need to move beyond notions that equate professionalism with absolute specialization. They will need to work with other experts across disciplinary boundaries as well as engage with lay stakeholders in inclusive fact–value deliberation, collaborative joint fact-finding, and decision–making processes. This will require them to draw upon four domains of knowledge: episteme (formal knowledge), phronesis (practical wisdom), techne (practiced technique), and praxis (thoughtful practice). Principled pragmatism provides a lens for synthesizing these four

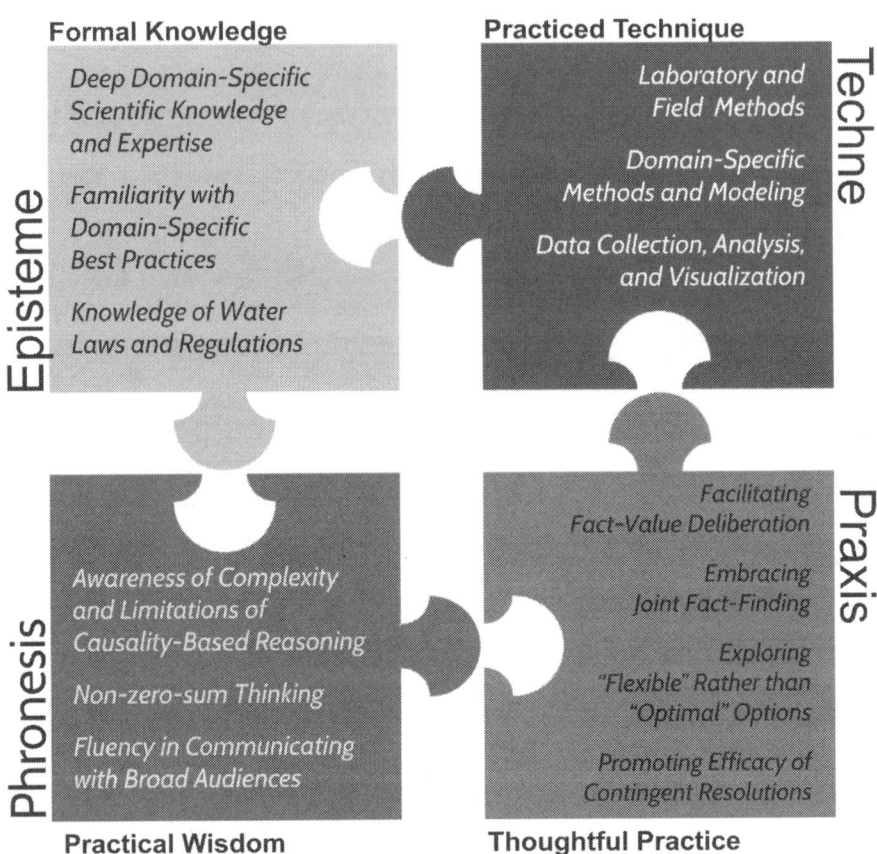

Figure 4.1 The practice of Water Diplomacy requires the synthesis and application of all four domains of knowledge through the lens of principled pragmatism.

domains into outcomes that are actionable, equitable, and sustainable. See Figure 4.1 for a graphical summary.

Episteme

Episteme represents formal knowledge. Water professionals will need a deep understanding of their discipline's scientific knowledge and best practices as well as an understanding about the formal constraints (e.g., laws and regulations) that limit the available options for the problem at hand.

Phronesis

Phronesis represents prudence and practical wisdom. For water professionals, applying phronesis when working on complex problems entails appreciating the irreducibility of complexity, recognizing the inadequacies of conventional causality-based reasoning, embracing non–zero-sum thinking, inventing – but not committing to – different options for intervention, and respecting the need to communicate effectively with broad audiences.

Techne

Techne represents practiced technique. It represents a fluency with the tools of the trade. For water professionals, this often means a familiarity with both laboratory and field methods, comfort with domain-specific methods and modeling tools, and an awareness of the available workflows for collecting data, analyzing it for insights, and visualizing those insights effectively.

Praxis

Praxis represents thoughtful practice. While techne represents the skills to produce scientific findings and materialize ideas, praxis represents the skills required for the operationalization of those tools and techniques in the messiness of the real world. For water professionals, this often means inviting and facilitating fact–value deliberation, embracing joint fact-finding without abandoning the importance of scientific expertise, seeking mutually beneficial options guided by the criteria of "flexibility" rather than "optimality", a commitment to finding resolutions that are contingent and do not require a particular realization of the future to be infallible, and adopting a conscious mode of operating in an inclusive joint fact-finding or decision-making process (e.g., as an Honest Broker, Humble Analyst, or Democratic Professional).

Principled pragmatism

We are not the first to introduce or promote the notion of "principled pragmatism." Indeed, it has been increasingly used and promoted by various voices in the international community. For example, Ban Ki-Moon wrote in an editorial, "if I were to sum up my view of the United Nations and its work today, it would be a spirit of principled pragmatism" (Ki-moon 2007). Two years later, Hillary Clinton, serving as the Secretary of State for the United States, said in a speech that "principled pragmatism informs our approach on human rights" and is a key element of "putting our principles into action" (Smith 2009). While early-20th-century reflections on this notion seem to see it as "paradoxical" (Heclo and Madsen 1987) and "enigmatic" (Esping-Andersen 1988), in the 21st century, it has been widely embraced and was even described as a key attribute of the first official European Union Global Strategy in 2016 (European Commission 2016, p. 16).[5] Indeed, principled pragmatism has been identified in the literature as an operational strategy for a wide variety of activities, ranging from a means of ensuring the inclusion of feminist principles in international peace agreements (Bell and McNicholl 2019) to the preparation of witnesses for the International Criminal Court (Jackson and Brunger 2015).

Our view of principled pragmatism

There has been significant growth in the use of the phrase "principled pragmatism" in the last decade and especially in the last three years since its adoption as part of the European Union Global Strategy in 2016 (see Figure 4.2). While we don't claim to have the only legitimate use of the phrase, we are also aware that this growth in usage has necessarily come with a diversity of meanings, which we don't have the space to fully compare and contrast here. In many cases, the phrase is also being used as a buzzword without any substantive discussion of its meaning. Our use of the phrase must therefore not be seen as an open endorsement of how it is applied elsewhere. At the same time, we acknowledge that we are part of a burgeoning global conversation about the importance of attending to both principles and pragmatism. In this section, we clarify our use of the expression and its implications for professionals engaged in complex water issues.

Pragmatic compromises don't entail a compromise of values

Water professionals engaged in complex water issues will need to engage in processes in which knowledge and solutions are co-produced in concert with a broad range of stakeholders with a pluralistic set of values. Coming

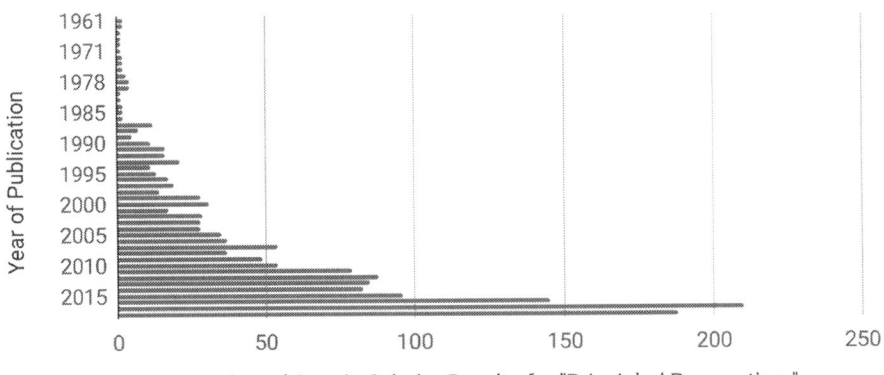

Figure 4.2 There has been significant growth in the use of the phrase "principled pragmatism," especially within the last decade. The figure represents the number of Google Scholar results for the exact phrase "principled pragmatism" grouped by year of publication.

Data source: Google Scholar

to compromises and contingent resolutions is an integral part of this process. However, the pragmatic process of arriving at compromise – a settlement of differences in interests – doesn't entail the compromise of one's guiding principles. Indeed, with respect to negotiating resolutions to complex water problems, "compromise over interests is possible and actionable while compromising principles is not" (Islam 2016).

With respect to the notion of principled pragmatism, at least one author suggested "perhaps the adjective and the noun should be reversed" (Packer 1963, p. 594). However, we think the original word order is correct. We aren't suggesting anyone should adopt or acknowledge only those principles that are pragmatic in nature but rather that they apply pragmatism in the realization of those principles in practice. Scholars have framed the policy making of former United States president Abraham Lincoln in terms of principled pragmatism. While he demonstrated a commitment to four principles – universal rights, the rule of law, the Union, and a popular government – he frequently reordered the priority of those principles in determining his political compromises as context changed. If he had committed only actions that promoted all of his principles simultaneously, it is unlikely he would have been able to make substantial progress on any of them (Siemers 2004).

Likewise, principled pragmatists must be aware of the context-dependent nature of their work and take care to prioritize principles in ways that allow them to be achieved in practice. Water professionals must understand how context dictates which interventions represent acceptable pragmatic compromises (in interests) and unacceptable compromises (in values). In the case of water management in Bangladesh, deal making over the quantities of water allocated for agriculture versus aquaculture may be acceptable, but such deal making is out of the question when it compromises the sustainability of the Sundarbans mangrove forest or equitable access of water to the local community (Islam 2017). Deal making is often an essential part of reaching effective contingent and negotiated resolutions, but making deals that don't appreciate norms and values often leads to outcomes that are unsustainable and unequitable. Instead, through principled pragmatism, we must remain committed to actionable options that are "grounded in translating global norms in terms of local understanding and the capacity to act on them" (Islam 2017).

Embrace other ways of knowing, but value your expertise

We've discussed how professionals engaged in complex water issues will be called to participate in inclusive and collaborative processes of fact–value deliberation, joint fact-finding, and decision making. While "a multi-dimensional, interdisciplinary, integrative response to this increased complexity seems to have the best chance of creating lasting change," such responses will often be "destabilizing for the traditional power structure" (Meek et al. 2001, p. 132). Indeed, such processes are destabilizing to the traditional vision of water professionals as expert "trustees" charged with making critical decisions on the basis of their interpretation of the public good (Pierce et al. 1976, p. 15). Today, this commonly takes the form of executive rule making by bureaucratic agencies (e.g., the United States Environmental Protection Agency), where the level of stakeholder interaction is limited to providing written responses to public comments on the proposed rules (Ozawa 1991, p. 20).

In order for professionals to be effective partners in the inclusive and collaborative processes we have been describing, they must work with experts from other disciplines as well as stakeholders with local expertise. This necessarily entails embracing the existence of other ways of understanding the problem being addressed – and, indeed, different ways of understanding the world – which can be uncomfortable for professionals trained in the "self-referential" practice of their own discipline (Tabulawa 2017, p. 16). However, acknowledging the legitimacy of multiple viewpoints is too often seen as "a threat to intellectual certainties, on the one hand, and to moral seriousness on the other" (Lukes 2008, p. 1). In reality, there is room for both principles and pragmatism in inclusive collaborations on complex water issues.

In acknowledging the principles of others, one does not have to sacrifice their own principles. In acknowledging that no one has exclusive and privileged access to reality, one does not need to abandon their expertise. Indeed, we would argue that in the case of complex water issues, to artificially deflate the relevance of one's expertise to the collective process is an abdication of professional responsibility. Professionals need to acknowledge the importance of their contributions as one of many members of a collaborative process that requires the diverse synthesis of both disciplinary expertise and local knowledge. One needs to remain humble and respect the flattened power structure of the collaborative setting but need not be "timid" (Andrews 2002, p. 41). Indeed, experts engaged to work on complex water issues have an obligation not to sit back but to engage fully in the collaborative process. Full participation by professionals requires not only the engagement of their formal knowledge (episteme) but also their practical wisdom (phronesis), their practiced techniques (techne), and their thoughtful practice (praxis).

Portraits of principled pragmatism in professional practice

In this section, we provide three portraits from the recent literature on professional practice that demonstrate how water professionals can draw upon all four domains of knowledge in a way that brings both principles and pragmatism to bear on complex water issues (i.e., engage in the practice of Water Diplomacy). While we continue to consider inclusive joint fact-finding and decision-making processes in the abstract (we don't have space for detailed treatments here), we will consider more concretely the roles that water professionals can expect to play in these processes.[6] We hope that water professionals can use these examples to better understand how they can put principled pragmatism into practice.

The Honest Broker

Our characterization of the *Honest Broker* comes from Pielke (2007). The Honest Broker serves the role of generating actionable ideas for a collective decision-making process. In particular, the Honest Broker has the goal of increasing the number of options available to stakeholders, especially when the circumstances are beset with deep uncertainties and a lack of consensus on shared values. The Honest Broker is contrasted with the professional Issue Advocate, who intentionally seeks to reduce the scope of choice available to stakeholders in such contexts by promoting particular options. While there is no fundamental fault to be placed with professionals who explicitly and consciously promote particular options or ideas, the risk we see for Principled Pragmatists is that they may accidentally fall into what Pielke refers to as "stealth issue advocacy" (Pielke 2007, p. 7). In these circumstances, professionals may believe they are eliminating and promoting options on the basis of value-neutral expert judgments when in reality their choices are guided primarily by their values and implicit biases.

The Honest Broker mode of practice circumvents these issues with a commitment to expand the scope of available options and not to take a stance or play a role in the decision between these alternatives. In Pielke's words, the Honest Broker "seeks explicitly to integrate scientific knowledge with stakeholder concerns in the form of alternative courses of action" that "may appeal to a wide range of interests" (Pielke 2007, p. 17). The "honest" aspect of the Honest Broker is that they attempt to present these options as objectively as possible while clarifying as necessary the potential costs and benefits, as well as the level of scientific consensus and characteristics of the uncertainties involved. In formulating portfolios of actionable options, Honest Brokers must engage with stakeholders to make sure the portfolio represents the diversity of values and interests within the group. Honest Brokers will often work together in interdisciplinary collaboration in order to obtain their objectives, as the "honest brokering of policy alternatives is often best achieved through a collection of experts working together with a range of views, experiences, and knowledge" (Pielke 2007, p. 3).

The Humble Analyst

Our vision of the Humble Analyst comes from Andrews (2002). Unlike Honest Brokers, who are primarily engaged to generate solutions and are not generally part of the problem-definition or decision-making phase, Humble Analysts are engaged to work throughout the entire process of fact–value deliberation, problem definition, data collection, analysis, modeling, option generation, and decision making. This mode of working is entirely collaborative and is based on an understanding that actionable knowledge often needs to be co-produced in complex contexts. As such, Humble Analysts "accept a supporting rather than a starring role for professional scientific inquiry" (Andrews 2002, p. 30).

This framing of the expert as a co-producer, colleague, and collaborator "tries to take better advantage of what the expert offers" (Andrews 2002, p. 10). Rather than attempting to appear neutral, the Humble Analyst is recognized as a stakeholder with values and interests just like any other. This approach "lays bare the basis for scientific disagreement, lets facts and values remain intertwined, and accepts that strict neutrality is too much to ask of most experts" (Andrews 2002, p. 10). The case studies presented by Andrews exemplify how Humble Analysts must draw upon all four domains of knowledge in order to effectively engage with others throughout the entire duration of the collaborative process. Andrews suggests that Humble Analysts are engaged professionally in a variety of contexts today, as the practice of "joint fact-finding has enjoyed increasing use in firms, local planning boards, regulatory agencies, courtrooms, legislative chambers, and international treaty negotiations" (Andrews 2002, p. 7).[7]

The Democratic Professional

Our description of the Democratic Professional is adapted from Dzur (2008). Like the Humble Analyst, the Democratic Professional may be engaged as an equal collaborator throughout the collective process of joint fact-finding and decision making. However, the motivation for this engagement as an equal collaborator is not based on the premise that knowledge must be co-produced but that in such settings, professionals need to democratize their power and authority through "task sharing." The intended effect of this is threefold: it serves as an expression of democratic equality and moral agency that promotes cooperation; it prevents technocratic capture of an otherwise inclusive and consensus-oriented process; and it lends legitimacy to the professional's role and status as a disciplinary expert by exposing collaborators to the intricacies of the way they work and practice.

The contrasting example to the Democratic Professional is the Technocratic Professional, who tries to "depoliticize issues by translating social problems into questions that can be solved by using the methods in which they are trained and over which they have greater command than the lay public" (Dzur 2008, p. 87). Rather than accepting the wickedness of the problem, the Technocratic Professional, "armed with systems theory and policy analysis and focusing on means, costs, and benefits of pregiven [sic] ends ... spends little time on the fundamental substantive value questions that concern laypeople" (Dzur 2008, p. 87). When it comes to water management, the opportunities for Democratic Professionalism are most apparent as part of a long-term adaptive management strategy (i.e., after initial commitments and contingent resolutions have been reached through consensus). In these cases, if professionals do not actively reaffirm the role of stakeholders, then there is significant risk that the newly formed governance institutions will default to a form of technocratic management that does not reflect the original outcomes of the consensus procedure. By making an ongoing commitment to engage lay stakeholders in the ongoing monitoring, analysis, and reporting on an intervention, Democratic Professionals are acting with attention to both principles and pragmatism.

Conclusion

Inclusive stakeholder processes are increasingly viewed as an essential component of arriving at resolutions to complex water problems (cf. Morrison 2003). In this chapter, we have argued that the water professionals of tomorrow will need to have command over four domains of knowledge to be effective in these environments: formal knowledge (episteme), prudence and practical wisdom (phronesis), practiced technique (techne), and thoughtful practice (praxis). This emphasis on four domains of knowledge stands in contrast with the structure of traditional education and training programs that too often equate professionalism with absolute specialization. We argue that the notion of *principled pragmatism* can guide professionals towards an effective integration and application of knowledge and skills from all four domains in working toward negotiated solutions to complex water problems (i.e., toward what we call the practice of Water Diplomacy).

In a colloquial sense, pragmatism is often taken to suggest practical, opportunistic, and expedient approaches at the expense of principles. We assert that professionals working on complex water issues require a more nuanced view of pragmatism that understands how it can serve as a mechanism for operationalizing principles rather than circumventing them. To realize this view, professionals will need to be attentive to the problem context to understand how principles can be prioritized for effective action. Context also dictates when interventions represent acceptable deal making that represents a compromise of interests and unacceptable deal making that represents a compromise of values.

Professionals need to embrace the pluralism of the new modes of collaboration in which they will find themselves without neglecting the value of their own expertise. By acknowledging multiple ways of knowing, water professionals acknowledge coming to actionable outcomes on complex water issues is "impossible through single disciplinary domains or models decontextualized from operational realities" (Fernandes and Philippi 2017, p. 11) and avoid the mistake of seeking only "technical solutions that sidestep the problematic social and political questions" (Fischer 1990, p. 23). However, participation in an inclusive process of fact–value deliberation, joint fact-finding, and decision making does not mean the professional must devalue their expertise, which is essential to arriving at actionable outcomes. Ultimately, arriving at resolutions to complex water issues requires "dialogue among conceptual disciplines of different types of knowledge and between nonscientific worldviews" (Fernandes and Philippi 2017, p. 11), and professionals must not exclude their voices from that dialogue.

We examined three archetypes from the literature that we believe demonstrate various ways in which principled pragmatism can be operationalized by water professionals. As Honest Brokers, they may be contracted to act as a neutral third-party source of creative policy and technology options. As Humble Analysts, they may be engaged as expert stakeholders who seek to co-produce knowledge throughout a consensus-driven process from problem formulation all the way to the decision-making process. As Democratic Professionals, they may arrange for "task sharing" to ensure the continued engagement of stakeholders in long-term collaborative adaptive management processes. All of these modes of professional practice represent principled pragmatism in action. They reflect an understanding that we need to move beyond the duality of "being pragmatic" or "being ideological" towards a method of practice that emphasizes how to connect ideas to action given the constraints and capacities of the problem context.

This is undoubtedly hard work. It requires a willingness to move beyond the comfort of disciplinary authority and step into the swampy lowlands of wicked socioenvironmental problems. However, such work is essential, as Donald Schön reminds us, since "in the swamp are the problems of greatest human concern" (Schön 1983, p. 42). Working in the swamp requires us to "muddle through" the messiness of the lowlands; we must relax our idealizations about professional practice that are constrained to situations that are ordered, logical, and rational (Fortun and Bernstein 1998). However, as we have argued, dispensing with this idealization about the contexts of professional work does not entail dispensing with scientific methods in the search for pragmatic action.

In terms of measuring our progress toward the symbolic aspiration of attaining equitable and sustainable water management, we can measure *where we are now* by our distance from *where we would like to be*; or we can measure *how far we have come* from *where we have been*. Neither measurement is perfect: we need both. By integrating both principles and pragmatism, our ideal Water Diplomat attempts to synthesize symbolic aspirations with a realistic assessment of the complex realities of many coupled natural and human systems. On this synthesis, we believe a new generation of Water Diplomats will play an important role in framing, formulating, and facilitating an inclusive and collaborative process along a trajectory from a world of seemingly infinite possibilities to an actionable subset of implementable ideas.

Notes

1 In this chapter, we will utilize these four categories as a diagnostic tool to understand the breadth of ways of knowing that need to be integrated in professional practice on complex water issues. In doing so, we are not endorsing any normative narrative that implies how sets of skills ought to be differentially valued (Hadorn et al. 2008, pp. 20–23).

2 Much has been written about the need to integrate multiple domains of knowledge into professional life. Indeed, this is at the heart of the vision of a modern interdisciplinary professional (Hadorn et al. 2008). However, despite this enthusiasm, it remains more or less aspirational in many education and training programs that were originally structured to encourage specialization in one domain of knowledge (Kim 2019).

3 Equity and sustainability are the guiding principles adopted by the Water Diplomacy Framework. For an in-depth discussion of how these principles are applied in stakeholder contexts, see Choudhury and Islam (2018).

4 These three modes of professional practice show some examples of how Water Diplomats may be asked to engage in stakeholder-centered joint fact-finding and collective decision making, but they do not constitute an exhaustive list.

5 This is not to say the concept is without criticism. Many of these objections outright reject the notion of any sort of pragmatism as a strategy when it comes to issues such as human rights. In our experience and understanding of complex issues, pragmatism is required to operationalize many principles in practice. Others have pointed to the use of the phrase as a sleight of hand that has allowed entities (especially corporations) to adopt nonspecific but warm-sounding language while allowing them to avoid specific societal obligations and responsibilities (Jägers and van Genugten 2011). On our use of the phrase, professionals are expected to take on additional societal obligations and responsibilities.

6 For concrete examples of joint inclusive fact-finding and consensual decision-making processes, we refer readers to Andrews (2002), Ozawa (1991), and Susskind et al. (1999).

7 We heed the advice of Andrews (2002, p. 3) in warning readers "joint-fact finding is a loaded phrase implying that facts exist and merely need to be found." The specific is used here to be consistent with the language used elsewhere to describe methodologies for the co-production of knowledge (e.g., Susskind et al. 1999).

References

Andrews, C.J., 2002. *Humble Analysis: The Practice of Joint Fact-finding.* Westport: Greenwood Publishing Group.

Bell, C., and McNicholl, K., 2019. Principled pragmatism and the 'inclusion project': Implementing a gender perspective in peace agreements. *feminists@law*, 9 (1).

Choudhury, E., and Islam, S., 2018. *Complexity of Transboundary Water Disputes: Enabling Conditions for Negotiating Contingent Resolutions.* New York: Anthem Press.

Clark, W.C., and Dickson, N.M., 2003. Sustainability science: The emerging research program. *Proceedings of the National Academy of Sciences of the United States of America,* 100 (14), 8059–8061.

Dzur, A.W., 2008. *Democratic Professionalism: Citizen Participation and the Reconstruction of Professional Ethics, Identity, and Practice.* University Park: Penn State Press.

Esping-Andersen, G., 1988. Reviews. *Economic and Industrial Democracy,* 9 (1), 141–142.

European Commission, 2016. *Shared Vision, Common Action: A Stronger Europe, A Global Strategy for the European Union's Foreign and Security Policy.* Brussels: European External Action Service.

Ferlie, E., Fitzgerald, L., McGivern, G., Dopson, S., and Bennett, C., 2011. Public policy networks and 'wicked problems': A nascent solution? *Public Administration,* 89 (2), 307–324.

Fernandes, V., and Philippi, A., Jr., 2017. Sustainability Sciences. *In:* R. Frodeman, ed. *The Oxford Handbook of Interdisciplinarity.* 2nd ed. Oxford: Oxford University Press.

Fischer, F., 1990. *Technocracy and the Politics of Expertise.* Newbury Park: SAGE Publications.

Fortun, M., and Bernstein, H.J., 1998. *Muddling Through: Pursuing Science and Truths in the 21st Century.* Washington, DC: Counterpoint.

Hadorn, G.H., Biber-Klemm, S., Grossenbacher-Mansuy, W., Hoffmann-Riem, H., Joye, D., Pohl, C., Wiesmann, U., and Zemp, E., 2008. The emergence of transdisciplinarity as a form of research. *In:* G.H. Hadorn, H. Hoffmann-Riem, S. Biber-Klemm, W.

Grossenbacher-Mansuy, D. Joye, C. Pohl, U. Wiesmann, and E. Zemp, eds. *Handbook of Transdisciplinary Research*. Dordrecht, The Netherlands: Springer, 19–39.

Heclo, H., and Madsen, H., 1987. *Policy and Politics in Sweden: Principled Pragmatism*. Philadelphia: Temple University Press.

Islam, S., 2016. A Lincoln for today: What would Lincoln do for Flint on his 207th birthday? *LinkedIn Pulse*. Available from: www.linkedin.com/pulse/lincoln-today-what-would-do-flint-his-207th-birthday-shafiqul/

Islam, S., 2017. Addressing water scarcity. *The Daily Star*. Available from: www.thedailystar.net/environment-and-climate-action/addressing-water-scarcity-1367101

Jackson, J.D., and Brunger, Y.M., 2015. Witness preparation in the ICC: An opportunity for principled pragmatism. *Journal of International Criminal Justice*, 13 (3), 601–624.

Jägers, N., and van Genugten, W., 2011. Corporations and human rights: Moving beyond 'principled pragmatism' to 'ruggie-plus'. *SSRN Electronic Journal*. https://doi.org/10.2139/ssrn.1844203

Kim, J.-H., 2019. *Exploring Ways to Develop Reflective Engineers: Toward Phronesis-Centered Engineering Education*. 2019 ASEE Annual Conference & Exposition.

Ki-moon, B., 2007. The spirit of principled pragmatism. *The Economist*.

Lukes, S., 2008. *Moral Relativism*. New York: Picador.

Meek, J., Wentworth, J., and Sebberson, D., 2001. The practice of interdisciplinarity: Complex conditions and the potential of interdisciplinary theory. *Issues in Interdisciplinary Studies*, 19 (1), 123–132.

Mitcham, C., and Schatzberg, E., 2009. Defining technology and the engineering sciences. *In*: A. Meijers, ed. *Philosophy of Technology and Engineering Sciences*. Amsterdam: North-Holland, 27–63.

Morgan, D.F., 1988. Administrative phronesis: Discretion and the problem of administrative legitimacy in our constitutional system. *Dialogue*, 11 (1),10–44.

Morgan, D.F., and Kass, H.D., 1989. Constitutional stewardship, phronesis and the American administrative ethos. *Dialogue*, 12 (1), 17–60.

Morrison, K., 2003. Stakeholder involvement in water management: Necessity or luxury? *Water Science and Technology: A Journal of the International Association on Water Pollution Research*, 47 (6), 43–51.

Ozawa, C.P., 1991. *Recasting Science: Consensual Procedures in Public Policy Making*. Boulder: Westview Press.

Packer, H.L., 1963. The model penal code and beyond. *Columbia Law Review*, 63 (4), 594–607.

Pielke, R.A., Jr., 2007. *The Honest Broker: Making Sense of Science in Policy and Politics*. Cambridge: Cambridge University Press.

Pierce, J.C., and Doerksen, H.R., 1976. *Water Politics and Public Involvement*. Ann Arbor, MI: Ann Arbor Science Publishers.

Rittel, H.W.J., and Webber, M.M., 1973. Dilemmas in a general theory of planning. *Policy Sciences*, 4 (2), 155–169.

Schön, D.A., 1983. *The Reflective Practitioner: How Professionals Think in Action*. New York: Basic Books.

Shrivastava, P., 1986. Is strategic management ideological? *Journal of Management*, 12 (3), 363–377.

Siemers, D.J., 2004. Principled pragmatism: Abraham Lincoln's method of political analysis. *Presidential Studies Quarterly*, 34 (4), 804–827.

Smith, B., 2009. 'Principled pragmatism' on human rights. *POLITICO*, 14 Dec.

Sullivan, D.L., 1991. The epideictic rhetoric of science. *Journal of Business and Technical Communication*, 5 (3), 229–245.

Susskind, L.E., McKearnen, S., and Jennifer, T.-L., 1999. *The Consensus Building Handbook: A Comprehensive Guide to Reaching Agreement*. Thousand Oaks: SAGE Publications.

Tabulawa, R., 2017. Interdisciplinarity, neoliberalism and academic identities. *Journal of Education*, 69, 11–42.

Wagner, J., 1998. A subtle tyranny: Rediscovering the purpose of the liberal arts. *Interchange*, 29 (3), 327–344.

Problem-driven interdisciplinary collaboration in action

Case studies from the Tufts Water Diplomacy program

5 Operationalizing problem-driven interdisciplinary collaboration

An overview of case studies from the Tufts Water Diplomacy program

Kevin M. Smith and Shafiqul Islam

Our case studies in context

In this section, we turn to concrete examples of research conducted as part of the Tufts Water Diplomacy program. These case studies exemplify the complex challenges faced by water professionals working on socioenvironmental issues. Moreover, they demonstrate how problem-driven interdisciplinary work can be facilitated by shared language (e.g., the Water Diplomacy Framework) and can be conducted with attention to both principles and pragmatism. The following chapter descriptions should not be taken as comprehensive summaries or as reflective of the authors' endorsement of the editors' arguments. Each chapter stands on its own and forms its own conclusions about the relevancy of our claims. Nevertheless, we have endeavored to arrange them to build progressively upon the themes explored earlier in the book.

As we argued in Chapter 2, the failure to recognize the inherent nature of the system (i.e., simple, complicated, or complex) can lead to interventions that are bound to fail. While narrow expertise, standards, and best practices can be effective in managing problems that arise from simple and complicated systems, they are an inadequate response when confronted with a complex system arising from a coupling of natural and human systems. In Chapter 6, Michal Russo and Laura Read highlight such a mismatch in the typical approach to floodplain delineation. In its most common usage, the word "floodplain" simply refers to the land adjacent to a body of water that is prone to flooding on some interval, be it months, years, or decades. Most hydrologists add a few nuances to this description and use "return periods" in probabilistic rather than anecdotal terms. However, on the basis of either usage, the notion of "delineating a floodplain" is ostensibly the same: it is the process of estimating the geographic extent of flooding that is expected to occur with a given frequency or probability.

On the traditional view, floodplain delineation is a purely procedural and analytical process that ought to be left in the hands of experts. Why, then, do so many floodplain maps fail to predict postflood property damage? While the authors identify a number of contributing factors from a technical perspective, they also suggest that resolving these technical challenges is not enough. The authors argue that floodplain delineation is too often framed as simply an exercise in representing observed reality, when in practice the process of delineation always entails a simultaneous exercise of social power to construct reality. After Foucault, the authors suggest that the process of floodplain delineation "is not a mere reflection of reality"; rather, "[delineation] also constructs its own reality."

Traditional techno-centric procedural practices fail to recognize not only the presence of sociopolitical dynamics inherent in flooding but also their coupling to and inseparability from its natural dynamics. In place of standards and best practices, the authors focus on developing actionable insights that can make floodplain delineation (1) more useful and reflective of those it is intended to serve, (2) more explicit in its coupling of environmental and social dynamics, and (3) more responsive to inevitable but unforeseeable social and environmental changes. What emerges is not a theory of how to do floodplain delineation *the right way* but a proposal for "Flood Diplomacy" as a shared conceptual framework for interdisciplinary collaboration that recognizes floodplain delineation as a complex and adaptive process that needs to involve fact–value deliberation among affected parties.

Failures to recognize system complexity do not only have consequences for planning, they can also cripple critical operations. In Chapter 7, Michael Ritter documents failures to respond to the ongoing cholera epidemic in Haiti. Ritter argues that a failure to recognize the nature of complexity of the situation led to failures at multiple scales.

At the policy level, a disconnect between the political attractiveness of establishing the goal of "cholera elimination" and the more feasible and scientifically creditable strategy of "cholera containment" created a rift between decision makers and the responding organizations. Ultimately, this disagreement forced commitments to technical strategies that were not appropriate given the high degree of uncertainty. At the sector level, attempts to establish a "sustainable" supply chain of household water treatment and safe storage products were beleaguered by a lack of established consensus on what constituted affordable, equitable, or effective adoption. At the organization level, programs that failed to adopt participatory approaches to identify, develop, and manage local staff were often unable to achieve sufficient levels of trust to be effective.

Complexity demands effective stakeholder engagement, but Ritter's analysis identifies three barriers that limited stakeholder engagement in response to Haiti's cholera epidemic. Without the requisite levels of trust, dynamic leadership, and stable resources, Ritter suggests achieving effective and sustained stakeholder engagement was fraught with practical difficulties. These same barriers that hindered stakeholder engagement also promoted deficiencies in the institutional capacities of responding organizations that ultimately limited the effectiveness of their day-to-day operations. Ritter concludes with actionable insights about how these barriers might be reduced or eliminated.

In Chapter 8, Sixt et al. explore the importance of requisite institutional capacities in more detail. The authors provide a case study of groundwater governance in the High Plains Aquifer in the United States, which is one of the largest groundwater systems in the world. Through a detailed analysis of state-level governance institutions in Kansas, Nebraska, and Texas, the authors hope to establish whether the capacity exists to support the type of collaborative and adaptive management advocated for by the Water Diplomacy Framework.

Sixt et al. distinguish between the micro-level capacity factors (e.g., the resource, leadership, and trust issues present in the response to cholera in Haiti) and the macro-level capacity factors engrained in centuries-old governance structures (e.g., rights of stakeholders to organize, to make collective decisions, and to establish boundaries and limits on resource extraction). Building on the work of Elinor Ostrom and others, the authors rank each state according to how well it meets six macro design principles for sustainable resource governance. No state meets all of the design principles, but the analysis yields important and actionable guidance (e.g., establishment of institutional alternatives to lawsuits) that can help remedy the deficits.

Such design principles, the authors argue, are indicators of requisite institutional capacities for sustainable groundwater governance. However, effective long-term governance of water resources requires not only institutional capacity but innovative policy and technology options that make interstate cooperation mutually attractive. In Chapter 9, Agustín Botteron demonstrates how the Water Diplomacy Framework's conceptualization of water as a flexible resource can be a starting point for such innovations.

Botteron's analysis is aimed at identifying creative options in the Eastern Nile Basin, especially those that can foster cooperation between Egypt, Ethiopia, Sudan, and South Sudan. Overall, the total inflow of water into the Eastern Nile Basin amounts to about 895 billion cubic meters (BCM) per year. The author notes this figure in sharp contrast to the 84 BCM per year that are currently the focus of ongoing disputes between the basin countries.

The large gap between the total inflow and the debated volume suggests that there is significant room for exploring mutually beneficial solutions. The author demonstrates this by examining the impact of improved irrigation efficiency, additional rainwater capture, and cropping changes in the region. On the basis of the scenarios considered for this research, Botteron estimated increases of the effective water availability in the Eastern Nile Basin of up to 11 BCM per year. The additional volume that can be achieved through cooperation on these strategies reduces the pressure of each state to pursue zero-sum thinking.

While states continue to be the only parties to agreements about transboundary waters, this reframing of water as a flexible resource also makes space for the consideration of nontraditional "actors" such as ecosystem functions and services. In Chapter 10, van Rees et al. argue that traditional decision-making approaches too often lead to the "reduction of complex ecological phenomena to simple metrics or hard rules that may constrain available solutions." Rather than seeing the environment as a monolithic consumer of water, an effective accounting of ecological complexity recognizes the critical role of diverse eco-system functions and services.

The authors explore how this recognition (or its absence) impacts the framing of complex water management challenges in three separate case studies. The first case documents how water management decisions in the absence of ecological considerations led to the ongoing decline of the Caroni Swamp in Trinidad – a decline that, in human terms, has led to increased incidence of severe flooding and has threatened the swamp's $1 million annual ecotourism draw (Trinidad and Tobago dollars). The second case considers coffee cultivation in Santa Maria, Costa Rica. In this example, the authors explore how an acknowledg-ment of ecological services allows agroforestry to be seen as an approach to coffee cultivation that increases the benefits available to all stakeholders. The third case explores how a comprehensive consideration of ecological phenom-ena identified opportunities to simultaneously protect culturally important birds, improve reef health, and reduce groundwater demand in Oʻahu, Hawaiʻi.

On this exploration of different linked processes and different ways of accounting, "the complexities of ecological and hydrological systems do not always lead to greater difficulties in management, but can present unanticipated, synergistic solutions to seemingly separate problems." Frameworks that explic-itly identify ecological phenomena as stakeholders in their own right – such as the authors' proposed "ecological stakeholder analog" concept – represent a principled and pragmatic mechanism for addressing the complexity of coupled natural and human systems.

This attention to both principles and pragmatism is imperative when address-ing complex water management challenges, as we explored in Chapter 4. This opens the door to an accommodation of both numbers and narratives, some-thing that is especially important in cases that require collaboration and joint fact-finding between indigenous and nonindigenous groups. In Chapter 11, Laura Corlin details how these important Water Diplomacy ideas can help address the complexity inherent in securing access to safe drinking water across the Navajo Nation.

The problems with water access in the Navajo Nation are inseparable from centuries of colonization and oppression. Past exploitation of the Navajo by federal and state agencies in the United States has led to fractured trust and deep and persistent power asymmetries. The Navajo government retains the authority for all water provisioning decisions in the Navajo Nation – a right and a responsibility that is a critical component of tribal sovereignty. Never-theless, collaboration with federal and state agencies is the only practical mech-anism for ensuring the Navajo get the water allocations owed to them under their existing agreements and settlements.

Corlin explores the nuances of this fraught relationship and then identifies opportunities to build trust and social capital, as well as to achieve equitable redistributions of power. Through the lens of Water Diplomacy, the author is able to explain why past efforts to negotiate solutions have failed and how more inclusive approaches could achieve sustainable, equitable, and mutually beneficial outcomes. Acknowledging the dynamic nature of coupled natural and human systems, Corlin points to the importance of developing "contingent solutions that meet the current needs of all stakeholders while offering opportunities to adapt to changing conditions over time."

Indeed, contingent approaches are a cornerstone of a collaborative and adaptive response to complex water management challenges. By adopting a contingent mode of planning, collaborative management institutions are better equipped to endure the stresses imposed by the emergent and unpredictable behaviors of coupled natural and human systems. In such systems, surprises are the rule rather than the exception, as Palash et al. remind us in Chapter 12. The authors provide several examples of such surprises arising from a coupling of natural and human systems within the southwest Bangladesh delta.

Perhaps the most dramatic of these examples was a well-intentioned policy shift that attempted to reduce exposure to microbiologically contaminated shallow wells and surface water sources. This internationally recognized campaign to install deep tube wells and encourage their use was initially declared a success, as microbial disease burden was reduced as expected. However, many of these tube wells actually tapped into reservoirs contaminated with high levels of naturally occurring arsenic. This exposure went unnoticed for decades, until seemingly all at once there was a recognition that more than 20 million people were suffering from the effects of arsenic exposure. Since the contaminated water was also used for irrigation in many instances, it has also contaminated crops and soils.

In this way, the policy promoting tube wells has extended arsenic exposure to people and ecosystems that were never anticipated. In addition to examining this and other surprises which have arisen in the southwest Bangladesh delta, Palash et al. emphasize the need to acknowledge that emergence – like arsenic poisoning – will remain unanticipated at the beginning of the planning and implementation phase; consequently, one must be willing to consider and adopt interventions appropriate for particular contexts. Because it is nearly impossible to design and execute permanent solutions to complex water problems like those in the southwest Bangladesh delta, one needs to seek and find contingent approaches to resolve these water problems with changing circumstances.

When considered together, these chapters offer a comprehensive commentary on the changing nature of professional practice and academic research on contemporary water issues. It starts with the importance of recognizing the inherent nature of the system (i.e., simple, complicated, or complex). There is also a growing awareness that the solutions that work for simple and complicated systems usually will not work for complex water problems arising from coupled natural and human systems. Not all water systems are complex; for simple and complicated systems, the notion of cause and effect, best practices, and prescriptive advice based on expert opinions usually work well. Complex systems by their very nature resist simple diagnosis. And in light of this recognition, the tacit goal of establishing cause-and-effect relationships that is present in many disciplines is abandoned in favor of the interdisciplinary activity of identifying the capacities and constraints of organizations and institutions at both the micro and macro scales.[1] This diverse and inclusive mode of thinking is carried further forward with the goal of establishing creative technology and policy options that view water as a flexible resource and appreciate the value of linked options and services (e.g., ecosystem services). All of this is done with attention to both principles and pragmatism as water professionals look for ways to build inclusive collaborative management processes that embrace equity and sustainability. Importantly, water professionals need to approach these problems with humility and recognize that complex coupled natural and human systems exhibit emergent phenomena that can manifest as unintended detrimental impacts on the health and prosperity of people and ecosystems.

Note

1 Choudhury and Islam (2018) proposed a minimal set of "enabling conditions" required to initiate, design, and implement a resilient negotiated process to resolve transboundary water management issues. These are an active recognition of interdependence, the presence of mutual value creation, and an adaptive regime of governance.

Reference

Choudhury, E., and Islam, S., 2018. *Complexity of Transboundary Water Disputes: Enabling Conditions for Negotiating Contingent Resolutions*. New York: Anthem Press.

6 Flood diplomacy

The hydrological, technical, and sociopolitical challenges of delineating usable floodplain boundaries

Michal Russo and Laura Read

Introduction

Floods are frequently cited as one of the most expensive natural hazards both in the US and globally (Highfield et al. 2013). The economic impacts of flooding have been increasing rapidly over the last half century, and future projections by leading global and national experts indicate that these trends will continue to escalate (Chang and Franczyk 2008; Ntelekos et al. 2010; Gall et al. 2011). A rich literature debates the causes and drivers of these increases, including development, (White and Greer 2006), water infrastructure (Brody et al. 2007), human practices that degrade the eco-system (e.g., deforestation, wetland filling; Villarini and Smith 2010), and climate change (Milly et al. 2002; Merz et al. 2012; Arnell and Gosling 2016). Still others postulate that vulnerability has increased, attributed to population growth and wealth concentration within exposed areas (Pielke and Downton 2000). Despite the evolving theories on how human-water dynamics add complexity to floodplain hydrology (Di Baldassarre et al. 2013), the criteria under which the federal government delineates flood hazards have not changed significantly in the last two decades (see Ntelekos et al. 2010 for a review of the National Flood Insurance Program).

More than a century ago, scientists established that a floodplain has unique hydrologic and ecosystem properties and that development within the floodplain can increase both the probability of flooding and vulnerability to damages (Leopold 1994). Hydrologically, the floodplain is a land surface with a certain set of characteristics that increase the likelihood of inundation during a flow event. The natural floodplain is an important part of the river ecosystem, where the periodic pulsing of high flows creates rich habitat for many species. As populations and property encroach on the floodplain, society's perception of the floodplain has become associated with the concept of hazards. In response, in 1968 the Federal Emergency Management Agency (FEMA) and the National Flood Insurance Program (NFIP) set national policy aimed at reducing impacts through development restrictions and insurance programs.

FEMA and the NFIP developed the floodplain concept based on the conventional framework whereby flood risk is characterized by historic streamflow records. Through this process, they standardized the floodplain as the

"100-year flood." This process served as a means of systematically identifying those populations and properties exposed to a certain level of risk to riverine, and in some areas, coastal storm surges (Anderson 1974). The 100-year flood concept refers to the annual exceedance probability (p) of experiencing a certain magnitude streamflow every year ($p = 0.01$) and assumes stationarity, i.e., that the probability of this certain event occurring does not change from year to year (Olsen 2006). This method relies on an expected relationship between streamflow magnitudes and flood losses to set US flood insurance policies. As a result, the term *floodplain* quickly became interchangeable with those areas having a 1% or greater chance of annual flood. For communities electing to be covered by FEMA's flood insurance program, mandatory requirements and insurance standards were applied to any property within the defined floodplain.

Over the past half century, the technology, including both data and methods used to determine the location of the 100-year floodplain, has improved significantly. When the NFIP adopted the concept of the 100-year flood, both flood damage data and computational capability were extremely limited (Schaake and Fiering 1967). Today, more sophisticated automated approaches match cross sectional data with historic floodplain data, topographic maps, and soil maps to produce water surface elevations for various riverine flood flows (and their recurrence intervals; FEMA 2015; FEMA 2017). In recent decades, advances in floodplain delineation include use of high-resolution topography, integration of hydraulic models and geographic information systems (GIS) to automate processes, and general improvements in computing speeds (Noman et al. 2001). These advancements, while notable, have functioned to sustain the operationalization of the 100-year concept and its underlying relationship between recurrence intervals and damage payments. However, if we step back to consider the initial goal of reducing flood impacts, there is mounting evidence that this relationship is fraught with limitations and misconceptions. Simply more accurate mapping may not suffice to capture all the complex risks that are needed to accurately depict the floodplain.

Technical challenges to floodplain delineation

Researchers have written at length on technical limitations to FEMA's flood hazard delineations (Burby 2001; Ntelekos et al. 2010). Major challenges center on (1) poor-quality or old data, (2) nonstationarity, (3) standardized delineations, and (4) nonintegrated models.

Poor-quality or old data

Maps are developed based on records from historic stream gauges and topographic data. Datasets can be crude or incomplete if data points are few or elevation maps are of poor resolution (Carolan 2007). These challenges can be especially problematic in relatively flat coastal areas, where one foot of vertical distance can translate into a mile of horizontal distance (Colby and Dobson 2010; Maidment 2010). Maps also have to be continually updated to reflect changing conditions (Burby 2001). Land cover change, for example, can dramatically alter the boundary of the floodplain. In 2003, FEMA sought to re-map 92% of the US with updated data through the Map Modernization Program.[1] Before then, only 21% of the maps met FEMA's own quality standards (National Academy of Sciences 2009).

Nonstationarity

Traditional mapping assumes stationarity; however, recent evidence confirms that floods can exhibit non-stationarity − i.e., annual risk is changing due to human influences such as land-use change, urban development, or climate change (Villarini et al. 2009). Nonstationarity changes the shape of the distribution of the average return period of peak floods, suggesting that a new probability distribution should be derived for characterizing such systems if a sufficient record shows trends in the annual maximum flood series (Salas and Obeysekera 2014; Read and Vogel 2015). As of the revised US National Flood Frequency Guidelines Bulletin 17C, nonstationarity is recognized as a consideration in the delineation process (England et al. 2019), though guidelines on accounting for its influence remain unspecified.

Standardized delineations

Some rivers migrate, sometimes miles, along their borders while other water bodies naturally or artificially restricted by permanent fixtures may fluctuate only minimally. Hydrologists Smith (2000), Murphy (1958), and Reuss (1993) argue that this heterogeneity "makes it impossible to set definitions for a designated flood that are universally applicable" (Smith 2000, p. 255). As a means of establishing fair and equitable allocation of protection, FEMA has standardized the floodplain at the national rather than regional level. However, this generalization may have eroded local hydrological nuances and dynamics, leveraging shared common denominators over locally specific models and datasets.

Nonintegrated models

Delineations are also challenged for not including or integrating system components including basement flooding, stormwater runoff, and infrastructure failures. Yet in urban areas, these mechanisms often dominate the sources of flood impacts (White 2010). Coastal and riverine systems are also not integrated, leading to major oversights in the coastal communities where tidal inundation and storm surges may interact with runoff and riverine flooding.

Sociopolitical challenges to floodplain delineations

While flood risk maps are delineated according to the physical environment (mainly topography), it has long been acknowledged that social institutions drive flood risks (Galloway et al. 2006). The technical challenges to integrating human behavior in flood models are numerous (Pahl-Wostl 2006). In addition, multiple complications arise due to the misapplication and miscommunication of floodplain delineations associated with social institutions (Highfield et al. 2013; Morss et al. 2005; Wachinger et al. 2013; Bubeck et al. 2012). Technical updates to the delineation process have done little to accommodate these less tangible mechanisms. The literature generally converges on three mechanisms by which current delineations lead to a misrepresentation of flood risks by social institutions: (1) (re)development subsidies, (2) inappropriate time horizons, and (3) binary delineations.

(Re)development subsidies

Flood insurance is aimed at reducing flood hazards as well as supporting property owners affected by flood damages. These two pursuits can sometimes have conflicting strategies and outcomes. By providing federally subsidized flood insurance, FEMA is providing financial support to homeowners who are affected by flooding. Often, this support comes in the form of payments to rebuild damaged homes at the very location they were damaged. This support is sustained as a social good despite our collective knowledge that development in the floodplain raises flood risk (Burby 2001). This practice sets up the notion that if you flood, the US government will bail you out. Updates to the regulation are aiming to reduce the excessive costs of structures that redevelop after flood damages. However, current protocol remains to rebuild in place (Moore 2017).

Inappropriate time horizons

Traditional communication of flood risk (e.g., 100-year flood) to an average homeowner is mismatched with the risk faced over an average mortgage planning horizon (Read and Vogel 2015). For example, when communicating risk with regard to an engineering infrastructure, risk is conveyed as the reliability to protect against an "event" over the given design life, like a 50-year planning horizon. Risk is expressed in terms of a failure due to at least one event occurring within that time period. If a homeowner was given a similar message, "the risk of experiencing at least one flood-flow 'event' is x% over the 30-year mortgage period," perhaps the relationship between an individual and their perception of flood risk would change.

Binary delineations

Floodplain delineation is binary such that properties are either "inside" or "outside" the floodplains. This convention, created for efficiency of regulation, does not reflect the physical properties of a floodplain which are continuous. This convention creates a false notion of safety even feet outside of the special flood hazard area (Blanchard-Boehm et al. 2001). FEMA's Hazus program estimates the damages to be 1.5 times larger directly outside of the 100-year floodplain (Galloway et al. 2006). Due to lower costs and fewer regulations (i.e., mandated insurance), there are incentives to develop immediately outside the delineated floodplain boundary. This has an unintended consequence of supporting increased population densities in this zone. Greater densities result in higher costs of damages and vulnerabilities from flooding (Patterson and Doyle 2009). In addition, the interaction between a static construction of risk and the dynamic nature of a floodplain means that what was once "outside" can become "inside." Communities are exposed to dynamics, such as increases in impervious urbanization, wetland filling, and deforestation (Patterson and Doyle 2009), as well as increased extreme precipitation, more frequent and higher-magnitude hurricanes, rising sea levels, and earlier snowmelt associated with climatic changes (see Easterling et al. 2000, references within).

Studies show a large portion of flood impacts occur outside of the floodplain

Over the years, researchers have been noting increasing evidence of impacts occurring outside of the delineated floodplain. Wescoat and White (2003, p. 153) described their personal account, based on decades of research, that the "majority of flood losses are occurring outside of the floodplain." In the preceding two decades, a series of reports estimated that between 23 to 40% of flooding occurs outside of the delineated floodplain (Federal Interagency Floodplain Management Task Force 1992; Dixon et al. 2006; Galloway et al. 2006; Highfield et al. 2013). Brody and Highfield (2011) estimated that based on a national sample of 450 jurisdictions over a decade, approximately 25% of insurance claims were located outside of the 100-year floodplain. This percentage increased even further in coastal jurisdictions due to high variability, where a study in Harris County, Texas, from 1978 to 2008 revealed that more than 47% of claims were located outside the 100-year floodplain boundary (Brody and Highfield 2011). A 2016 study outside of Houston looked at five consecutive tropical storms and showed approximately 80% of the 1,096 claims during the 11-year study period were located outside of the FEMA-derived 100-year floodplain (Blessing et al. 2017).

However, these numbers also reflect an enormous sampling bias – in reality, the challenge may be much more extreme. While insurance coverage is mandatory for mortgaged homes located within the 100-year floodplain, there is no requirement for insurance outside of this binary boundary. As a consequence, less than 1% of the households outside of the delineated floodplain have flood insurance policies (Dixon et al. 2006). Looking at a finer resolution, a recent study by the Center for Neighborhood Technology (Festing et al. 2013) showed that in Cook County, Illinois, home to Chicago, there is no correlation between zip codes with any land within a floodplain and aggregate damage payouts. A follow-up statewide study showed that 92% of urban flood-damaged properties were outside of FEMA's delineated floodplain (Cossio et al. 2015). The proprietary methods by which flood damage losses and flood event extent data is collected has led to a dearth of national evidence to systematically support or refute the claims made in Illinois. Due to the interdependence of social and hydrological flood drivers, it is also challenging to determine which of the many critiques of floodplain delineation is causing these discrepancies.

The implications of an inaccurate delineation

While flood insurance is only one of many flood management actions that communities can take, once approved, FEMA's 100-year flood delineation serves as the key spatial indicator for determining flood risk in the United States (Highfield et al. 2013). NFIP's flood maps have become de facto regulatory and administrative documents, used to inform a range of policies and development from infrastructure siting to open-space corridor planning. When floodplain delineations fail to represent hazards that exist outside of the floodplain boundary, they (1) support building in harm's way, (2) reduce preparedness and awareness, (3) increase the vulnerability of ecological systems, (4) exacerbate unfair allocation of financial burdens, and (5) result in trust issues between the public and the federal government.

Building in harm's way

Residential development in the floodplain is regulated with specific standards, for example raising the base elevation. Public, commercial, industrial, and institutional developments have higher standards in the floodplain to ensure public health and safety. Improper delineation can result in communities planning development of critical facilities – for example schools, nursing homes, wastewater plants,[2] fire houses – in harm's way, i.e., within the floodplain boundary (Galloway et al. 2006).

Preparedness and awareness

Residents living outside of the delineated floodplain are less likely to be prepared for flooding by purchasing flood insurance, implementing structural accommodations (Gilbert F. White National Flood Policy Forum 2004), or integrating nonstructural adaptations such as awareness of evacuation routes or moving mementos out of the basement. Emergency managers are also less prepared to assist vulnerable populations (e.g., elderly, infirm, low income) when flooding occurs outside of the delineated floodplain.

Ecological vulnerability

Damages from floods outside the floodplain can also have substantial long-term ecological impacts and reduce the effectiveness of the natural environment in protecting against floods (Brody et al. 2013). Large flood disasters typically prioritize aid to support public health and economic stability, deprioritizing ecological impacts (Newbery et al. 2010); waste goes untreated into water bodies to relieve sewer backups (Nilsen et al. 2011); and minimum base flows are overlooked to provide water in extended droughts (Carter et al. 2009). However, when floods occur in unexpected magnitudes and locations, it is more likely that safeguards to handle such events are not in place, potentially resulting in ecological disasters (e.g., toxic spills, water quality issues, and stress on sensitive lands). In urbanized areas where pressures on ecological systems are already high, the increased uncertainty associated with inaccurate flood delineation may result in surpassing thresholds and permanent loss of critical ecosystem functions (Ernstson et al. 2010).

Unfair allocation of financial burdens

Premiums are based on the relationship between flood damages and expected inundation risks. If delineations underestimate flood risks, then premiums place an undue burden on those homeowners with a mandatory flood insurance policy within delineated areas.

Trust issues

Residents largely base their flood risk perceptions on recent evidence (Atreya and Ferreira 2015). If repeatedly the areas affected do not correlate with the areas paying premiums, map delineations are viewed as inaccurate, unfair, or irrelevant by the populations they serve. As a result, trust is eroded. Notions of mistrust can spread from residents to public service providers and ultimately result in substantial inefficiencies (Samaddar et al. 2012; Wachinger et al. 2013).

As the proceeding section suggests, there are multiple avenues by which flood delineation could be improved upon, with substantial and overlapping implications to flood prone communities. Based on the literature, we hypothesize that floodplain delineation is not strictly a reflection of physical drivers including climate, hydrology, and landform but also of social drivers including technology and information, politics and regulation, and worldview and risk tolerance. Further, in addition to informing flood management actions such as construction and preparedness and response, flood delineation shapes risk perception – i.e., how individuals understand flood risk including their beliefs and expectations about future risks. Importantly, these perceptions have direct influence on flood management and therefore flood risks and impacts. Consequently, challenges with floodplain delineation, stemming from ineffective or erroneous regulation and mapping elements, not only shape flood management but can also harm risk perceptions, magnifying problems with the implementation of flood management strategies. In the next section, we test these notions "on the ground" by examining a case study in a city where a recent devastating flood had a large impact outside of the delineated floodplain.

A case study in Warwick, Rhode Island

To investigate flooding outside of the delineated floodplain – including both the causes and impacts – we traveled to Warwick, Rhode Island. In 2010, the city experienced a massive flood that caused $79 million in direct damages (City of Warwick 2011). The flood was categorized as a 200-year flood, so it is not surprising that areas outside of the 100-year floodplain were affected. However, the flood event was seen as a surprise because the damages occurred largely along the Pawtuxet River, which was not even in Warwick's Hazard Mitigation Plan. Up to the 2010 flood, community flood risks were perceived as largely a coastal issue. Technical limitations such as poorly calibrated models or coarse and older datasets might have contributed to this oversight, but what about sociopolitical limitations? Did technical experts, decision makers, and affected parties perceive risks differently? Did these differences shape their responses and therefore contribute to (e.g., amplify) flood damages? Our study seeks to understand – in a city with a recent, significant, and unexpected flood event – *how different are the interpretations of flood risks between stakeholders, and what role did they play in flood risk mitigation behavior on an individual and institutional level?*

A background on Warwick's flood risks: flood mapping and the 2010 flood impacts

The City of Warwick, Rhode Island, is located along Narragansett Bay, south of Providence (see Figure 6.1). The city consists of 35 square miles of land area and 39 miles of coastline and hosts a population of about 86,000 (City of Warwick 2011). Warwick is located at the intersection of the Narragansett Bay and Pawtuxet River watersheds and is heavily developed (see Figure 6.2). According to the updated 2015 Warwick Hazard Mitigation Plan, the flood risk in Warwick is two-fold – (1) riverine influence from the Pawtuxet River which runs along the northern border of Warwick and drains east, and (2) tidal surge potential from the coast, i.e., Narragansett Bay. Prior to 2010 the city had seen major flooding in association with major coastal storms – namely the 1938, 1954, and 1991 hurricanes (see Table 6.1). In 1925, the Scituate Reservoir was built along the Pawtuxet River, securing a water source for the watershed as well as regulating flood levels in the river. Developments were built along the river. However, after decades of flooding regularly they were evacuated. A few developments remain in low-lying areas of the city (both along the river and along the coast), and flood fairly regularly (e.g., Conimicut Point, Elmwood Avenue, and the Industrial Park area). Warwick's coastal and riverine systems are connected such that during high tide or surge events, water flowing down the Pawtuxet gets constrained, and water levels can rise rapidly.

A review by the National Research Council in 2009 pointed out a number of potential issues with some of the methods FEMA uses to estimate inundation resulting from storm surges and waves. Spaulding et al. (2017) noted that

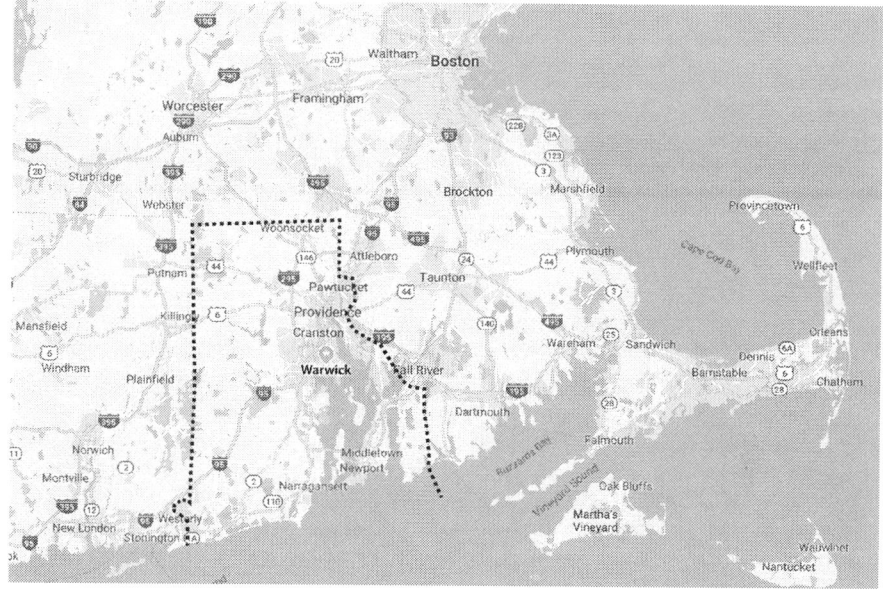

Figure 6.1 Context map of Warwick, RI.

Data source: Google Maps

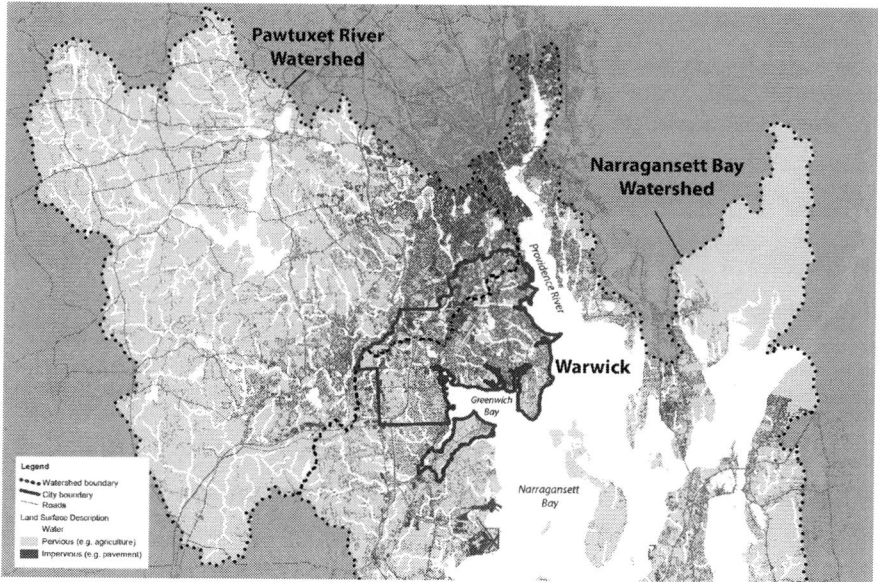

Figure 6.2 Land use map of watershed.

Data source: Rhode Island Geographic Information System

Table 6.1 Historic Damages of Coastal Flooding

Date	Disaster	Damage (in millions USD)
1938	Storm of '38	$306
1954	Hurricane Carol	$461
1991	Hurricane Bob	$115
2010	Severe Storms and Floods	$79

Data source: Warwick Hazard Mitigation Plan (City of Warwick 2011)

Figure 6.3 Photo of 2010 wastewater treatment facility flooding.

Source: Rhode Island Emergency Management Agency

several of these issues applied to Warwick, primarily the use of a 1D model instead of a 2D coupled model to represent the coastal–riverine system and that in remapping in 2013, FEMA did not follow the NRC's recommendations. Further, prior to the 2010 flood event, Warwick's flood maps as well as Hazard Mitigation Plan did not highlight risks along the Pawtuxet, focusing instead on coastal risks. However, the 2010 flood overtopped the Pawtuxet River by 3 feet, leading to unprecedented flooding affecting many of Warwick's low-lying neighborhoods, as well as several critical facilities, roads, public facilities, and regional employment centers: (1) the wastewater treatment plant flooded, resulting in more than $14 million in damages (see Figure 6.3), (2) the local mall was heavily damaged, and (3) the city closed down schools, the airport, and the hospital for a few days as clean water and energy were resupplied. One of the biggest challenges of the 2010 flood, as CNN (2010) reported on April 2, was that "nobody was prepared." Even as the rain continued to fall, the wastewater treatment facility manager, the mall owner, residents, and even the mayor expected that they were safe. The pattern of precipitation was so novel that the National Weather Service model used to predict flood risks and release alerts underpredicted impacts. In addition, thousands of flooded homes were not in the floodplain and did not have flood insurance.

The occurrence of this hazardous event provided an opportune time to reevaluate the defined risks and determine whether changes had occurred or new information was available to warrant a significant update to the city's flood delineation and hazard mitigation plan. In February 2011, the Warwick Mitigation Plan was updated in accordance with FEMA regulations that plans must be revisited every 10 years for eligibility in certain FEMA funding programs. At that point, however, the plan was written as data from the 2010 flood event was still being collected. The plan only nominally addressed changing perceptions about flood risks in the City.

The resulting risk scores were reevaluated; however, only one risk score was recommended for adjustment: flood. "After the extreme storms and riverine flooding experienced by the City of Warwick in March 2010, *the area of impact for this hazard type was increased from 10 to 50 square miles. This was done to account for the fact that riverine flooding occurs in a different geographical area than previously identified coastal flood hazards.* Although a first impression is that the risk score for flooding should have increased substantially, this was not the case, since the risk of flooding in Warwick is already high due to tropical cyclones, nor'easters, and storm surge. The addition of riverine flooding was of relatively little impact to the overall flood risk score" (City of Warwick 2011, pp. 2–3; emphasis ours).

Following the 2010 flood event, the USGS conducted a detailed study of the flooding in the Pawtuxet River (Zarriello et al. 2014). The study compared new cross-section data points with updated hydraulic models including tidal zones to create a better estimate of flood risks. In addition, the 2015 FEMA flood insurance rate map (FIRM) was updated due to FEMA requirements, and at this time, several key revisions to the Warwick Hazard Mitigation Plan were implemented. These new maps utilized new data from the 2010 flood as well as Hurricane Sandy in 2012.

In terms of on-the-ground flood mitigation activities, not much has been done (by the accounts of state and local flood managers). The wastewater treatment plant has been retrofitted to the 500-year floodplain level, a handful of site-level projects were completed to remove road ends and reduce impervious surfaces, and multiple houses were raised or retrofitted with backflow valves. In 2011, the Pawtuxet Falls Dam was removed, but it was explicitly expected to *not* have an effect on flood levels.

Methods

In the spring of 2016, we conducted 25 interviews with 32 diverse stakeholders in and around Warwick. Like most coastal communities, Warwick's flood mitigation behavior is shaped by a network of public and private actors. For example, both the Rhode Island Emergency Management Agency (RIEMA) and the Rhode Island Coastal Resources Management Council (CRMC) administer FEMA's flood insurance rate maps as a means of regulating development. REIMA focuses on inland properties, while CRMC focuses on coastal properties. Our interview selection process aimed to capture the diversity of roles and relationships within this network. Subjects were selected based on their positions in relevant organizations and through snowball sampling targeting specific interests. We explicitly sought to increase diversity along three dimensions: (1) diverse interests *affected* by flooding (e.g., houses getting flooded, businesses damaged, habitat they are trying to protect threatened, infrastructure they manage shaped by flooding); (2) experts at agencies who *analyze* flooding (e.g., hydrologists at the state, federal or local level, as private consultants, academic researchers, or public works); and (3) decision makers who *affect* flood mitigation action and investment funds (e.g., mayor's office, regulatory agencies). We spoke to individuals representing federal, state, and local interests; coastal, riverine, and other affiliations; regulatory and advisory roles and those directly affected; as well as both technical experts and lay people (see Figure 6.4). For the purpose of brevity, acronyms and codenames are used to represent stakeholders in the remainder of this chapter (see Table 6.2). Stakeholders are classified by their operation level (local, state, or federal), hydrology (coastal, riverine, or both), roles (advisory

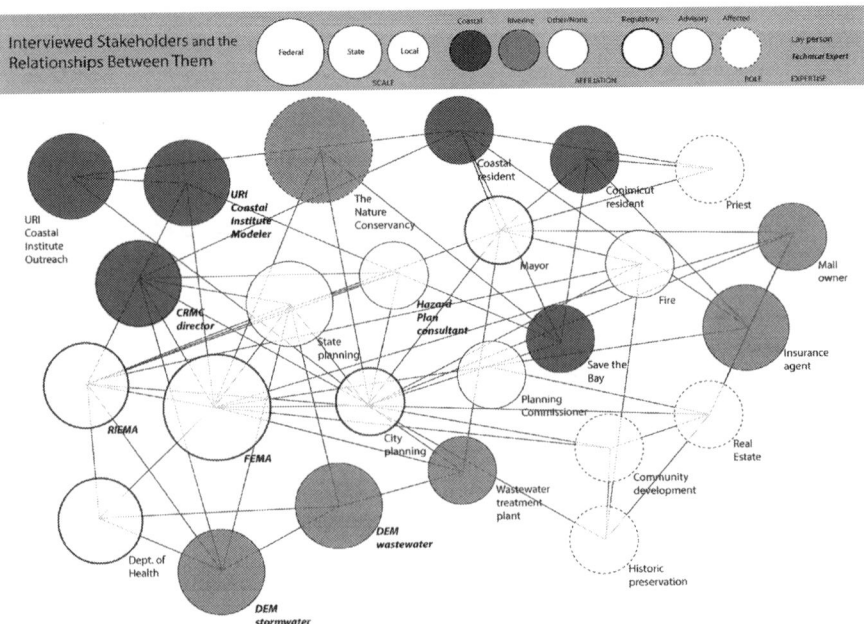

Figure 6.4 Interviewed stakeholders and the relationships between them.

Table 6.2 List of Stakeholder Code Names, Agencies, and Positions

Code Name (Abbreviation)	Agency, Position
Planning	Warwick City Planning
Mayor	Mayor of the City of Warwick
Community	Warwick Community Development
Fire	Warwick Fire Department
WWTF	Wastewater Treatment Facility
Mitigation	Private Consultant for Warwick Hazard Mitigation Plan
Commissioner (Commiss.)	City Planning Commissioner
Historic	Private river resident on historic preservation board
Coastal In	Private coastal resident, living in the floodplain
Costal Out	Private coastal resident, living out of the floodplain
Priest	Coastal resident who runs a congregation in Warwick
Realtor	Private realtor who serves on the City Planning Council
Mall	Private local business, the Warwick Mall
Insurance	Private insurance provider for state
RIEMA	RI Emergency Management Agency
CRMC	RI Coastal Resource Management Council
State Stormwater	RI Department of Environmental Management, Stormwater Division
State Wastewater	RI Department of Environmental Management, Wastewater Division
Health	RI Department of Health
Transportation (Transport)	RI State Planning Office
URI Models	University of Rhode Island Coastal Resources Center, modeler
URI Outreach	University of Rhode Island Coastal Resources Center, outreach
Save the Bay	A non-profit conservation organization
TNC	The Nature Conservancy
FEMA	Federal Emergency Management Agency

affected, manager), expertise (direct or affected), and interests in Table 6.3. Additional details on select agencies and their missions are included in Table 6.11 at the end of this chapter.

Interviews were conducted in people's offices, public spaces (e.g., coffee shops), and homes. Interviews generally lasted 1 hour and were audio recorded, transcribed, and coded in NVivo®, a qualitative data analysis software. The research team identified both categorical and emergent themes from the transcripts. The interviews were developed to reveal differences in their understanding as they pertain to the following four dimensions: (1) *Problem definition*: what counts as flood risk? (2) *Causes/Drivers*: what is causing Warwick's flood risk? (3) *Uncertainty & Knowledge*: what is known about flood risks, and by whom? (4) *Approach and Barriers*: what does a desirable future look like, and what ought to be done to reduce risks? See Box 6.1 for our interview questions. The interview protocol was largely informal and open ended, aimed at revealing the subject's framing as opposed to their specific confirmation of key attributes.

Table 6.3 Stakeholder Levels, Hydrology, Roles, Expertise, and Interests

Name Code	Level			Hydrology			Role				Expert		Interest
	Federal	State	Local	River	Coast	Other/None	Decision Making/ Regulatory	Advisory	Affected	Manager	Direct	Technical	
Planning			x	x	x		x	x	x	x	x	x	varied
Mayor			x	x	x		x	x	x	x	x	x	varied
Community			x	x				x		x	x		community
Fire			x			x				x	x		public safety
WWTF			x	x					x	x	x	x	infrastructure
Mitigation			x	x						x		x	hazards
Commissioner		x			x		x	x	x	x	x	x	varied
Historic			x	x				x	x	x	x	x	community preservation
Coastal In			x		x				x		x		residents
Coastal Out			x		x				x		x		residents
Priest			x		x			x	x		x		community
Realtor			x		x		x	x			x		business
Mall			x			x			x		x		business
Insurance		x				x		x				x	business
RIEMA		x		x			x	x		x		x	hazards
CRMC		x			x		x	x				x	varied
State Stormwater		x		x			x	x	x	x		x	infrastructure
State Wastewater		x		x			x	x	x	x		x	infrastructure
Health		x				x		x		x		x	public safety
Transportation		x		x				x	x	x	x	x	transportation
URI Models		x		x	x			x				x	ecosystem health
URI Outreach		x		x	x			x		x		x	ecosystem health
Save the Bay			x	x	x			x	x		x	x	ecosystem health
TNC	x					x		x			x	x	ecosystem health
FEMA	x			x	x		x					x	hazards

Box 6.1   Interview questions

1   What is your role in relationship to flooding in the City of Warwick?
2   What have been some of the major impacts of flooding in the City of Warwick over the last 5 years?
3   What do you think is causing these impacts?

   a   How do you expect impacts to change in the future?

4   How well do you think we understand future risk?
5   What does a desirable and realistic future for the city in terms of flood impacts look like?
6   What do we need to do to make this desirable future a reality?
7   What are the barriers preventing the city from making this future a reality?
8   We want to get a wide variety of perspectives. Based on your experience, who else do you think we should talk to? Do you know of any assessments or organizations we should look further into?

Results

Due to the small sample size, no statistical significance can be derived from these results. Instead, narratives are used in a more exploratory framework to reveal important and potentially overlooked sentiments and factors affecting flood risk management. Interviews revealed a diversity of perspectives in relation to the four flood risk dimensions. While we expected perspectives to fall largely along categorical lines – e.g., state managers have a different perspective than national or local managers, residents have a different perspective than technical experts, etc. – there was as much variability within groups as across groups. It is critical to stress that the majority of stakeholders we talked to could not be classified strictly within one category. Technical experts were often also affected parties (living in the areas), while some individuals served on both local and state boards.

Question 1: problem definition – what counts as flooding?

River vs. coastal: FEMA's flood delineation modeling approach separates flood risks due to coastal inundation from fluvial or river flooding. This separation is pragmatic – the models that predict hazards in each system are unique. However, in many coastal cities, the two hydrological systems are linked, such that high tide or "inflow" can severely limit relief from river overflow during extreme precipitation events. Multiple interviews ($n = 6$) suggested that a delineation of two systems through separate models and maps can limit hazard mitigation personnel from effectively predicting the extent of flood damage when both systems are at play. Coupled storm–surge and riverine flood models are still relatively uncommon but are gaining in popularity from the local-scale (Kew et al. 2013) to the global-scale (Ikeuchi et al. 2017). Research suggests that regions prone to storm surge and riverine flood risks may experience compounded flood damages (e.g., discussed for an event in the Netherlands: van den Hurk et al. 2015). This may have been the case for Warwick during the 2010 storm and has also been the case in significant flooding events in the United States (US) and globally (see the list in Wahl et al. 2015). In addition, while the majority of stakeholders interviewed mentioned both riverine and coastal flood risks, they had highly uneven knowledge of both systems (understanding far more about the dynamics of one than the other).

Intense vs. urban and nuisance flooding: In addition to river- and coastal-driven flooding, stakeholders described flood risks driven by poorly situated development and infrastructure. These flood risks, indicative of urban flooding, may account for a major discrepancy between the location of losses and traditional floodplain delineation (Cossio et al. 2015). Floodplain delineation is associated with infrequent flood events (by definition, 1% annual probability) that cause intense, substantial, or major impacts. However, some research is showing that more frequent but minor flood events may cumulatively be causing even more damage in some areas (Armstrong et al. 2012; Archfield et al. 2016). Urban flooding is often associated with these types of events (i.e., "every time it rains, my basement floods"), but along the coast, nuisance flooding is associated with high-tide events (Moftakhari et al. 2017; i.e., "sunny day flooding"). Recent studies show that nuisance flooding – which can compromise infrastructure such as storm drains and road closures – has increased along US coasts between 300% and 925% since 1960 (Sweet et al. 2014). In Warwick, both minor and major flooding are a challenge, and both tidal and urban drivers support frequent but minor floods. The vast majority of stakeholders interviewed (21 out of 24) mentioned issues associated with minor but frequent flooding in addition to intense flooding. However, interviewees revealed knowledge gaps when describing how the two types of floods are connected, both hydrologically and in terms of regulations.

Flood impacts: Interviewees also defined flood risks in terms of impacts. Here we found a substantial correlation between roles and impact identification. For example, while the Department of Health manager mentioned mental health issues, the wastewater treatment manager spent the vast majority of the interview discussing damage estimates for infrastructure (see Table 6.4). Less than half of the stakeholders mentioned ecological impacts including effects on coastal marshes and water quality changes. Not surprisingly, among those who mentioned it were stakeholders from Save the Bay, The Nature Conservancy, and DEM's Wastewater division. Even fewer stakeholders mentioned public or mental health issues (7/24). On the other hand, everyone mentioned damages to property, and the vast majority described impacts to people or the community. It is important to highlight these discrepancies, as flooding can affect people in dramatically different ways – but not all costs are well accounted

Table 6.4 Impacts of Flooding as Discussed by Stakeholders

Name Code	public safety (evacuation)	public health (mold)	mental health	vulnerable populations	character of place	people/ community	businesses	jobs/the economy	costs	infrastructure services/ WWTF	stormwater infrastructure	roads & access	properties/ houses	airport, schools, hospital	marsh/ natural sys./env. impacts	water quality
Planning	+				+											
RIEMA			+		+	+	+	+				+	+	+	+	+
Fire	+					+				+		+	+	+	+	
Priest	+					+	+	+	+				+			
Mayor	+	+	+					+	+			+	+		+	+
Realtor				+		+				+		+	+		+	+
CRMC	+	+				+	+ +			+ +		+	+			+
Mall					+	+	+			+	+	+	+			
Community		+				+	+	+								+
WWTF	+		+		+	+	+	+	+	+		+	+			
State Wastewater						+	+	+	+	+			+		+	
State Stormwater						+	+	+	+				+			
FEMA		+				+	+		+	+			+			
Insurance						+				+			+			
Health	+					+		+		+		+	+	+		+
Mitigation									+	+		+	+			
Commissioner							+		+	+			+			
Historic			+ +		+	+					+					
Coastal In		+	+	+		+				+		+				
URI Models				+		+				+		+				
Save the Bay						+				+	+	+		+	+	
TNC		+		+		+		+			+	+			+	
Transportation	+					+				+		+	+		+	+
URI Outreach		+				+				+		+	+		+	+
Coastal Out	+	+	+			+		+	+			+	+		+	+ +

Table 6.5 Drivers of Flooding, Past and Future (Legend)

Drivers of Flooding (Legend)	
A	preparedness/knowledge
B	development (impervious surfaces)
C	building in harm's way
D	water infrastructure (capacity, siting, maintenance)
E	loss of wetlands
F	riverbank overflow
G	river flow alterations/armoring/flood gates
H	groundwater
I	elevation/topography/shoreline exposure
J	erosion
K	natural cycles
L	location in watershed/upstream development
M	high tides/waves/tidal
N	surge
O	storms (nor'easters, hurricanes, coastal storms)
P	precipitation
Q	extreme events
R	changes in intensity/frequency of P
S	snow/precipitation timing
T	climate change
U	sea level rise

for within the current system, causing many stakeholders to feel excluded (French et al. 2010).

Question 2: causes & drivers: what is causing Warwick's flood risks?

Interviewees in Warwick were keenly aware of the influence of human development on both coastal inundation and riverine floods. However, there was great variation in the dynamics that interviewees described when asked about flood risk. Broadly, when interviewees were prompted to discuss the current drivers of flood risks in Warwick, they focused largely on development (impervious surfaces, filling in wetlands, building in harm's way, infrastructure, altering the river flow) as well as major storms. However, when discussing future risks, interviewees focused primarily on climatic changes, especially sea level rise and changes to the intensity and frequency of precipitation events (see Tables 6.5 and 6.6). The majority of stakeholders (19 out of 25, or 76%) believe the flood risk is changing (another 2 are unsure). More than 84% of stakeholders (21 out of 25) view sea level rise as a major driver of flood risk.

Question 3: uncertainty and knowledge: what is known, and by whom?

How well do we understand flood risks? Consistent with the literature, interviewees described flood risks to be heavily studied but complex phenomena. When asked to describe how well we, as a society, understand flooding, interviewees characterized system components as fairly well understood but

Table 6.6 Drivers of Flooding, Past and Future

	Are risks changing?	A	B	C	D	E	F	G	H	I	J	K	L	M	N	O	P	Q	R	S	T	U
Planning	Y	↕	↕						↓				↑	↓		↓			↓			↑
Mayor	Y	↓	↓	↓	↓	↓			↓	↓	↓			↓		↓	↓		↓		↕	↑
Comm.	Y	↕	↕	↕	↑									↓	↕	↕						↕
Fire	?	↓																				
WWTF	Y	↓	↓	↓	↓		↓			↓		↓		↓		↑			↓		↓	↓
Mitigation	Y	↑	↕	↑	↑			↓							↓	↑	↓		↑	↑	↑	↑
Commiss.	Y		↕	↕	↑	↓		↕		↕					↑	↑		↑	↑	↑	↑	↑
Historic	Y		↓	↑	↑			↓		↑		↓		↓		↓		↕	↑		↕	↓
Costal In	N		↓	↓	↑	↓	↓	↕		↑		↑			↑			↑				↑
Costal Out	N		↓	↓				↓				↓				↓		↓				↓
Priest	N/Y		↓	↑			↓					↑				↓		↑			↑	
Realtor	N			↓																		
Mall	?																					
Insurance	Y		↓	↓	↓			↓	↓	↓	↓		↓	↓	↑	↕	↑	↓	↓	↑	↕	↕
RIEMA	Y		↓	↕	↑	↑		↑		↑				↑		↑		↑	↑	↑	↑	↑
CRMC	Y		↓	↕												↓			↓			↕
State Stormwater	Y		↓		↑											↕			↑		↕	↕
State Wastewater	Y		↓		↑			↓	↓							↓			↑		↕	↑
Health	Y		↓	↓												↑	↑	↑				↑
Transport.	Y													↓		↕						↑
URI Models	Y														↑	↑			↕			↕
URI Outreach	Y		↓	↓							↑			↓	↕	↑	↑		↕			↑
Save the Bay	Y										↑			↓	↓	↓	↓					↑
TNC	Y	↕	↕	↓	↕	↕		↓		↑		↓		↓		↓		↓	↑		↑	↑
FEMA	Y	↓	↓	↓	↕	↕		↓	↓				↓				↑	↑	↑		↑	↑

This driver was an influence in the: past ↓ | future → | both ↔

our predictability to be low. As one manager said, "I think the science is pretty good, I don't think we understand it all, but it certainly explains what, in my experience, is happening." One of the repeated challenges stems from how well models can estimate the interaction between system components. When asked if flood models adequately capture social dynamics, the majority of respondents said no. Other model integration challenges included (1) river and coastal hydrology (see above) and the (2) impact of urban infrastructure on flood risks.

How well can we predict flooding? Several stakeholders discussed our limited ability to anticipate how fast climatic changes such as sea level rise will occur. Several individuals mentioned that past flood maps were inadequate in predicting the 2010 flood – either because the maps were wrong or because of the faulty forecasting system (National Weather Service). In addition, some stakeholders suggested that coastal flood risks are more predictable than river-based flooding, as storms off the Atlantic can be monitored days in advance. The reporting on coastal risks was also described as more widely understood by the general population. Riverine flooding was perceived as less predictable although in a better state today than in 2010. This finding is supported by literature that also reports that information collected during post-flood risk assessments can lead to improved system capacity for preparation for future floods (Simonovic and Carson 2003). In fact, after significant flood events, in-depth studies can improve hydrometeorological understanding and forecast ability (e.g., Colorado's 2013 extreme flood, Gochis et al. 2015; Slovenia's 2007 floods, Rusjan et al. 2009).

How much agreement is there on flood risks? Cross-agency disagreements were frequently cited by stakeholders. Warwick is typical of many coastal towns, where multiple agencies model flood risks for different reasons and with varying levels of sophistication; 42% of the interviews mentioned a lack of agreement between agencies with regards to floodplain delineation. In Warwick, two primary modelers are FEMA and Coastal Resources Management Council (CRMC), who work collaboratively with the University of Rhode Island's Coastal Institute and focus primarily on coastal dynamics. Their respective flood risk maps are not congruent (Spaulding et al. 2017). While CRMC's model integrates climatic change and future sea level rise scenarios, even if those variables are removed, their model still estimates the 100-year coastal floodplain as far larger than FEMA's model. Several stakeholders described complications in harmonizing these two models and the implications they have had on public communication of risk. While some stakeholders described one set of data as being more usable than the other, the perceived disagreement was seen as a major hurdle in getting the decision makers to act.

How usable is the flood risk data? According to Cash et al.'s (2003) study, knowledge is usable when it is deemed credible, relevant, and legitimate by end users. Here, *credible* refers to how authoritative, believable, and trusted the information is; *relevant* refers to how appropriate the information is to specific decisions being made; and *legitimate* refers to how fair an information-producing process is and whether it considers appropriate values, concerns, and perspectives of different actors. Based on the set of interviews we conducted, StormTools, a modeling program developed by the University of Rhode Island's Coastal Institute in collaboration with RI's CRMC, was seen as very useful. While there was some positive praise for FEMA among those interviewed, more than half of the stakeholders were critical of FEMAs maps and the process they use to calculate risk. Criticisms included all three aspects of usability elements from Cash et al. In terms of credibility, we were surprised to hear that FEMA's credibility was challenged not only because prior maps failed to capture 2010 flood extent but due to specific model limitations. For example, one responder said, "FEMA uses some generalizations in terms of the sediment dynamics and the shore face, and their models seem to be either mid-Atlantic or Southern models because our sediment dynamics are totally different up here." FEMA's legitimacy was also heavily challenged. As one stakeholder noted, "we are down here and then FEMA comes down and they set up their little tent. They try to pretend that they are part of the solution when they are part of the problem." This interviewee referred to the subsidized flood insurance policies that allow redevelopment in the floodplain. However, others challenged FEMA's legitimacy based on the data-gathering process, citing that FEMA seems unwilling to discuss models and datasets in anything but a unidirectional flow: "you can provide FEMA with data, but they are not obliged to look at it or incorporate it in any way." Lastly, FEMA's maps were challenged in terms of their relevance – "FEMA's maps are historic maps, so I wouldn't base future planning on that." Further, FEMA's maps fail to capture nuisance and urban flood risks and therefore failed to deal with a large portion of impacts that the city is observing on a more continual basis.

How well do technical experts, decision makers and managers, and the public understand risk? When asked if flood risks are understood, stakeholders generally clarified – "by whom?" While the majority (71%) believe that technical experts understand flood risks, only 17% believe affected parties such as residents, business owners, and interest groups are aware of flood risks. Many clarified that awareness increased after the 2010 storm and Hurricane Sandy but that memories faded fast. There were also distinctions between those who were directly affected by flood (who generally have a greater awareness) and those who were not, and those who live along the coast and those who live along the river. About a third of the respondents believed that decision makers, politicians, and managers understood flood risks, about a third believe they did not, and about a third either did not know or simply did not discuss it. For many, there were subtle differences between (1) decision makers at the state versus local level, (2) within Warwick or elsewhere, or (3) whether they were elected officials or administrators. For several respondents, this lack of awareness was seen as a major barrier to productive flood management.

Approach and barriers: what does a desirable future look like?

Interviewees were asked a series of questions about what they thought an ideal future might look like in terms of flood risks for Warwick. Broadly, stakeholders did not expect flood risks to get better. When discussing risks, stakeholders generally framed the problem in terms of challenges and limitations as opposed to a focused vision for the future. That said, within the discussion of what is not currently happening, there was a rich narrative about what the city ought to do to manage flood risks and what specific barriers are currently preventing effective action.

Effective approaches to reduce Warwick's flood risks: Stakeholders discussed dozens of diverse ideas about how to minimize future losses. These actions ranged from small-scale (backflow valves) to regional cooperation or changes to federal policies. They further ranged from actions that only agencies can make (e.g., integrate models, build higher bridges) to action individuals (e.g., homeowners) can take. The vast majority of stakeholders (88%) suggested approaches that centered on restricting development in the floodplain. On the other hand, very few stakeholders focused on reducing risk through evacuation (12%) and changing behavior (8%; see Tables 6.7 and 6.8). Most stakeholders proposed more than one approach. We found no significant difference between the type of strategy proposed and stakeholder roles.

Barriers preventing effective action: In addition to approaches, interviewees were asked to describe the barriers that prevent Warwick from having the capacity to implement the above solutions. The top three barriers were (1) funding or lack of resources (whether at the household, city, or state level), (2) the federal insurance policy (specifically referring to subsidies allowing for redevelopment in the floodplain) and (3) politics (including disagreements or lack of coordination between agencies; see Tables 6.9 and 6.10). A large portion of stakeholders discussed barriers associated with information gaps including a lack of public perception and awareness, limited knowledge overall, or misleading data (i.e., misinformation). Additional barriers included a limited capacity associated with the legacy of past development as well as the City's location at the mouth of the watershed; demographic and behavioral limitations associated with mobility and access, emotional ties to place, and human behavior; and competing interests at the city and state levels resulting in poor standards, limited funding for water infrastructure improvements, and lack of staff.

Discussion – three key findings from the interviews

Diversity of understandings, commonality in goals

Our interviews revealed a diversity of perspectives with regard to what flood risks are, which is not surprising given that we intentionally sought that diversity. However, despite this diversity, sentiments about risk varied by backgrounds and preferences more than values or beliefs about what ought to be done. The most prevalent example of this was the responses regarding redevelopment in the floodplain. Stakeholders representing domains as diverse as FEMA, the city, historic preservation, The Nature Conservancy, and the Department of Environmental Management all stressed the importance of minimizing development and restricting redevelopment in the

Table 6.7 Approaches to Minimize Future Flood Risks (Legend)

Approaches (Legend)
A **floodplain restrictions:** don't rebuild in the floodplain; deter people from (re)building in floodplains; buyouts; retreat; move repetitive-loss properties
B **building restrictions:** building codes; stricter or more enforced regulations; flood construction standards
C **comprehensive planning:** land swaps and conservation easements; zoning
D **coordination:** cohesive regulations at multiple scales; across agencies and municipalities; look upstream; unified approach; regionalization; public cooperation; watershed basis; leadership
E **improve natural function:** remove impervious surface; install seashell driveways; rain gardens; sustainable development; invest in open space; let the natural system take land back; connected hydrology; transition away from control toward accommodation of natural flows; improve ecosystem functions; enhance natural buffer; avoid cutting down trees; vegetate riparian areas; remove dams; remove nonessential infrastructure; enhance habitats; living shorelines; protect coastal wetlands
F **water infrastructure improvements:** finish sewer system extension; improve wastewater infrastructure; improve drainage; increase stormwater retention; capital improvements to public infrastructure; invest in pumps; increase stormwater capacity; stormwater infrastructure maintenance; increase capacity in reservoir; raise nutrient level standards
G **engineering solutions:** harden; armor; raise levee elevate buildings; mitigate; invest in a hurricane barrier; fill in basements
H **evacuation and response:** emergency response
I **household preparedness:** backup power; subpumps; Plan B; Code Red; roof tie-downs; flood vents; flood proof; back flow valve preventers; wet proof; retrofits; Fortify
J **education and communication:** raise awareness; communicate risks; engage community; outreach; information sharing
K **information and innovation:** better information (models); risk assessment; models of future risk; maps; build off community knowledge and creativity
L **change behavior:** stop global warming; change culture; act globally; enhance energy efficiency
M **programs:** flood insurance; community rating system; program support; monitor

floodplain. Thus, participants saw this as the best way to increase resilience of the city to future flood impacts. However, there was a wide range of perspectives in terms of why Warwick is still allowing development in the floodplain and how development should be regulated in the future. While some focused on FEMA, others focused on the city, courts, or funding restrictions. An important element of this particular challenge is that each party saw the challenge as lying outside of their domain of control.

Usability issues with FEMA's flood delineation

While we expected that FEMA's flood delineation was controversial, we were surprised to hear just how extensively – in terms of both who and what – the

Table 6.8 Approaches to Minimize Future Flood Risks

| | Approaches | | | | | | | | | | | | |
	A	B	C	D	E	F	G	H	I	J	K	L	M
Planning	+		+	+	+		+	+		+			
Mayor					+	+	+		+		+		
Community	+		+			+	+		+				+
Fire	+		+	+			+		+	+			
WWTF	+	+	+	+	+	+	+	+		+		+	+
Mitigation	+		+				+		+	+	+		
Commissioner	+		+	+			+			+			
Historic	+	+					+						
Coastal In	+						+						+
Coastal Out	+												
Priest	+	+				+	+						+
Realtor		+	+				+		+		+		
Mall				+								+	+
Insurance	+	+	+		+	+	+		+	+			+
RIEMA	+		+	+	+		+		+	+	+		
CRMC	+		+				+		+	+	+		
State Stormwater	+			+	+	+							
State Wastewater	+		+				+		+	+			
Health	+		+	+	+				+	+			
Transportation	+		+	+				+	+				
URI Models	+	+									+		
URI Outreach	+		+		+					+	+		
Save the Bay	+		+		+	+				+			
TNC	+	+	+	+	+	+				+	+		
FEMA	+	+	+				+		+	+	+		+
Total (%)	88	32	52	52	40	44	64	12	44	56	44	8	28

challenges range. We selected Warwick because the 2010 flood impacts were incongruent with FEMA floodplain delineations. However, the interviews revealed a challenging of FEMA's maps that goes far beyond inaccuracies. As a whole, FEMA's flood delineation – in terms of both product and process – was not viewed as usable, with enormous implications for risk perception and capacity building. FEMA's maps still exerted significant power – they were well known by everyone interviewed and perceived as having broad-reaching influence over flood risk management decisions. However, stakeholders consistently saw flood risks as not captured by the model, proactive and strategic actions as not supported by regulation, and opportunities for dialogue and coordination dismissed. In response, they padded their risk perceptions with alternative formal and informal (accurate and inaccurate) decision support tools. These divergences seem to have resulted in a lack of shared understanding, which has resulted in substantial inefficiencies and poor group dynamics.

Opportunity to capture coupled system dynamics

Stakeholders discussed three ways in which flood risk dynamics are coupled – flood risks are both (1) coastal and riverine, (2) physical and social, and

Table 6.9 Barriers to Effective Risk Management (Legend)

Knowledge and Information	1	public perception and awareness
	2	limited knowledge
	3	misleading data/misinformation
Politics and Regulation	4	federal insurance policy
	5	(re)building in harm's way
	6	Supreme Court
	7	regulatory environment
	8	politics
Physical Environment	9	a legacy of development
	10	issue is regional/watershed scale
	11	infrastructure improvements/capacity
Resources and Priorities	12	resources (money, time, staff)
	13	competing priorities
People and Culture	14	demographics (elderly, mobility, low income)
	15	cultural change/human behavior/emotion

(3) frequent and infrequent. These couplings reflect an opportunity to more effectively capture the city's flood risks through the use of more sophisticated models and delineation processes. By their nature, complex systems are not *reducible* – that is, we cannot break the system apart and study components in isolation. For flood risks, this means that when our coastal models don't integrate with our riverine models, we can potentially miss a lot more than that strip of tidal mixing. We can overlook an unprecedented flood event. It may also mean that flood impacts can't be decoupled from flood delineation, since behavioral responses to frequent low-order flooding shape future flood risks in the watershed.

Conclusions – tying our work to the Water Diplomacy Framework

In 2009, the National Academy of Sciences (NAS) published a report aimed at synthesizing the current state of flood risks. In support of the billion dollars in funding for FEMA to digitize and update their maps, they affirmed that "the key to anticipating, preparing for, and insuring against flooding is summed up in one word: *maps*" (National Academy of Sciences 2009). While our findings suggest that Warwick would have been able to reduce impacts if the 2010 maps had more accurately represented riverine flood risks, it is also clear that investments in only the technical aspects of flood delineation will fail to fundamentally improve cities' capacity to withstand and recover from flood events. In 1984, Gilbert White suggested that investments are best placed in *mitigation*. "Better hazard maps, more refined forecasts, and more efficient emergency operations will be important but they will not necessarily reduce damages, and they neglect the measures that might assure sound use of hazardous areas" (White

Table 6.10 Barriers to Effective Risk Management

| | Barriers | | | | | | | | | | | | | | |
| | Knowledge and Information | | | Politics and Regulation | | | | | Physical Environment | | | Resources and Priorities | | People and Culture | |
	1	2	3	4	5	6	7	8	9	10	11	12	13	14	15
Planning	+	+		+	+	+	+	+	+			+	+		
Mayor				+		+			+			+			
Community	+											+		+	
Fire															
WWTF															+
Mitigation								+				+			
Commissioner										+					+
Historic				+	+			+							
Coastal In				+	+		+	+							
Coastal Out					+		+								
Priest				+		+		+				+			
Realtor				+									+		
Mall		+						+							
Insurance				+		+						+			
RIEMA				+								+			
CRMC			+									+			
State Stormwater	+			+	+						+	+			
State Wastewater								+				+		+	+
Health			+			+						+			
Transportation				+											
URI Models		+										+			
URI Outreach	+	+	+					+				+	+		
Save the Bay										+		+		+	+
TNC		+						+				+			+
FEMA								+							

1994, p. 1240). Again, our findings agree: investments in mitigation – in the form of flood walls, elevating structures, installing pumps, etc. – would have reduced Warwick's flood impacts and are likely to provide cost-effective savings into the future. However, focusing on either (or both) better maps and mitigation would still leave cities like Warwick poorly prepared when the next flood comes; holistic actions that integrate technical aspects with societal preparedness and risk awareness are sorely needed.

Based on our case study in Warwick and knowledge of the literature, we propose a solution based not on maps or mitigation but rather a ***floodplain diplomacy***. Building on the Water Diplomacy Framework, we present a set of insights and recommendations for overcoming current limitations focusing on meaningful improvements in *capturing coupled system dynamics*, prioritizing usability through multiparty *knowledge production and decision making*, and developing a delineation process that can adapt to changing conditions (see Boxes 6.2 and 6.3). The essence of floodplain diplomacy is to improve how society interacts, regulates, and perceives the natural and engineered floodplain system. This approach involves acknowledging the limitations of the current FEMA delineation methodology and building on our improved understanding of flooding, complex systems science, and the importance of diverse knowledge production. To support an improved adaptive capacity to anticipate, prepare for, and respond to future flooding, we pair technological advancement with a focus on relationship building and a recognition of the limits of predictive models.

Box 6.2   Key insights

Flood Risks Are Emergent Phenomena

- Flood risks stem from coupled drivers (e.g., social and physical, coastal and riverine).
- Flood risk dynamics are driven by lower-order interactions and should not be externally controlled.
- Interactions are led by nonlinearity, and feedbacks results in irreducible uncertainty.

Flood Knowledge Is Power

- By whom, where, and how a floodplain is delineated is a form of power.
- Flood perception shapes flood risk.
- Flood knowledge is distributed across actors.
- Diverse knowledge is legitimate and useful.

Floodplain Delineation Is Messy

- Floodplains are constantly changing.
- Floodplains are locally unique.
- Floodplain boundaries are soft (not dichotomous).

Box 6.3   Recommendations

Leverage Distributed Knowledge Toward a More Usable Flood Risk Delineation

- Actively seek out and recruit diverse stakeholder representatives.
- Ensure a fair process through third-party neutral facilitation.
- Support open co-production (joint fact-finding) and decision making (consensus-based assessment).
- Evaluate products and process to affirm credibility, legitimacy, as deemed by knowledge users.

Expand Current Models to Capture Coupled System Dynamics

- Identify multiple system components: e.g., social and physical, coastal and riverine, rare/intense and frequent/nuisance, local and global.
- Integrate components and their relationships/feedbacks.
- Characterize future risks and uncertainty.

Restructure the Process of Delineation to Adapt to Changing Conditions

- Design a delineation system that is responsive and adaptive (expect change).
- Support systems for real-time/seamless data updates.
- Maintain a regular process for input and revisions.

Key insights

Flood risks are emergent phenomena

Flood risks stem from emergent phenomena – i.e., risks arise from interactions among fine or lower-order entities (components) such that the higher-order entities exhibit properties the lower-order entities do not exhibit. When technical experts ignore, dismiss, or oversimplify these attributes, we mislead communities and dramatically misrepresent hazards. Foremost, the system cannot be taken apart and understood by component parts in isolation (i.e., reduced). When predictive models decouple or exclude system components, outcomes can be erroneous. For example, flood risks stem from both social and physical drivers and both coastal and riverine mechanisms. Evidence of violating this principle was evident in the interviews from Warwick in which many participants cited the National Flood Insurance Program and broader "politics" in failing to encompass the dynamics of the total system. Second, the system is internally controlled by the actions of low-order entities, e.g., homeowners, agency directors, and nonprofit organizations. The actions, and therefore the understanding and beliefs, of these entities are essential to directing and predicting future conditions. As with other complex systems, an external command-and-control approach results in fragility or collapse (Gunderson and Holling 2002). A final characteristic of flood risks as a complex system is that the component feedbacks substantially amplify and dampen system changes and, in doing so, create nonlinear shifts in conditions. For example, top-down policies and insurance regulations have a ripple effect on local residents and the physical characteristic of the floodplain. These attributes have enormous implications for estimating future conditions and therefore how we delineate and manage flood risks.

Floodplain delineation is messy

The current system of delineation promotes an order that does not reflect reality. Institutional networks further function to reaffirm this order, unintentionally alienating affected stakeholders. Current policy and regulations do not reflect the complexity of the delineation process. Floodplains are dynamic, non-stationary systems due to urbanization and climate drivers, yet the resources to reevaluate the risks as they change simply do not occur. Floodplains are also locally unique – so algorithms for estimating boundaries in one area may not be effective in another region. Third, floodplain boundaries are soft (not binary or dichotomous), and thus a gradient or continuous variable approach could be more effective at capturing flood risks.

Flood knowledge is power

In policy making and in science, there is great privilege in interpretation, or in this case delineation. S/he who defines what is "in" and what is "out" has the power to leverage restrictions, financial costs, and social influence on his/her subjects (Foucault 2012). The biggest challenge in floodplain delineation may stem from the historic assumptions in which communicating flood risks is conceived as a one-way channel. Experts finely tuned their message to poorly informed decision makers and the general public (i.e., loading dock model; Cash et al. 2006). Problematically, this approach fails to capture distributed knowledge. Cities like Warwick understand flooding through a wide variety of direct and technical expertise. The missions of agencies like CRMC and DOH let them explore flood risks from unique perspectives that are not afforded to agencies like FEMA. Residents, nonprofit organizations, business owners, realtors, insurance brokers, and decision makers also possess irreplaceable and distributed knowledge about flood risks. By limiting floodplain delineation to a one-directional channel of communication, cities like Warwick fail to capitalize on the vast knowledge available to them. This limited knowledge base also influences the ability to reform the behavior (i.e., desired risk management behavior) of colleagues and constituents. Increasingly, agencies are realizing that the core of the challenge is not more and better resources but rather enhanced support of relationships between knowledge holders and integrated decision making among stake-holders (Morss et al. 2005).

Recommendations

Based on the insights discussed above, we offer three sets of recommendations for future floodplain delineation process and policy.

Expand current models to capture coupled system dynamics

Flood delineation fails to integrate coupled system dynamics. The model suites employed to develop the floodplain, e.g., FEMA's Hazus model or proprietary models from the insurance companies, are inaccessible to the majority of stake-holders in floodplain management. Breaking open the black box on flood risk modeling may increase the representation of multiple system components related to social elements (perception, demographics, finances) and physical characteristics (coastal vs. riverine risks, urbanization, etc.) based on feedback from the community. We recommend that once the system components have been identified and cataloged, their dependencies, feedbacks, and interactions be mapped such that the impact of a policy could be traced from the top down to the bottom. Once these components are understood, it will be pos-sible and necessary to use this knowledge for predicting future risks and uncertainty.

Restructure the process of delineation to adapt to changing conditions

The basic floodplain delineation principles have remained unchanged for nearly half a century, yet advances in science have brought forth concepts of nonstationarity and the impacts of climate and human development on the floodplain. This highlights a need for designing a delineation system that is responsive and adaptive; in other words, we should expect change. Such a system design should include support systems to make use of near-real-time data, creating flexibility to use the best information available for mapping and determination of risk zones. With advances in computing resources and the development of the National Oceanic and Atmospheric Administration's National Water Model[3] to seamlessly produce operational flood forecasts across the US, these changes are more feasible than before.

Leverage distributed knowledge toward a more usable flood risk delineation

At the core, flood delineation fails because it does not integrate multiple view-points or distributed knowledge. Warwick, like many coastal cities in the US, possesses a diverse range of unique perspectives about what flood risks are and how they ought to be reduced. While these perspectives have varying

Table 6.11 Select Agencies and Their Missions

Agency	*Mission*, link
Warwick City Planning	To serve the needs of Warwick residents by providing planned, responsible and orderly growth and development of the city. *www.warwickri.gov/planning-department*
Mayor of the City of Warwick	To address requests and concerns from city residents and the business community while also assisting in a number of special projects. Additionally, the mayor develops and implements policies, procedures, and programs and is responsible for crafting the annual municipal budget. The office is responsible for representing and promoting Warwick, and the mayor and his staff work collaboratively with local, state, and federal officials, as well as civic, religious, and nonprofit organizations, to address areas of mutual concern throughout Warwick. *www.warwickri.gov/mayors-office*
Warwick Community Development	The primary purpose of the Community Development Office is to assist Warwick-based nonprofit organizations with meeting the needs of the City's low-moderate income individuals and to help revitalize low-moderate-income eligible neighborhoods through the use of CDBG funds. *www.warwickri.gov/community-development*
Warwick Fire Department	To provide the highest level of emergency services to mitigate incidents which arise from fire, medical, hazardous material, or environmental mishaps either on land or water. We dedicate ourselves to a lasting partnership with the community, to support a higher quality of life through public education, loss prevention and service response. *www.warwickri.gov/warwick-fire-department*
RI Emergency Management Agency	To reduce the loss of life and property for the whole community while ensuring that as a state we work together to build, sustain, and improve our capability to prepare for, protect against, respond to, recover from, and mitigate all natural, human-caused, and technological hazards. *www.riema.ri.gov/*
RI Coastal Resource Management Council	The Coastal Resources Management Council is a management agency with regulatory functions. Its primary responsibility is for the preservation, protection, development, and, where possible, the restoration of the coastal areas of the state via the implementation of its integrated and comprehensive coastal management plans and the issuance of permits for work with the coastal zone of the state. *www.crmc.ri.gov/*
RI Department of Health	To prevent disease and protect and promote the health and safety of the people of Rhode Island. *www.health.state.ri.us/*

(*Continued*)

Table 6.11 (Continued)

Agency	Mission, link
RI State Planning Office	A transportation policy-making body authorized by the federal government to ensure that expenditures for transportation projects and programs are based on a continuing, cooperative, and comprehensive planning process. *www.planning.ri.gov/planning-areas/transportation/*
URI Coastal Resources Center	Keeping our coasts healthy and productive by helping those who live, work and play along our coasts to engage in dialogue and promote the good governance that is essential to keeping this vitally important ecosystem healthy and productive. We do this by providing professional services that include policy development, planning, technical assistance, outreach, public/stakeholder meeting facilitation, practitioner workshops and courses, and social science research on the what and why of real problems facing today's coast. *www.crc.uri.edu/*
The Nature Conservancy	A nonprofit conservation organization in the United States that works with public and private partners to ensure our lands and waters are protected for future generations. *www.nature.org/*
FEMA	To support our citizens and first responders to ensure that as a nation we work together to build, sustain, and improve our capability to prepare for, protect against, respond to, recover from, and mitigate all hazards. *www.fema.gov/*

degrees of technical knowledge and salience, bringing the various voices together to negotiate flood boundaries can increase the usability of flood risk delineation products. Leveraging the various perspectives into an integrated product could not only reveal blind spots but also enhance the legitimacy of the product, thereby increasing its usability across individuals and agencies. Consensus-building processes can support the development of a negotiated floodplain delineation process through an inclusive stakeholder assessment process, an effective conversation facilitated with a neutral third party, and a mutual-gains approach (Islam and Susskind 2013). Long-term success depends on the existence of a regular process for input and revisions such that the public has a voice in the delineation process. Policies governed by the private insurance agency and FEMA without public feedback are at risk for being ineffective at reducing the risk they aim to protect against.

This chapter has provided a basis for understanding the complexity of floodplain delineation from a systematic, technical, and social perspective. We have presented an abbreviated history on the delineation process, a discussion on defining flood risk, and the reasons and consequences for misrepresenting the floodplain. While a significant section of this chapter discusses the faults of FEMA, we recognize that calling FEMA the scapegoat is not the solution. FEMA is not the sole responsible party for all of the aspects of flood risks and damages – the science and the social policies, etc. It is too complex a problem for one agency that has limited resources. However, the interviews from Warwick highlighted that at the very least, FEMA has a responsibility in bringing others to the table and allowing for more interaction between entities that have knowledge and need knowledge for preparing, mitigating, and responding to disasters.

We intend for this work to present a vision for improving floodplain delineation through a floodplain diplomacy framework. In this vision, the floodplain is not perfect, but the uncertainties are characterized and communicated. The delineation process is negotiated within the community, building on distributed knowledge and diverse interests. Thirdly, the process of delineation is open and adaptive, supporting a more responsive and future-oriented approach to capturing flood risks. Finally, we acknowledge that this study is limited in scope and that interviews from another flood event could have revealed a different set of insights. Additional postdisaster studies and a database to access them are needed to understand emergent and divergent trends among flood survivors and responders.

Notes

1 Redelineation of the floodplain can occur when better topographic data is available to match with existing cross-sections; however, normally, redelineation does not involve additional engineering studies, but instead uses the new topographic data to delineate a new floodplain using the original cross-sections and the automated delineation process (FEMA 2015).
2 Many critical facilities are sited within the floodplain either intentionally (e.g., pumping stations are more efficient if they are at lower elevations of the watershed) or prior to the use of floodplain maps, there are still modern facilities built in these blurred risk zones.
3 Information about the National Water Model is available from: http://water.noaa.gov/about/nwm

References

Anderson, D., 1974. The national flood insurance program: Problems and potential. *The Journal of Risk and Insurance*, 41 (4), 579–599.

Archfield, S.A., Hirsch, R.M., Viglione, A., and Blöschl, G., 2016. Fragmented patterns of flood change across the United States. *Geophysical Research Letters*, 43 (19), 10232–10239.

Armstrong, W.H., Collins, M.J., and Snyder, N.P., 2012. Increased frequency of low-magnitude floods in New England. *Journal of the American Water Resources Association*, 48 (2), 306–320.

Arnell, N.W., and Gosling, S.N., 2016. The impacts of climate change on river flood risk at the global scale. *Climatic Change*, 134 (3), 387–401.

Atreya, A., and Ferreira, S., 2015. Seeing is believing? Evidence from property prices in inundated areas. *Risk Analysis*, 35 (5), 828–848.

Blanchard-Boehm, R., Berry, K., and Showalter, P., 2001. Should flood insurance be mandatory? Insights in the wake of the 1997 new year's day flood in Reno-Sparks, Nevada. *Applied Geography*, 21 (3), 199–221.

Blessing, R., Sebastian, A., and Brody, S.D., 2017. Flood risk delineation in the U.S.: How much loss are we capturing? *Natural Hazards Review*, 18 (3), 1–10.

Brody, S.D., Blessing, R., Sebastian, A., and Bedient, P., 2013. Delineating the reality of flood risk and loss in Southeast Texas. *Natural Hazards Review*, 14 (2), 89–97.

Brody, S.D., and Highfield, W.E., 2011. *Evaluating the Effectiveness of the FEMA Community Rating System in Reducing Flood Losses*. Final Report for FEMA Mitigation Division Study, Phase I.

Brody, S.D., Highfield, W.E., Ryu, H.C., and Spanel-Weber, L., 2007. Examining the relationship between wetland alteration and watershed flooding in Texas and Florida. *Natural Hazards*, 40 (2), 413–428.

Bubeck, P., Botzen, W.J.W., and Aerts, J.C.J.H., 2012. A review of risk perceptions and other factors that influence flood mitigation behavior. *Risk Analysis*, 32 (9), 1481–1495.

Burby, R.J., 2001. Flood insurance and floodplain management: The US experience. *Global Environmental Change Part B: Environmental Hazards*, 3 (3), 111–122.

Carolan, M.S., 2007. One step forward, two steps back: Flood management policy in the United States. *Environmental Politics*, 16 (1), 36–51.

Carter, J.G., White, I., and Richards, J., 2009. Sustainability appraisal and flood risk management. *Environmental Impact Assessment Review*, 29 (1), 7–14.

Cash, D., Clark, W.C., Alcock, F., Dickson, N., Eckley, N., and Jäger, J., 2003. Salience, credibility, legitimacy and boundaries: Linking research, assessment and decision making. *SSRN Electronic Journal*.

Cash, D.W., Borck, J.C., and Patt, A.G., 2006. Countering the loading-dock approach to linking science and decision making: comparative analysis of El Niño/Southern Oscillation (ENSO) forecasting systems. *Science, Technology, & Human Values*, 31 (4), 465–494.

Chang, H., and Franczyk, J., 2008. Climate change, land-use change, and floods: Toward an integrated assessment. *Geography Compass*, 2 (5), 1549–1579.

City of Warwick, 2011. *City of Warwick, Rhode Island Local Multi-Hazard Mitigation Plan*. Available from: https://www.warwickri.gov/sites/warwickri/files/uploads/warwick_hazard_mitigation_plan.pdf

CNN, 2010. *Rhode Island Flooding: 'Nobody Was Prepared'*. 2nd April. Available from: www.cnn.com/2010/US/weather/04/01/northeast.flooding/index.html

Colby, J.D., and Dobson, J.G., 2010, February. Flood modeling in the coastal plains and mountains: Analysis of terrain resolution. *Natural Hazards Review*, 11 (1), 19–28.

Cossio, M.L.T., Giesen, L.F., Araya, G., Pérez-Cotapos, M.L.S., VERGARA, R.L., Manca, M., Tohme, R.A., Holmberg, S.D., Bressmann, T., Lirio, D.R., Román, J.S., Solís, R.G., Thakur, S., Rao, S.N., Modelado, E.L., La, A.D.E., Durante, C., Tradición, U.N.A., En, M., Espejo, E.L., Fuentes, D.E.L.A.S., Yucatán, U.A. De, Lenin, C.M., Cian, L.F., Douglas, M.J., Plata, L., and Héritier, F., 2015. Report for the urban flooding awareness act. *State of Illinois, Illinois Department of Natural Resources*, 2.

Di Baldassarre, G., Kooy, M., Kemerink, J.S., and Brandimarte, L., 2013. Towards understanding the dynamic behaviour of floodplains as human-water systems. *Hydrology and Earth System Sciences*, 17, 3235–3244.

Dixon, L., Clancy, N., Seabury, S.A., and Overton, A., 2006. *The National Flood Insurance Program's Market Penetration Rate: Estimate and Policy Implications*. RAND Infrastructure, Safety, and Environment and Institute for Civil Justice.

Easterling, D.R., Meehl, G.A., Parmesan, C., Changnon, S.A., Karl, T.R., and Mearns, L.O., 2000. Climate extremes: Observations, modeling, and impacts. *Science*, 289 (5487), 2068–2074.

England, J., 2019. *Bulletin 17C*. Available from: https://sites.google.com/a/alumni.colos tate.edu/jengland/bulletin-17c.

Ernstson, H., van der Leeuw, S.E., Redman, C.L., Meffert, D.J., Davis, G., Alfsen, C., and Elmqvist, T., 2010. Urban transitions: On urban resilience and human-dominated eco-systems. *Ambio*, 39 (8), 531–545.

Federal Interagency Floodplain Management Task Force, 1992. *Floodplain Management in the United States*. Available from: https://www.fema.gov/media-library-data/20130 726-1504-20490-0436/fema18.pdf

FEMA, 2017. *Delineating Flood Prone Areas*. Available from: https://training.fema.gov/ hiedu/docs/fmc/chapter%205%20-%20delineating%20flood-prone%20areas.pdf

FEMA, 2015. *Guidance for Flood Risk Analysis and Mapping: Redelination Guidance*. Available from: www.fema.gov/media-library-data/1449866809789-9c0d4332ad18d207f33c72c fe9dc3b1a/Riverine_Mapping_and_Floodplain_Guidance_Nov_2015.pdf

Festing, H., Copp, C., Sprague, H., Wolf, D., Shorofsky, B., and Nichols, K., 2013. *The Prevalence and Cost of Urban Flooding: A Case Study of Cook County, IL*. Available from: https://www.cnt.org/sites/default/files/publications/CNT_PrevalenceAndCostOfUr banFlooding2014.pdf

Foucault, M., 2012. *Discipline and Punish*. New York: Vintage Books.

French, S.P., Lee, D., and Anderson, K., 2010. Estimating the social and economic con-sequences of natural hazards: Fiscal impact example. *Natural Hazards Review*, 11 (5), 49–57.

Gall, M., Borden, K.A., Emrich, C.T., and Cutter, S.L., 2011. The unsustainable trend of natural hazard losses in the United States. *Sustainability*, 3 (11), 2157–2181.

Galloway, G.E., Baecher, G.B., Plasencia, D., Coulton, K.G., Louthain, J., and Bagha, M., 2006. *Assessing the Adequacy of the National Flood Insurance Program's 1 Percent Flood Standard*. Available from: https://www.fema.gov/media-library/assets/docu ments/9594

Gilbert F. White National Flood Policy Forum, 2004. *Reducing Flood Losses: Is the 1% Chance (100-year) Flood Standard Sufficient?* Available from: https://www.nrcs.usda.gov/ Internet/FSE_DOCUMENTS/16/nrcs143_009401.pdf

Gochis, D., Schumacher, R., Friedrich, K., Doesken, N., Kelsch, M., Sun, J., Ikeda, K., Lindsey, D., Wood, A., Dolan, B., Matrosov, S., Newman, A., Mahoney, K., Rutledge, S., Johnson, R., Kucera, P., Kennedy, P., Sempere-Torres, D., Steiner, M., Roberts, R., Wilson, J., Yu, W., Chandrasekar, V., Rasmussen, R., Anderson, A., Brown, B., Gochis, D., Schumacher, R., Friedrich, K., Doesken, N., Kelsch, M., Sun, J., Ikeda, K., Lindsey, D., Wood, A., Dolan, B., Matrosov, S., Newman, A., Mahoney, K., Rutledge, S., Johnson, R., Kucera, P., Kennedy, P., Sempere-Torres, D., Steiner, M., Roberts, R., Wilson, J., Yu, W., Chandrasekar, V., Rasmussen, R., Anderson, A., and Brown, B., 2015. The great Colorado flood of September 2013. *Bulletin of the American Meteorological Society*, 96 (9), 1461–1487.

Gunderson, L.H., and Holling, C.S., 2002. *Panarchy: Understanding Transformations in Human and Natural Systems*. Washington D.C.: Island Press.

Highfield, W.E., Norman, S.A., and Brody, S.D., 2013. Examining the 100-year floodplain as a metric of risk, loss, and household adjustment. *Risk Analysis*, 33 (2), 186–191.

Ikeuchi, H., Hirabayashi, Y., Yamazaki, D., Muis, S., Ward, P.J., Winsemius, H.C., Verlaan, M., and Kanae, S., 2017. Compound simulation of fluvial floods and storm surges in a global coupled river-coast flood model: Model development and its application to 2007 Cyclone Sidr in Bangladesh. *Journal of Advances in Modeling Earth Systems*, 9 (4), 1847–1862.

Islam, S., and Susskind, L.E., 2013. *Water Diplomacy: A Negotiated Approach to Managing Complex Water Networks*. New York: RFF Press.

Kew, S.F., Selten, F.M., Lenderink, G., and Hazeleger, W., 2013. The simultaneous occurrence of surge and discharge extremes for the Rhine delta. *Natural Hazards and Earth System Sciences*, 13 (8), 2017–2029.

Leopold, L.B., and Luna, B., 1994. *A View of the River*. Cambridge, MA: Harvard University Press.

Maidment, D. R., 2010. *Flood Map Accuracy*. Statement before the U.S. Senate. Available from: http://www.ce.utexas.edu/prof/maidment/giswr2011/docs/FloodplainMapping Senate.pdf

Merz, B., Vorogushyn, S., Uhlemann, S., Delgado, J., and Hundecha, Y., 2012. HESS opinions 'more efforts and scientific rigour are needed to attribute trends in flood time series'. *Hydrology and Earth System Sciences*, 16 (5), 1379–1387.

Milly, P.C.D., Wetherald, R.T., Dunne, K.A., and Delworth, T.L., 2002. Increasing risk of great floods in a changing climate. *Nature*, 415 (6871), 514–517.

Moftakhari, H.R., AghaKouchak, A., Sanders, B.F., and Matthew, R.A., 2017. Cumulative hazard: The case of nuisance flooding. *Earth's Future*, 5 (2), 214–223.

Moore, R., 2017. Seeking higher ground: How to break the cycle of repeated flooding with climate-smart flood insurance reforms. *NRDC*. Available from: www.nrdc.org/resources/seeking-higher-ground-how-break-cycle-repeated-flooding-climate-smart-flood-insurance [Accessed 22 Feb 2018].

Morss, R.E., Wilhelmi, O.V., Downton, M.W., and Gruntfest, E., 2005. Flood risk, uncertainty, and scientific information for decision making: Lessons from an interdisciplinary project. *Bulletin of the American Meteorological Society*, 86 (11), 1593–1601.

Murphy, F., 1958. *Regulating Floodplain Development*. Chicago: University of Chicago Press.

National Academy of Sciences, 2009. *Mapping the Zone: Improving Flood Map Accuracy*. Washington, D.C.: The National Academies Press.

Newbery, D., Echenique, M., Goddard, J., Heathwaite, L., Morris, J., Schultz, W., and Swanwick, C., 2010. *Land Use Futures: Making the Most of Land in the 21st Century*. London: Government Office for Science.

Nilsen, V., Lier, J.A., Bjerkholt, J.T., and Lindholm, O.G., 2011. Analysing urban floods and combined sewer overflows in a changing climate. *Journal of Water and Climate Change*, 2 (4), 260–271.

Noman, N.S., Nelson, E.J., and Zundel, A.K., 2001, December. Review of automated floodplain delineation from digital terrain models. *Journal of Water Resources Planning and Management*, 127 (6), 394–402.

Ntelekos, A.A., Oppenheimer, M., Smith, J.A., and Miller, A.J., 2010. Urbanization, climate change and flood policy in the United States. *Climatic Change*, 103 (3), 597–616.

Olsen, J.R., 2006. Climate change and floodplain management in the United States. *Climatic Change*, 76 (3–4), 407–426.

Pahl-Wostl, C., 2006. The importance of social learning in restoring the multifunctionality of rivers and floodplains. *Ecology and Society*, 11 (1), 10.

Patterson, L.A., and Doyle, M.W., 2009. Assessing effectiveness of national flood policy through spatiotemporal monitoring of socioeconomic exposure. *Journal of the American Water Resources Association*, 45 (1), 237–252.

Pielke, R.A.J., and Downton, M.W., 2000. Precipitation and damaging floods: Trends in the United States, 1932–97. *Journal of Climate*, 13 (20), 3625–3637.

Read, L.K., and Vogel, R.M., 2015. Reliability, return periods, and risk under nonstationarity. *Water Resources Research*, 51 (8), 6381–6398.

Reuss, M., 1993. *Water Resources People and Issues: Gilbert F. White*. Fort Belvoir, Virginia: United States Army Corps of Engineering Office of History.

Rusjan, S., Kobold, M., and Mikoš, M., 2009. Characteristics of the extreme rainfall event and consequent flash floods in W Slovenia in September 2007. *Natural Hazards and Earth System Sciences*, 9, 947–956.

Salas, J.D., and Obeysekera, J., 2014. Revisiting the concepts of return period and risk for nonstationary hydrologic extreme events. *Journal of Hydrologic Engineering*, 19 (3), 554–568.

Samaddar, S., Misra, B.A., and Tatano, H., 2012. *Flood Risk Awareness and Preparedness: The Role of Trust in Information Sources*. Proceedings 2012 IEEE International Conference on Systems, Man, and Cybernetics (SMC), pp. 3099–3104.

Schaake, J.C., and Fiering, M.B., 1967. Simulation of a national flood insurance fund. *Water Resources Research*, 3 (4), 913–929.

Simonovic, S.P., and Carson, R.W., 2003. Flooding in the Red River Basin: Lessons from post flood activities. *Natural Hazards*, 28 (2/3), 345–365.

Smith, D.I., 2000. Floodplain management: Problems, issues and opportunities. *In*: D. Parker, ed. *Floods*. New York: Routledge, 254–267.

Spaulding, M.L., Grilli, A., Damon, C., Fugate, G., Isaji, T., and Schambach, L., 2017. Application of state of the art modeling techniques to predict flooding and waves for a coastal area within a protected bay. *Journal of Marine Science and Engineering*, 5 (1), 14.

Sweet, W., Park, J., Marra, J., Zervas, C., and Gill, S., 2014. *Sea Level Rise and Nuisance Flood Frequency Changes around the United States*. NOAA Technical Report NOS CO-OPS 073, (June), 58.

Van Den Hurk, B., Van Meijgaard, E., De Valk, P., Van Heeringen, K.J., and Gooijer, J., 2015. Analysis of a compounding surge and precipitation event in the Netherlands. *Environmental Research Letters*, 10 (3).

Villarini, G., Serinaldi, F., Smith, J.A., and Krajewski, W.F., 2009. On the stationarity of annual flood peaks in the continental United States during the 20th century. *Water Resources Research*, 45 (8), 1–17.

Villarini, G., and Smith, J.A., 2010. Flood peak distributions for the eastern United States. *Water Resources Research*, 46 (6), 1–18.

Wachinger, G., Renn, O., Begg, C., and Kuhlicke, C., 2013. The risk perception paradox-implications for governance and communication of natural hazards. *Risk Analysis*, 33 (6), 1049–1065.

Wahl, T., Jain, S., Bender, J., Meyers, S.D., and Luther, M.E., 2015. Increasing risk of compound flooding from storm surge and rainfall for major US cities. *Nature Climate Change*, 5 (12), 1093–1097.

Wescoat, J.I., and White, G.F., 2003. *Water for Life*. Cambridge: Cambridge University Press.

White, G.F., 1994. A perspective on reducing losses from natural hazards. *Bulletin of the American Meteorological Society*, 75 (7), 1237–1240.

White, I., 2010. *Water in the City: Risk, Resilience and Planning for a Sustainable Future*. New York: Routledge.

White, M.D., and Greer, K.A., 2006. The effects of watershed urbanization on the stream hydrology and riparian vegetation of Los Peñasquitos Creek, California. *Landscape and Urban Planning*, 74 (2), 125–138.

Zarriello, P.J., Olson, S.A., Flynn, R.H., Strauch, K.R., and Murphy, E.A., 2014. *Simulated and Observed 2010 Floodwater Elevations in Selected River Reaches in the Pawtuxet River Basin*. Rhode Island Scientific Investigations Report 2013–5192.

7 Cholera in Haiti

Why many efforts have failed and how we can do better

Michael Ritter

Introduction

Cholera is an infectious disease caused by ingesting food or water that is contaminated with the bacterium *Vibrio cholerae*. The first case of cholera in Haiti in more than a century was confirmed in October 2010, and this quickly led to the largest single country cholera outbreak in modern history (CDC 2010). In the first seven years of the outbreak, more than 815,000 people (about 7% of the population) were infected with cholera, and a reported 9,735 died (MSPP 2017).

Authors of several scientific investigations concluded that the most likely source of the epidemic was entry of contaminated waste from a United Nations (UN) peacekeeping camp into the Artibonite River (Chin et al. 2011; Hendriksen et al. 2011; Piarroux et al. 2011; Frerichs et al. 2012). The introduction of the microbe was a necessary but not sufficient condition for its rapid spread throughout the country, which was enabled by many factors such as widespread use of river and irrigation water for household and agricultural purposes, the lack of immunity of the Haitian population to cholera, the migration of infected individuals, the potency of the South Asian type *V. cholera* strain, and deficiencies in water and sanitation infrastructure (Cravioto et al. 2011). Poor access to safe drinking water and sanitation was a problem long before the cholera outbreak began, as public health experts identified outbreaks of waterborne diseases in Haiti shortly after microbes like dysentery were discovered (Behrmann et al. 1900). An estimated 65% of the population had access to an improved drinking water source in 2017, showing little progress from the 62% of the population with improved water in 1990 (UNICEF 2019).

These low levels of safe water provision have been enabled by a lack of consistent, sustained investment in the sector. Prior to the cholera outbreak, most government resources that were allocated to water infrastructure were focused on Port-au-Prince and a few secondary cities (Gelting et al. 2013). Historically, even the projects that did receive government funding were hindered by political instability, rent-seeking, and corruption (Fass 1988). Currently, the government agencies directly implicated in the reduction of waterborne disease are the Ministry of Public Health and Population (MSPP) and the

National Directorate for Potable Water and Sanitation (DINEPA), the latter of which was created by Parliament in 2009.

In the five years after the cholera outbreak had started, only 1% of DINEPA's budget was funded by the Haitian government, with the remainder coming from foreign donors (DINEPA 2014; USAID 2014). The international community has a long history of investments in Haiti, but its impact has been both helpful and harmful. Foreign involvement began with colonization, brutal slavery, and a debt imposed by France after Haiti's independence. Multilateral institutions such as the UN and the Inter-American Development Bank (IDB) have financed projects and taken leadership roles in the sector since the early 20th century (Dewitt 1987). During the US occupation of Haiti from 1915 to 1934, the US constructed water infrastructure throughout the country (Gelting et al. 2013). Since official US aid to Haiti began in 1943, the size of commitments has fluctuated dramatically based on US political interests. This is perhaps best illustrated by the US' blocking of an IDB loan for water infrastructure in the early 2000s (Smith Fawzi et al. 2009).

An analysis of the causes of poor water and sanitation must take into account how this history is interconnected with diverse issues and how feedback loops have trapped the country in its current situation. The absence of a sustainable source of sector financing described above is one of the primary causes of low levels of safe water provision. The resulting resource uncertainty hinders government agencies' ability to attract and retain competent and committed staff. This environment enables indifference and corruption on the part of public officials and results in frequent turnover that harms institutional memory and effectiveness. The inability of the government to ensure provision of safe water damaged trust among other stakeholders like citizens and the private sector. This skepticism about the government's ability to fulfill this role has made it difficult for the government to enforce policy such as water tariffs (where infrastructure does exist; Gelting et al. 2013). Foreign donors' perception of corruption has led foreign donors to divert funding from the government (Ramachandran and Walz 2015). This increases the strain on government resources, fueling the cycle of ineffectiveness and giving citizens and donors further justification to withhold funds from the government. Currently, the majority of foreign aid is channeled through NGOs, which typically operate on short-term budgets with specific project deliverables (Ramachandran and Walz 2015). Given the diversity of goals and incentives among NGOs and lack of strong sector leadership, alignment toward long-term strategies is difficult, further impeding progress beyond the present situation.

This is the fragmented environment in which responses to the cholera outbreak emerged: a mix of stakeholders with multiple objectives competing for limited resources in a climate beset by resource uncertainty and distrust. Progress has been both incomplete and impressive given the challenges. To understand the conditions that have enabled effective responses, this chapter uses cases at three scales: high-level strategic policy coalitions, a working group within the household water treatment and storage sector, and a social enterprise NGO operating within this sector. First, complexity theory is applied to understand how stakeholders at each level defined the problem and the degree of complexity of these problems. Second, cases of effective and ineffective stakeholder engagement at each level are explored. The complexities of this environment, especially those related to resource uncertainty and low levels of trust, explain much of why the cholera response in Haiti has struggled to meet high-level goals like eliminating cholera. Rather than choosing more rational or more "correct" technical solutions to the problem, future cholera response efforts can be improved by adapting stakeholder engagement strategies to situations characterized by complexity, resource uncertainty, and low levels of trust.

Problem definitions and complexity

Many stakeholders operating at multiple levels have attempted to address cholera in Haiti, but they have not all defined the problem in the same way. In addition, even within a single institution, definitions of problems and goals have been dynamic. This diversity of problem definitions among stakeholders and time periods is a source of complexity, and acknowledging this is critical for analyzing the response to cholera. In this section, the analysis tracks the evolution of problem definitions among stakeholders at three levels during the first eight years of the cholera outbreak in Haiti.

Policy level: increasing consensus on cholera response strategy despite high uncertainty

The highest level of stakeholders included the Haitian government, major donors (including foreign governments and UN agencies), and select international NGOs and scientific researchers. As stakeholders at this level debated how to respond to cholera, one of the key questions was whether the goal should be elimination or containment of the disease. The scientific community produced mathematical models of short-term cholera transmission, and those that attempted to influence policy gave conflicting recommendations (Andrews and Basu 2011; Bertuzzo et al. 2011; Chao et al. 2011; Tuite et al. 2011). Despite scientific uncertainty about the feasible time frame and interventions required to eliminate cholera from the island, policy makers pushed ahead to set actionable goals accompanied by funding requests. Slightly over one year into the outbreak, a coalition of major stakeholders rallied around the goal of elimination by 2022, including the Haitian government, UN agencies like UNICEF and the World Health Organization (WHO), and the United States' Centers for Disease Control and Prevention (CDC; Pan American Health Organization 2012; Periago et al. 2012).

Once the call to action had been announced, decision makers at the policy level shifted to the task of developing a technical plan that would provide the highest probability of achieving elimination (DINEPA 2013). The key questions at this phase became how to attract the necessary funding and how to allocate limited resources between technical interventions. Answers to these questions were still highly uncertain and were connected to many problems (these are described in the stakeholder engagement section later in this chapter) on which consensus was still not yet achieved.

Sector level: increasing complexity of problems over time

Exploring the evolution of problem definitions in one subsector is instructive because it contrasts the policy-level example with one in which the decision-making process grew more complex over time. One sub-sector focused on household water treatment and safe storage (HWTS), which is a set of technologies that includes consumables such as chlorine and durables such as filters, that have been shown to improve drinking water quality (Clasen and Haller 2008) and reduce diarrheal disease (Quick et al. 1999; Quick et al. 2002; Luby et al. 2004; Fewtrell et al. 2005; Clasen et al. 2006; Clasen et al. 2007; Clasen 2015). Microbiological contamination between the water source and the point of use is well documented in developing countries (Gundry et al. 2004), highlighting the need for treatment in the home even in cases in which the water source is microbiologically safe (Clasen and Bastable 2003). Despite the health benefits of HWTS products, demand remains low. An estimated 33% of households in countries without reliable access to safe water self-report treating their drinking water, and socioeconomic and urban/rural disparities persist (Rosa and Clasen 2010).

In Haiti, water is piped onto the premises of only 10% of households (4.5% in rural areas; WHO and UNICEF 2015). Prior to the cholera outbreak, HWTS products were distributed in Haiti exclusively by the private sector, mostly by NGOs through subsidized models. While self-reported HWTS use (76%) is higher than similar countries (Cayemittes et al. 2013), chlorine (the most common HWTS method in Haiti) was present and confirmed in only 13% of households in rural areas of the Artibonite region (Patrick et al. 2013) and only 26% of all rural households in a national survey (PSI 2012). Due to these low levels of baseline HWTS use, the quick inactivation of *V. cholerae* by chlorine, and the lower cost and shorter time frame of HWTS interventions relative to larger infrastructure projects, HWTS was viewed as a key component of cholera response even in the earliest strategy documents (DINEPA 2010).

The early phases of the strategy included delivery of free HWTS products to communities directly affected by cholera outbreaks. While there were many challenges in achieving this goal, it was a simple problem when viewed through the lens of complexity theory. Emergency responders used known solutions to the problem of delivering HWTS to distinct geographical areas during short time frames (Lantagne and Clasen 2012). Cholera rates and geographical hotspots are well defined; HWTS technologies that inactivate the cholera bacteria are known (Wolfe et al. 2018); and interventions exist that successfully train people to use these technologies correctly, at least in the short run (Evans et al. 2014).

Subsequent phases sought to create networks of access to HWTS products at the national level, which is a more complicated problem. Establishing a sustainable supply chain requires not just recruiting and training individuals to distribute products but also retaining and incentivizing these agents on a long-term basis. Creating ongoing demand requires not just educating households about how to use products but also convincing them to use a portion of their budget to pay for HWTS products (Dupas 2011; Martin et al. 2018). Reaching national scale requires models that can adapt to population segments that are difficult to reach. For example, in remote mountain communities, ability to pay for HWTS products is lower, while the costs of delivering them are higher. Education levels are lower, while the microbiological profile of water sources may necessitate HWTS technologies that require users to perform multiple steps in order to treat water correctly. Solutions to these problems exist, but they require moving beyond simple technology provision to comprehensive approaches that require expertise in business models, behavior change, logistics, finance, and other fields.

The national strategy called for an eventual transition out of the acute emergency phase to implementation of national networks to supply HWTS products at an affordable price without government subsidy (DINEPA 2010). The inclusion of the dual goals of affordability and sustainability introduced a major source of complexity. The emphasis on accessible prices coupled with the removal of government subsidies implied a vision of sustainable and equitable access, but strategy documents did little to clarify what would be considered a success. There is little consensus in the scientific literature on definitions of "sustained adoption" of HWTS products (Martin et al. 2018), let alone clarity with respect to questions about what percentage of the population must use HWTS products and for what period of time in order to eliminate cholera. Even for individual stakeholders that set their own target levels of HWTS adoption among well-defined populations and time frames, little is known about how to achieve consistent use over periods of multiple years. While promising models exist (Harshfield et al. 2012), there are no clear examples of HWTS suppliers in Haiti that could supply large portions of the population without subsidies, casting doubt on whether achieving the goal was even possible.

One could argue that when viewed at the scale of the entire sector, achieving sustainable and affordable access to HWTS is a wicked problem. Given the diversity of definitions of sustainability and equity among a wide range of stakeholders operating in the HWTS sector, the policy problem is not definitively described, which implies that solutions cannot be objectively categorized as correct or false or optimal (Rittel and Webber 1973). Rather, multiple stakeholders are left to compete for limited and common resources in an attempt to achieve multiple objectives that are, at best, roughly aligned with an unclear national strategy and, at worst, irrelevant or even in conflict with the strategy.

Organizational level: increasing clarity of the problem in a social enterprise NGO

Organizations operating within the HWTS sector dealt with similar challenges with respect to defining the problem and goals. One NGO that illustrates these challenges well is Deep Springs International (DSI). DSI was chosen as a case study because it provides insights into multiple stakeholder engagement strategies and because the author's role as program director during the first six years of the cholera response provides a unique perspective on the internal debates, challenges, and lessons learned. DSI is a nonprofit founded in the US that seeks "to improve public health while creating jobs through an integrated and sustainable safe water program that can be scaled throughout under-served communities in developing countries" (DSI 2017). As a social enterprise, DSI aims to develop business models that can simultaneously sustain themselves and reduce waterborne disease among vulnerable populations.

In 2008, DSI took over management responsibilities of a pilot project in Haiti that produced chlorine solution locally and distributed it for households to treat their own drinking water. The pilot project covered much but not all of its costs through revenue from chlorine sales to households in a rural, low-income community. DSI aimed to increase the financial sustainability of this model and scale it up throughout Haiti. In 2010, DSI became involved in the emergency response to the earthquake and cholera outbreak. DSI expanded from a network of about 30 chlorine sales agents serving about 4,000 households in one region to a network of about 250 sales agents that had contact with more than 50,000 households in multiple regions.

Throughout the expansion process, DSI's definition of the problem evolved. DSI was always committed to both health impact and sustainability, but DSI leadership held diverse views on how to define these terms and how much to emphasize each. For some, success would require developing a full cost recovery model for distributing HWTS products, and subsidies were only useful in the short run. Others sought to make marginal improvements on the sustainability of existing HWTS models and were open to long-term subsidies if health impacts were achieved and donor funds could be obtained.

The earthquake and cholera outbreak forced DSI to adapt to a new environment and contributed to an eventual clarification of its goals. DSI leadership coalesced around a tighter understanding of the balance between its dual goals and expanded its use of quantitative indicators to measure its impact. The focus of discussions among leadership shifted from debates about its fundamental mission toward identifying technical strategies that were most likely to increase its chosen indicators.

Evaluating stakeholder engagement methods in the cholera response

The ways in which each stakeholder defined the problem created unique conditions in which they participated in the cholera response. In this section, I use

Hurlbert and Gupta's "split ladder of participation" as a lens for understanding stakeholder engagement at each of these three levels (Hurlbert and Gupta 2015). On their vertical axis, Hurlbert and Gupta incorporate Arnstein's notion (Arnstein 1969) of various degrees of participation ranging from low to high. They add the notion of trust along the horizontal axis, which results in four quadrants that correspond to different types of problems and learning. I use the split ladder primarily as an evaluation tool for critiquing whether the types of participation used were appropriate for the given context and to make recommendations for how stakeholder engagement can be done more effectively.

Policy level: strategic technical decisions without scientific certainty

In the early phases of the cholera outbreak, the Water, Sanitation, and Hygiene (WASH) Cluster coordinated humanitarian actors in the WASH sector. Established in response to the earthquake in January 2010, the WASH Cluster aimed to identify gaps in the emergency response and to direct resources where they were needed. Any institution engaged in WASH activities was invited to attend meetings that were convened by UNICEF and the Haitian government (DINEPA), the joint leaders of the WASH cluster in Haiti. The cluster was limited in its ability to develop and mobilize stakeholders toward a unifying strategy, partly because of low levels of certainty and consensus on fundamental questions about the problem being addressed. Individuals involved in overall coordination efforts reported that the "task of coordinating and integrating diverse constituencies is more difficult than it would appear on programmatic documents. This is not necessarily because of the (technical) lack of coordinating mechanisms, but because of (political) differences in priorities and strategic choices" (Zanotti 2010).

In parallel to regular WASH cluster meetings, high-level donors and policy makers discussed what goals to pursue and how to allocate resources. The goal-setting process involved scientists interacting with individuals and institutions that held financial and political power, like bilateral donors, DINEPA, and UN agencies. As described above, achieving consensus on the goal of cholera elimination by 2022 was primarily a political process, and there was some degree of disconnect between the scientific and policy-making discussions. Even after the call to action was announced, the scientific debate over the feasibility of elimination continued in the academic community. One peer-reviewed publication suggested that the probability of elimination in the short term was "of the order of 1%" (Bertuzzo et al. 2014), while another pointed to the lack of evidence of an environmental reservoir of *V. cholera* in Haiti as evidence for the feasibility of elimination (Baron et al. 2013).

The role of vaccines and water and sanitation interventions

One topic that generated substantial disagreement and high levels of participation by NGOs, scientific researchers, donors, and the Haitian government was whether oral cholera vaccine (OCV) should be prioritized as part of the intervention package. Opponents expressed concerns about diversion of resources from water and sanitation, arguing that the strengthening of water infrastructure is what led to elimination of cholera in other Latin American countries. Decisions to exclude vaccines from early stages were based on "limited vaccine availability, complex logistical and operational challenges of multidose regimen, and obstacles to conducting campaign in setting with population displacement and civil unrest" (Date et al. 2011). The Ministry of Health (MSPP) initially opposed vaccination unless the campaign would start with 1 million doses and commit to reach 6 million, citing concerns about resentment from populations that didn't receive it (Cyranoski 2011). Despite remaining uncertainty, consensus emerged among the key decision makers that OCV should be part of a complementary package (Ivers et al. 2010; Adams 2012; Ivers et al. 2012; Ivers 2017). Several studies have demonstrated feasibility and effectiveness of OCV in Haiti (Ivers et al. 2013; Rouzier et al. 2013; Ivers et al. 2015; Tohme et al. 2015; Severe et al. 2016), but debate continues with respect to the validity of some of these studies (Rebaudet et al. 2016), and even proponents admit that OCV is not sufficient to eliminate cholera (Matias et al. 2017).

Sector level: attempting high participation of HWTS stakeholders despite disagreements

In the aftermath of the earthquake in Haiti, the WASH cluster organized smaller "technical working groups" in order to bring together stakeholders that were working on specific problems within the larger WASH sector. The HWTS working group was composed of representatives from government, international and local NGOs, the private sector, technical advisors, and donors. It focused on providing strategic direction to implementing organizations on topics that arose during the emergency response. Guidance was primarily communicated during working group meetings and policy briefs (Cluster 2011).

Even in the early months of the emergency response, stakeholders acknowledged the need for a coherent, long-term strategy. Most stakeholders saw this as a government function but also recognized that it had little existing capacity to take on this type of leadership. Donors agreed to fund DINEPA to recruit a local HWTS specialist and a neutral party to provide technical assistance throughout the process. Initial attitudes among NGOs were mixed, but the dynamism of the individual was one reason that trust grew and attendance at monthly stakeholder meetings was eventually fairly high. This process led to the identification of key topics on which DINEPA planned to issue regulations (DINEPA 2013).

However, momentum was lost after the position was left unfilled for more than a year. Communication between NGOs declined in the interim, and the majority of NGO representatives had turned over before a replacement was hired. While data on stakeholders and draft policy documents were available from the first round of engagement, the process of building trust and consensus had to restart. This second round of stakeholder engagement identified more than 60 organizations and involved 32 interviews with people working on HWTS in Haiti (The Water Institute at UNC 2015). By the end of the funding cycle that allowed a neutral party to assist DINEPA in this capacity, several workshops were conducted and policy documents drafted, but there was little in the way of formal guidance that DINEPA could enforce. A third round of engagement occurred after this and focused on finalizing strategy documents and obtaining signatures from key stakeholders.

Throughout this process, levels of trust and participation fluctuated, and outcomes were mixed. Two topics are helpful in understanding the appropriateness of stakeholder engagement techniques that were used. Both topics have caused conflict between stakeholders, and both remain incompletely resolved.

Product certification

Identifying and certifying appropriate HWTS products is an example in which there were relatively high levels of participation but low levels of trust. Multiple NGOs convened workshops to educate other stakeholders about technology options, and DINEPA offered space for NGOs and companies to present products and share lessons learned about technology. External consultants synthesized findings and presented research. This allowed for a vibrant discussion of what technologies are effective and under what conditions. However, some NGOs and companies were suspicious of which products the government would decide to approve and whether it would actively promote certain brands. This suspicion was furthered by reports that a product that was new to the Haitian market received rapid certification and wide distribution in a region where a politician had ties to the company promoting the product.

Education and power delegation were the main forms of engagement used to make progress on product certification. External consultants developed evidence-based guidelines on product certification and conducted laboratory product evaluations (Murray et al. 2015). Early strategy documents explicitly stated that DINEPA would not promote specific brands, and DINEPA made efforts to communicate the strategy and certification process to all stakeholders to help work through trust issues. The regulatory role was delegated to the Ministry of Health after consultants trained officials on product certification standards and how to integrate them.

Pricing policy

Pricing was another policy issue characterized by generally low levels of trust. In the early weeks of the cholera response, NGOs expressed concern that some people, especially the most vulnerable groups, would not be able to afford HWTS products on the market. Private companies and social enterprises expressed concerns that free distributions by NGOs could destroy markets and develop a dependency mindset among the population. The government articulated the importance of affordability and the necessity of free distributions in "emergency settings" while recognizing the need for sustainable financing and supply chains.

Early working group meetings allowed stakeholders to discuss these issues, and early strategy documents set dates for transitions away from free products (mid-2011), defined who would determine if a situation after this date required free distributions (DINEPA), and limited the amount that could be distributed (a 2-month supply of HWTS products). When many actors did not follow these guidelines, working groups continued to be a forum for debate about definitions of sustainability and equity, but meetings did not change the fact that some stakeholders' values were still at odds. In 2015, facilitators reopened discussions on how to handle free distributions but then relegated pricing policy to the "lessons learned" appendix in a strategy document, concluding, "there is little consensus on the issue and this remains a challenge to be addressed" (The Water Institute at UNC 2015). Some stakeholders felt that this decreased clarity on HWTS pricing and that the facilitated discussions did more to placate participants than advance policy.

Organizational level: adaptive personnel management in diverse trust environments

In the transition from emergency response to development programs, DSI maintained staff in three regions: (1) the pilot site in the northwest department, (2) the earthquake-affected areas; and (3) small projects in multiple communities in the north department. Many of the strategic variables were similar across program sites, such as frequency of follow-up household visits conducted by sales agents, ratios of sales agents to households, product branding, advertising channels, and promotional messages. However, differences between program sites with respect to history and institutional culture led to differences in levels of trust and participation of staff. In this section, lessons about management in the context of different staff cultures is presented from the author's insider perspective as program director and with input from DSI's local manager.

The pilot project was characterized by high levels of trust and participation of staff. Prior to DSI's involvement, the project director used a participatory approach to identify and develop local project staff. Local staff was involved in many strategic project decisions, and relatively high levels of mutual trust existed between the project director and staff. DSI continued and built on this management style in many ways, including conducting all meetings with staff in Haitian Creole.

Trust and participation were mixed in the program site that began in earthquake-affected areas. Initial activities were funded through short-term grants from partner NGOs. Relative to the pilot project, DSI focused more on short-term results and rapid recruitment of local staff, using a more top-down management style. Within a year of the earthquake, a majority of the international NGOs that had responded to the emergency had left or significantly downsized their operations, which created perceived uncertainty among local staff about their long-term employment prospects with DSI. Trust increased as DSI solicited greater levels of participation in decision making from local staff, but the process was not linear and was often characterized by conflict.

Low levels of trust and participation characterized a third program site that DSI took over shortly before the earthquake. Prior to DSI management, local staff operated in an environment with little supervision. DSI maintained most of the same staff but implemented changes with respect to compensation and expectations. Trust and participation levels were slow to develop and remained lower than in the other program sites.

As the country transitioned from the acute cholera response to focusing more on long-term strategies, DSI implemented a series of management practices designed to bring the organization closer to its dual goals of sustainability and health impact. To understand how different levels of trust and participation affected outcomes in different program sites, it is useful to consider three initiatives (joint learning, delegation of decision-making power, and performance-based incentives) that involved participation of local staff. While the three initiatives may be considered "best practices" in management, the cases illustrate that the effectiveness of these initiatives was contingent on levels of trust.

Joint learning

DSI used many different strategies to promote its household chlorination product, but little was known about which strategies were most effective. Historically, promotion techniques focused on community meetings, radio spots, branded signs, and sponsorship of community events. As DSI transitioned out of the emergency response phase, the project director announced a "Year of Innovation" during which new promotional strategies would be encouraged. Each regional team was given a budget for promotion, which was to be used for experimentation based on new ideas from staff. Emphasis was placed on learning rather than success; it was stressed that the DSI leadership team would be pleased with regional teams that tried novel marketing techniques, even if they failed to increase chlorine sales.

The team with previously high levels of trust and participation embraced this approach. One field supervisor commented that he was previously afraid to suggest ideas that might fail, but the experiment was empowering. The majority of the truly new ideas across all program sites came from this region. The team organized community education sessions in new locations and product expositions near rural marketplaces and commissioned the painting of promotional murals. Staff satisfaction with the process was high, and they were open to evaluations of the effectiveness of these techniques. The promotion strategy in the following year incorporated the lessons learned by increasing allocation of resources toward techniques that appeared to work and decreasing resources for those that did not.

In contrast, the site that began as an emergency response project pursued a promotion strategy that was very similar to those used previously. At the end of the year, when an evaluation of sales data failed to show evidence of any positive impact of the promotion activities, the data seemed to have a "back-fire effect" on local staff's beliefs, similar to that documented in political science literature (Nyhan and Reifler 2010). Rather than leading to a more informed team, as was intended, local staff dug into their prior beliefs that the techniques that had been used for several years worked, despite evidence to the contrary. Tension resulted, with some local supervisors doubting whether the organization had ever been willing to continue funding effective promotion techniques after the experimental year. To them, the experiment was perceived as one of placation, and it had a negative impact on the planning process in the following year.

Delegation of decision-making power

This joint learning effort was related to another management initiative, empowering local staff to take increased levels of responsibility for strategic decisions. One element of all programs was household visits conducted by chlorine sales agents. DSI embarked on an initiative to transfer power downward to regional site managers. Each regional team was given a budget and the responsibility for strategic decisions such as how frequently to conduct visits and how to structure contracts with the agents that conducted visits, including setting compensation structures and expectations.

In the site that began as an emergency response, this transition was initially met with resistance. When given the budget figure, the regional managers negotiated with DSI leadership for more resources, requesting a higher budget so that visits could be continued at an even higher rate than before. Upon realizing that the total budget was nonnegotiable, the team still didn't entirely trust that decision-making power was being transferred to them, and they sought to craft a proposal that they thought DSI leadership wanted. Through a series of discussions over time, supervisors grew to trust the organization and feel empowered to make decisions. Future rounds of budgeting were much more in line with truly delegated power, and supervisors proposed their own ways of incentivizing and managing sales agents.

Accountability and performance-based incentives

As DSI transferred more decision-making power to its field supervisors, it also sought to reward them with additional compensation for high performance. DSI began incorporating performance-based incentives into the contracts of field supervisors and sales agents. This led to different results in each of the three program sites.

In the site with high levels of trust and participation, the process itself was well received. Staff pushed back on the targets, complaining that they were too high. However, they agreed with the ways in which their performance was to be measured. Over the following years, they began to meet and exceed the thresholds necessary for additional compensation and eventually became a leading region among programs.

In the site where DSI took over an existing network of employees, the introduction of incentives was met with more intense skepticism. Sales agents in particular did not agree with the system and expressed their objections to participating. Very few even attempted to achieve the targets. Rather than increasing accountability and performance, this management initiative seemed to lead to lower levels of engagement of sales agents.

While supervisors and sales agents in the emergency response site did not react in this way, another source of conflict arose as a result of the implementation of incentives. Most supervisors were optimistic about their capacities to gain additional income and embraced the idea, but intense competition developed between them. In some cases, this expanded the reach of the program into new areas, while other locations became battlegrounds for supervisors who treated it as a zero-sum game. The overall outcomes of the process were mixed. Total sales increased, and the magnitude of the change was large enough that incentives likely explain at least a portion of the increase. However, this came at the cost of increased conflict within the team, which is a problem that persisted after the introduction of performance-based incentives.

Synthesis: insights for stakeholder engagement in complex water problems

Several interconnected themes emerge from these cases that may apply to a broader set of complex water problems. The following issues seem to act as enabling conditions for effective stakeholder engagement in this context:

- *High levels of trust*: In the examples of management practices in a social enterprise NGO, trust was necessary in order for "best practices" to be effective. When trust was low, staff did not participate fully in and benefit from learning initiatives, they did not fully take on decision-making power, and they did not accept measures to increase accountability. At the sector level, one reason for the lack of further progress on regulatory issues was low trust between stakeholders.
- *Dynamic leadership:* Leadership helped overcome challenges posed by uncertainty, partly by engendering trust among stakeholders. At the policy level, dynamic individuals were able to drive consensus toward particular positions despite scientific uncertainty. In the HWTS sector, engagement of NGOs fluctuated, and the greatest levels of engagement seemed to occur in the presence of strong leaders at DINEPA even when future resources were not guaranteed.

- *Stable resources:* The conditions above seem well suited for operating in contexts with resource uncertainty, but relationships between these enabling conditions are reciprocal. The HWTS sector case illustrates how fluctuating resources can hinder development of institutional leadership, and weak leadership can in turn fail to attract more donor funds. Resource uncertainty also increases the difficulty of establishing trust among stakeholders, though the examples of the high trust DSI program site show how trust can improve resource utilization even if budgets are not large. This theme has been identified as important to success in other contexts as well; in a study of enabling conditions for effective HWTS implementation, organizational resource availability was identified as important for sustaining and scaling water projects.

(Ojomo et al. 2015)

Based on these themes, several recommendations for improving stakeholder engagement can be made:

- *Stakeholder engagement techniques should be adapted to the current levels of participation among stakeholders.* Leaders at every level should evaluate if and what forms of stakeholder engagement may be most effective. Facilitators should seek to understand the diversity of problem definitions, set clear goals for the process, and align the types of engagement with the goals and nature of the problem. Frameworks such as the "split ladder of participation" may be helpful.
- *Engagement should be contextualized to the levels of trust.* The examples of organizational initiatives that succeeded only where trust was high support the idea that management in complex situations must be adaptive rather than a straightforward implementation of "best practices." Leaders at every level should evaluate current levels of trust. In contexts where trust is well-established, traditional management practices and higher levels of participation can be pursued. In contexts without high overall levels of trust, leaders should employ a different set of processes aimed at building trust.
- *Seek and support dynamic leaders who can adapt to complexity.* Incorporating the recommendations above requires much more than technical skills. Leaders who understand how to build consensus and trust can overcome barriers related to uncertainty. As developing and retaining leaders requires resources, this is connected to the ability to reduce resource uncertainty. Strategic plans at any level should budget for and prioritize dynamic leaders.

How interdisciplinarity and the Water Diplomacy Framework add value

The cases and lessons highlight several ways in which an interdisciplinary approach and the Water Diplomacy Framework provide a better theoretical lens and offer a greater chance at achieving practical resolutions on this problem.

- *What types of technical analyses are useful?* Traditional approaches tend to make assumptions about shared values and build these assumptions into technical models. The Water Diplomacy Framework takes a step back and highlights the importance of diagnosing the degree of consensus on a problem before constructing models. These cases illustrate that consensus may exist on one part of the problem but not another and that some values that are assumed in most models may not actually be shared among all stakeholders. This has implications for the types of technical models that are useful. Acknowledging a broader set of problem definitions and starting points implies the need for a broader set of analytic tools. For example, if a HWTS supplier's definition of sustainability requires full cost recovery, then an analysis of profitability may be sufficient to answer questions about how to structure its programs and marketing. However, if definitions of sustainability are relaxed to allow donor funds to be incorporated, models must take into account epidemiologic data and incorporate additional approaches, such as cost-effectiveness analysis (Ritter et al. 2017).

- *What role do technical analyses play in response to problems?* More comprehensive models that integrate multiple disciplines may still not be sufficient for responding to the problem at hand. The Water Diplomacy Framework views technical analyses as a starting point in negotiated resolutions rather than an endpoint in solving the problem. Models should be viewed as tools for discussion between and within stakeholders. In the example of the HWTS supplier deciding how to structure its marketing activities, models can actually serve as a tool for internal discussions that clarify the goals and values within an institution. They can also be useful for translating ideas and values between stakeholders.

- *How should technical analyses be applied in practice?* Traditional approaches may rush to apply findings from one context to another. The Water Diplomacy Framework views management as an adaptive process. Rather than assuming that the techniques that worked at one scale or context are broadly applicable, practitioners must account for the complexities of their context and undergo continuous evaluation. Cases like the ones described above are most useful for providing examples of how marginal improvements can be made in the process of engaging stakeholders and seeking resolutions to complex problems.

References

Adams, P., 2012. Haiti prepares for cholera vaccination but concerns remain. *The Lancet*, 379, 16.

Andrews, J.R., and Basu, S., 2011. Transmission dynamics and control of cholera in Haiti: An epidemic model. *The Lancet*, 377, 1248–1255.

Arnstein, S.R., 1969. A ladder of citizen participation. *Journal of the American Institute of Planners*, 35, 216–224.

Baron, S., Lesne, J., Moore, S., Rossignol, E., Rebaudet, S., Gazin, P., Barrais, R., Magloire, R., Boncy, J., and Piarroux, R., 2013. No evidence of significant levels of toxigenic V. cholerae O1 in the Haitian aquatic environment during the 2012 rainy season. *PLoS Currents*, 5.

Behrmann, T., Salomon, P., and Hudicourt, L., 1900. Haiti: Dysentery prevalent in Nippes. *Public Health Reports (1876–1970),* 15, 497–499.

Bertuzzo, E., Finger, F., Mari, L., Gatto, M., and Rinaldo, A., 2014. On the probability of extinction of the Haiti cholera epidemic. *Stochastic Environmental Research and Risk Assessment,* 30 (8), 2043–2055.

Bertuzzo, E., Mari, L., Righetto, L., Gatto, M., Casagrandi, R., Blokesch, M., Rodriguez-Iturbe, I., and Rinaldo, A., 2011. Prediction of the spatial evolution and effects of control measures for the unfolding Haiti cholera outbreak. *Geophysical Research Letters,* 38, 1–5.

Cayemittes, M., Busangu, M.F., Bizimana, J.D.D., Barrere, B., Severe, V., Cayemittes, V., and Charles, E., 2013. *Enquête Mortalité, Morbidité et Utilisation des Services EMMUS-V.* Calverton: Institut Haitien de l'Enfance, Petion-Ville, Haiti and MEASURE.

CDC, 2010. Update: Cholera outbreak – Haiti, 2010. *MMWR,* 59, 1473–1479.

Chao, D.L., Halloran, E., and Longini, J., M., 2011. Vaccination strategies for epidemic cholera in Haiti with implications for the developing world. *PNAS,* 108, 7081–7085.

Chin, C.S., Sorenson, J., Harris, J.B., Robins, W.P., Charles, R.C., Jean-Charles, R.R., Bullard, J., Webster, D.R., Kasarskis, A., Peluso, P., Paxinos, E.E., Yamaichi, Y., Calderwood, S.B., Mekalanos, J.J., Schadt, E.E., and Waldor, M.K., 2011. The origin of the Haitian Cholera outbreak strain. *New England Journal of Medicine,* 364 (1), 33–42.

Clasen, T., 2015. Household water treatment and safe storage to prevent diarrheal disease in developing countries. *Current Environmental Health Reports,* 2, 69–74.

Clasen, T., and Bastable, A., 2003. Faecal contamination of drinking water during collection and household storage: The need to extend protection to the point of use. *Journal of Water and Health,* 1 (3), 109–115.

Clasen, T., and Haller, L., 2008. *Water Quality Interventions to Prevent Diarrhoea: Cost and Cost-Effectiveness.* Geneva, Switzerland.

Clasen, T., Roberts, I., Rabie, T., Schmidt, W.P., and Cairncross, S., 2006. Interventions to improve water quality for preventing diarrhea. *Cochrane Database of Systematic Reviews,* CD004794 (3), 1–201.

Clasen, T., Schmidt, W.P., Rabie, T., Roberts, I., and Cairncross, S., 2007. Interventions to improve water quality for preventing diarrhoea: Systematic review and meta-analysis. *BMJ,* 334, 782.

Cluster, W., 2011 *Cadre pour la promotion de l'hygiene et de l'assainissement: Résultats de l'Atelier des enseignements en matière de promotion de l'hygiène en Haïti en 2010.* Available from: https://www.pseau.org/outils/ouvrages/cadre_promotion_hygiene_haiti_atelier2011.pdf

Cravioto, A., Lanata, C.F., Lantagne, D., and Nair, G.B., 2011. *Final Report of the Independent Panel of Experts on the Cholera Outbreak in Haiti.* Available from: https://reliefweb.int/sites/reliefweb.int/files/resources/Full_Report_525.pdf

Cyranoski, D., 2011. Cholera vaccine plan splits experts. *Nature,* 469, 273–274.

Date, K.A., Vicari, A., Hyde, T.B., Mintz, E., Danovaro-Holliday, M.C., Henry, A., Tappero, J.W., Roels, T.H., Abrams, J., Burkholder, B.T., Ruiz-Matus, C., Andrus, J., and Dietz, V., 2011. Considerations for oral cholera vaccine use during outbreak after earthquake in Haiti, 2010–2011. *Emerging Infectious Diseases,* 17, 2105–2112.

Deep Springs International, 2017. *Mission and Vision.* www.deepspringsinternational.org/AboutUs/MissionVision.aspx

Dewitt, R.P., 1987. Policy directions in international lending, 1961–1984: The case of the Inter-American Bank. *Journal of Developing Areas,* 21, 277–284.

DINEPA, 2010. *Stratégie Nationale de Réponses à l'Epidémie de Choléra.* Available from: https://www.pseau.org/outils/ouvrages/dinepa_strategie_nationale_cholera_v1_0.pdf

DINEPA, 2013. *National Plan for the Elimination of Cholera in Haiti, 2013–2022.* Available from: https://reliefweb.int/report/haiti/national-plan-elimination-cholera-haiti-2013-2022

DINEPA, 2014. *Présentation et Bilan 2012–2014.*

Dupas, P., 2011. Health behavior in developing countries. *Annual Review of Economics*, 3 (1), 425–449.

Evans, W.D., Pattanayak, S.K., Young, S., Buszin, J., Rai, S., and Bihm, J.W., 2014. Social marketing of water and sanitation products: A systematic review of peer-reviewed literature. *Social Science & Medicine*, 110, 18–25.

Fass, S.M., 1988. *Political Economy in Haiti: The Drama of Survival.* Brunswick: New Transaction Books.

Fewtrell, L., Kaufmann, R.B., Kay, D., Enanoria, W., Haller, L., and Colford, J.M., 2005. Water, sanitation, and hygiene interventions to reduce diarrhoea in less developed countries: A systematic review and meta-analysis. *The Lancet Infectious Diseases*, 5, 42–52.

Frerichs, R.R., Keim, P.S., Barrais, R., and Piarroux, R., 2012. Nepalese origin of cholera epidemic in Haiti. *Clinical Microbiology and Infection*, 18, E158–E163.

Gelting, R., Bliss, K., Patrick, M., Lockhart, G., and Handzel, T., 2013. Water, sanitation and hygiene in Haiti: Past, present, and future. *American Journal of Tropical Medicine and Hygiene*, 89, 665–670.

Gundry, S., Wright, J., and Conroy, R.M., 2004. Household drinking water in developing countries: A systematic review of microbiological contamination between source and point-of-use. *Tropical Medicine & International Health*, 9, 106–117.

Harshfield, E., Lantagne, D., Turbes, A., and Null, C., 2012. Evaluating the sustained health impact of household chlorination of drinking water in rural Haiti. *American Journal of Tropical Medicine and Hygiene*, 87, 786–795.

Hendriksen, R.S., Price, L.B., Schupp, J.M., Gillece, J.D., Kaas, R.S., Engelthaler, D.M., Bortolaia, V., Pearson, T., Waters, A.E., Upadhyay, B.P., Shrestha, S.D., Adhikari, S., Shakya, G., Keim, P.S., and Aarestrup, F.M., 2011. Population genetics of Vibrio cholerae from Nepal in 2010: Evidence on the origin of the Haitian outbreak. *mBio*, 2. https://doi.org/10.1128/mBio.00157-11

Hurlbert, M., and Gupta, J., 2015. The split ladder of participation: A diagnostic, strategic, and evaluation tool to assess when participation is necessary. *Environmental Science & Policy*, 50, 100–113.

Ivers, L.C., 2017. Eliminating cholera transmission in Haiti. *The New England Journal of Medicine*, 376, 101–103.

Ivers, L.C., Farmer, P.E., Almazor, C.P., and Léandre, F., 2010. Five complementary interventions to slow cholera: Haiti. *The Lancet*, 376, 2048–2051.

Ivers, L.C., Farmer, P.E., and Pape, W.J., 2012. Oral cholera vaccine and integrated cholera control in Haiti. *The Lancet*, 379, 2026–2028.

Ivers, L.C., Hilaire, I.J., Teng, J.E., Almazor, C.P., Jerome, J.G., Ternier, R., Boncy, J., Buteau, J., Murray, M.B., Harris, J.B., and Franke, M.F., 2015. Effectiveness of reactive oral cholera vaccination in rural Haiti: A case-control study and bias-indicator analysis. *The Lancet Global Health*, 3, e162–e168.

Ivers, L.C., Teng, J.E., Lascher, J., Raymond, M., Weigel, J., Victor, N., Jerome, J.G., Hilaire, I.J., Almazor, C.P., Ternier, R., Cadet, J., Francois, J., Guillaume, F.D., and Farmer, P.E., 2013. Use of oral cholera vaccine in Haiti: A rural demonstration project. *American Journal of Tropical Medicine and Hygiene*, 89, 617–624.

Lantagne, D., and Clasen, T., 2012. Use of household water treatment and safe storage methods in acute emergency response: Case study results from Nepal, Indonesia, Kenya, and Haiti. *Environmental Science & Technology*, 46 (20), 11352–11360.

Luby, S., Agboatwalla, M., Hoekstra, R.M., Rahbar, M.H., Billhimer, W., and Keswick, B., 2004. Delayed effectiveness of home-based interventions in reducing childhood diarrhea, Karachi, Pakistan. *American Journal of Tropical Medicine and Hygiene*, 71, 420–427.

Martin, N.A., Hulland, K.R.S., Dreibelbis, R., Sultana, F., and Winch, P.J., 2018. Sustained adoption of water, sanitation, and hygiene interventions: Systematic review. *Tropical Medicine & International Health*, 23 (2), 122–135.

Matias, W.R., Teng, J.E., Hilaire, I.J., Harris, J.B., Franke, M.F., and Ivers, L.C., 2017. Household and individual risk factors for cholera among cholera vaccine recipients in rural Haiti. *American Journal of Tropical Medicine and Hygiene*, 97 (2), 436–442.

MSPP, 2017. *Rapport du Réseau National de Surveillance: 47ème semaine épidémiologique 2017*.

Murray, A., Pierre-Louis, J., Joseph, F., Sylvain, G., Patrick, M., and Lantagne, D., 2015. Need for certification of household water treatment products: Examples from Haiti. *Tropical Medicine & International Health*, 20, 462–470.

Nyhan, B., and Reifler, J., 2010. When corrections fail: The persistence of political misperceptions. *Political Behavior*, 32, 303–330.

Ojomo, E., Elliott, M., Goodyear, L., Forson, M., and Bartram, J., 2015. Sustainability and scale-up of household water treatment and safe storage practices: Enablers and barriers to effective implementation. *International Journal of Hygiene and Environmental Health*, 218, 704–713.

Pan American Health Organization, 2012. *Call to Action: A Cholera-Free Hispaniola: Moving from Cholera Control to Cholera Elimination Through Essential Investments in Water, Sanitation, and Hygiene Infrastructure*. Available from: https://en.calameo.com/books/00027 0407e351b51f4654

Patrick, M., Berendes, D., Murphy, J., Bertrand, F., Husain, F., and Handzel, T., 2013. Access to safe water in rural Artibonite, Haiti 16 months after the onset of the cholera epidemic. *The American Journal of Tropical Medicine and Hygiene*, 89, 647–653.

Periago, M.R., Frieden, T.R., Tappero, J.W., De Cock, K.M., Aasen, B., and Andrus, J.K., 2012. Elimination of cholera transmission in Haiti and the Dominican Republic. *The Lancet*, 379, e12–e13.

Piarroux, R., Barrais, R., Faucher, B., Haus, R., Piarroux, M., Gaudart, J., Magloire, R., and Raoult, D., 2011. Understanding the cholera epidemic, Haiti. *Emerging Infectious Diseases*, 17, 1161–1168.

PSI, 2012. *Enquete de suivi sur les déterminants de l'utilisation des produits de traitement de l'eau au niveau des ménages ayant des enfants de moins 5 ans en Haiti*. Washington, DC: Population Services International.

Quick, R., Kimura, A., Thevos, A.K., Tembo, M., Shamputa, I., Hutwagner, L., and Mintz, E., 2002. Diarrhea prevention through household-level water disinfection and safe storage in Zambia. *American Journal of Tropical Medicine and Hygiene*, 66, 584–589.

Quick, R., Venczel, L.V., Mintz, E., Soleto, L., Aparicio, J., Gironaz, M., Hutwagner, L., Greene, K.D., Bopp, C., Maloney, K., Chavez, D., Sobsey, M.D., and Tauxe, R.V., 1999. Diarrhoea prevention in Bolivia through point-of-use water treatment and safe storage: A promising new strategy. *Epidemiology & Infection*, 1222, 83–90.

Ramachandran, V., and Walz, J., 2015. Haiti: Where has all the money gone? *Journal of Haitian Studies*, 21, 26–65.

Rebaudet, S., Gaudart, J., Abedi, A.A., and Piarroux, R., 2016. Questioning the effectiveness of oral cholera vaccine in Port-au-Prince slums. *American Journal of Tropical Medicine and Hygiene*, 95, 493–494.

Rittel, H.W.J., and Webber, M.W., 1973. Dilemmas in a general theory of planning. *Policy Sciences*, 4, 155–169.

Ritter, M., Camille, E., Velcine, C., Guillaume, R.K., and Lantagne, D., 2017. Optimizing household chlorination marketing strategies: A randomized controlled trial on the effect of price and promotion on adoption in Haiti. *American Journal of Tropical Medicine and Hygiene*, 97, 271–280.

Rosa, G., and Clasen, T., 2010. Estimating the scope of household water treatment in low- and medium-income countries. *American Journal of Tropical Medicine and Hygiene*, 82, 289–300.

Rouzier, V., Severe, K., Juste, M.A., Peck, M., Perodin, C., Severe, P., Deschamps, M.M., Verdier, R.I., Prince, S., Francois, J., Cadet, J.R., Guillaume, F.D., Wright, P.F., and Pape, J.W., 2013. Cholera vaccination in urban Haiti. *American Journal of Tropical Medicine and Hygiene*, 89, 671–681.

Severe, K., Rouzier, V., Anglade, S.B., Bertil, C., Joseph, P., Deroncelay, A., Mabou, M.M., Wright, P.F., Guillaume, F.D., and Pape, J.W., 2016. Effectiveness of oral cholera vaccine in Haiti: 37-month follow-up. *American Journal of Tropical Medicine and Hygiene*, 94 (5), 1136–1142.

Smith Fawzi, M.K., Grout, M., Jean, J., Johnstone, M., Klasing, A., Lyon, E., Satterthwaite, M., Shoranick, T., and Varma, M.K., 2009. Wòch nan Soley: The Denial of the right to water in Haiti. *Health and Human Rights*, 10 (2), 67–89.

Tohme, R.A., Francois, J., Wannemuehler, K., Iyengar, P., Dismer, A., Adrien, P., Hyde, T.B., Marston, B.J., Date, K., Mintz, E., and Katz, M.A., 2015. Oral cholera vaccine coverage, barriers to vaccination, and adverse events following vaccination, Haiti, 2013. *Emerging Infectious Diseases*, 21, 984–991.

Tuite, A.R., Tien, J.H., Eisenberg, M.C., Earn, D., Ma, J., and Fisman, D.N., 2011. Cholera epidemic in Haiti, 2010: Using a transmission model to explain spatial spread of disease and identify optimal control interventions. *Annals of Internal Medicine*, 154, 593–601.

UNICEF, 2019. *Progress on Household Drinking Water, Sanitation, and Hygiene, 2000–2017.* Geneva, Switzerland.

USAID, 2014. *Water, Sanitation, and Hygiene Sector Status and Trends Assessment in Haiti.* Available from: https://www.globalwaters.org/node/119

The Water Institute at UNC, 2015. *Household Water Treatment and Safe Storage in Haiti: Report of Stakeholders' Consultations.*

WHO and UNICEF, 2015. *Progress on Sanitation and Drinking Water: 2015 Update and MDG Assessment.* Geneva, Switzerland.

Wolfe, M., Kaur, M., Yates, T., Woodin, M., and Lantagne, D., 2018. A systematic review and meta-analysis of the association between water, sanitation, and hygiene exposures and cholera in case-control studies. *American Journal of Tropical Medicine and Hygiene*, 99, 534–545.

Zanotti, L., 2010. Cacophonies of aid, failed state building and NGOs in Haiti: Setting the stage for disaster, envisioning the future. *Third World Quarterly*, 31, 755–771.

8 Water Diplomacy at the macro scale

Agricultural groundwater governance in the High Plains Aquifer region of the United States

Gregory N. Sixt, Ashley C. McCarthy, Kent E. Portney, and Timothy S. Griffin

Introduction

The High Plains Aquifer (HPA) in the United States, also known as the Ogallala Aquifer,[1] is one of the largest aquifer systems in the world, underlying more than 450,000 km^2 (174,050 mi^2) of eight states, from South Dakota to Texas (see Figure 8.1; Sophocleous 2011). Crop sales from lands irrigated from the HPA generated more than $7 billion in 2007 (Gollehon and Winston 2013). The aquifer supports approximately 20 percent of the corn, wheat, cotton, and cattle production in the United States and provides drinking water for more than 80% of the people who live within its boundaries (Dennehy et al. 2002; Sophocleous 2011; USDA-NRCS 2016). It is the most intensively used aquifer in the country, accounting for about 30% of total withdrawals from all aquifers used for irrigation (Sophocleous 2011; USDA-NRCS 2013).

Across most of the HPA region, declining groundwater quantity is a significant problem. Natural recharge of the aquifer from precipitation is low, and current groundwater abstraction exceeds the rate of recharge in many areas (Peck 2007; USDA-NRCS 2013). Recoverable water in storage in the aquifer declined 9% from before the widespread development of irrigation began in the 1950s to 2015 (McGuire 2017). However, this figure, while alarming, represents an average decline in groundwater levels across an immense land area and masks significant variability. For example, while stored groundwater has declined by 29% in the HPA region of Texas, it has only declined between 0.5% and 1% in Nebraska (Stanton et al. 2011; Bleed and Babbitt 2015).

There are a variety of reasons for this geographic variability, such as differences in precipitation patterns and underlying hydrogeology, which impact percolation of surface water to the aquifer. In addition, different agricultural practices and production systems and, perhaps most importantly, differences in governance institutions and policies that regulate how groundwater is used are causes of variability. The states in the HPA region have pursued

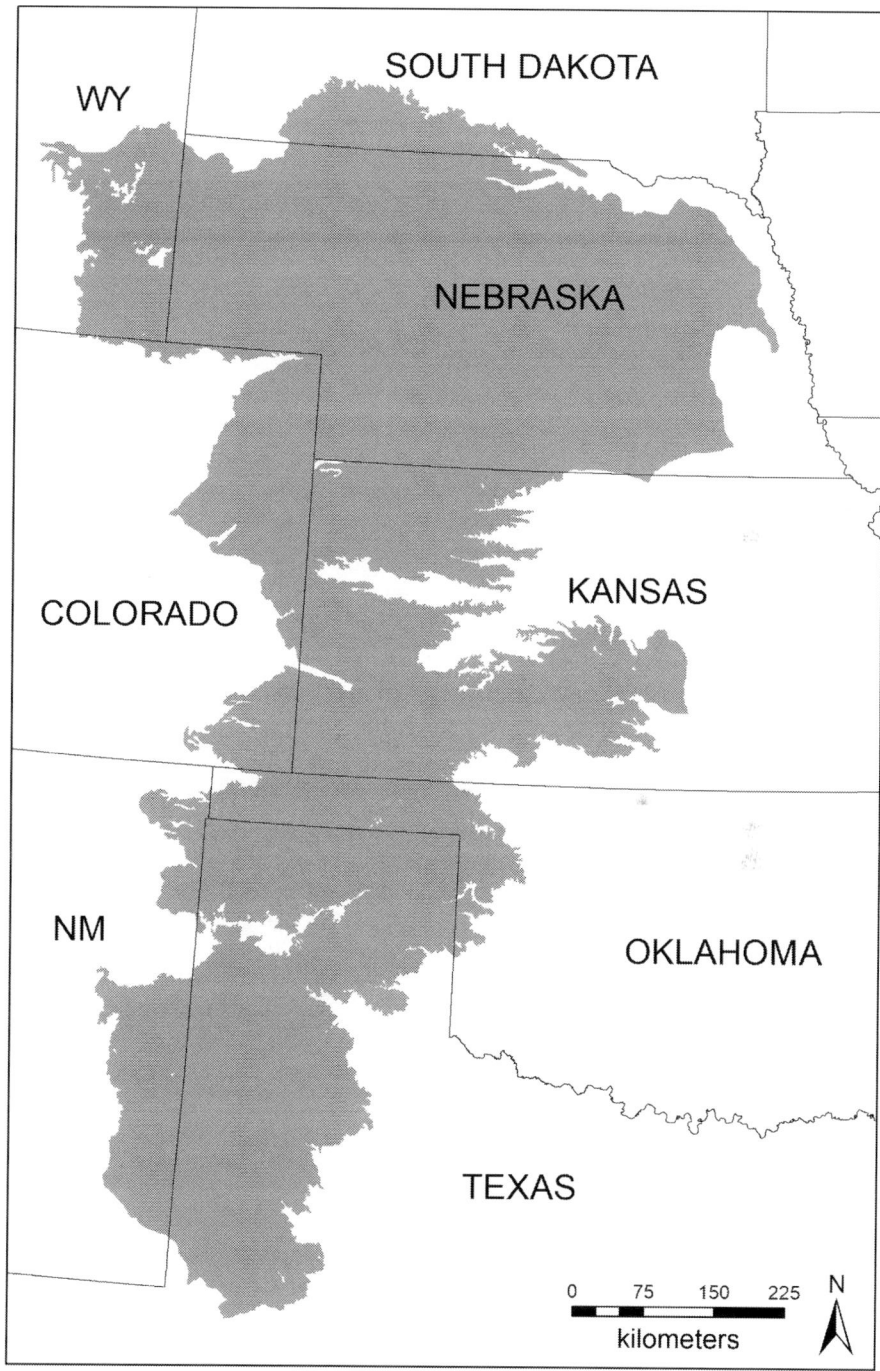

Figure 8.1 Boundary of the High Plains Aquifer.

Data sources: U.S. Geological Survey, U.S. Census Bureau

Table 8.1 High Plains Aquifer Area, Distribution, Irrigated Area, and Water Level by State. Modified and updated from Sophocleous (2011).

	Kansas	*Nebraska*	*Texas*	*Total*
HPA area in state[1] (km^2) (Total extent of HPA 450,788 km^2)	78,995	164,853	91,815	335,663
% of total HPA area in state	17.5	36.6	20.4	74.5
% of HPA water underlying state	10.0	65.0	12.0	87.0
Hectares irrigated using HPA water in 2013[2]	987,433	2,901,596	1,432,587	5,321,616
Change in water level from predevelopment to 2015 (m)[3]*	−8.0	−0.3	−12.5	N/A

Notes:
*Area Weighted Average;
[1]Gutentag et al. (1984),
[2]Gollehon and Winston (2013),
[3]McGuire (2017)

different approaches to groundwater governance with varying degrees of success toward sustainability.

This chapter has two primary objectives: (1) to expand the concept of Water Diplomacy to encompass groundwater governance and (2) to assess where and to what extent Water Diplomacy principles are in use for groundwater governance in the HPA region of the United States. Here, we look at the states of Kansas, Nebraska, and Texas because they collectively overlay 74.5% of the area of the HPA (Table 8.1), are all highly productive agricultural states, and their different regulatory systems offer a useful case study for identifying the extent to which U.S. state-level governance institutions possess the principles needed for Water Diplomacy solutions in the context of groundwater governance.

The chapter first discusses the concept of Water Diplomacy and its application to the governance of groundwater, generally understood to represent a common–pool resource (CPR). This is followed by a description of governance of CPRs and an introduction to Elinor Ostrom's design principles for governing sustainable resources (Ostrom 1990). We then provide a brief background on groundwater governance in the United States, followed by a discussion of groundwater governance institutions in each of the three states on which this study is based: Kansas, Nebraska, and Texas. We close with an analysis of the extent to which these three state-level institutions might be said to capture significant aspects of the principles of Water Diplomacy and finally a discussion on the factors that influence these variations.

Water Diplomacy in the context of groundwater governance

Water Diplomacy (WD) is a novel framework within the developing field of interdisciplinary water management. Broadly speaking, WD is a process engaged in by state or nonstate actors to address, resolve, or avoid conflicts over access to water and water-related issues. It can also represent a skill set possessed by an individual or group of individuals (e.g., negotiators or mediators) exercised in dealing with conflicts over water on behalf of state or nonstate actors. As an emerging interdisciplinary field, the definition, the disciplinary foci, and the tools it utilizes are still evolving, and an increasing body of literature is in the process of refining what constitutes WD (e.g., Islam and Madani 2017; Islam and Susskind 2013; Susskind and Islam 2012; van Rees and Reed 2015).

In their effort to advance the concept and practice of WD, Islam and Susskind (2013) identified six key principles that are thought to represent necessary conditions for achieving successful negotiated water results. These are:

1   Appropriate stakeholders and stakeholder interactions are identified and adequately represented.
2   Stakeholders engage in joint fact-finding to develop a shared understanding of their resource system and how key variables impact the system.
3   Relevant parties need to create value for stakeholders through mutual-gains, non–zero-sum policy solutions.
4   Informal problem-solving processes are linked to formal decision-making processes and authorities at a higher governance levels (i.e., at the national level or individual state level in federalist systems).
5   Policy solutions are developed within the context of collaborative adaptive management.
6   Management of water systems is improved through capacity building and societal learning among individuals, organizations, and networks.

WD as described by Islam and Susskind (2013) has been applied primarily to cross-national transboundary water negotiations involving national policy makers (e.g., Berndtsson et al. 2017; Brady et al. 2015; Choudhury 2017) or to situations in which a history of conflict over water is already present (e.g., Moazezi et al. 2017) or both (e.g., Huntjens 2017). Largely absent from the literature (an exception being Koebele 2015) is WD analysis that looks at water governance institutions at the subnational level (e.g., water governance at the interstate or individual state levels in the United States). This area of analysis is important and merits additional investigation. However, WD manifests differently at this level and thus requires different tools for analysis. This chapter represents one of the first forays into expanding WD analysis to the water governance context.

We argue that WD must be viewed through two different lenses, the micro level and the macro level. The micro level primarily encompasses water negotiations in which the focus is on utilizing trained negotiators, or on training people to be negotiators, in the six key principles central to the WD Framework described by Islam and Susskind (2013). The macro level of WD is more nebulous in that it attempts to take a higher-level look across complex water conflicts or governance situations to identify key factors that lead to robust governance institutions that produce sustainable water management outcomes. While the micro and macro levels of WD may seem independent of each other, the fact is that the policies, laws, histories, and values reflected at the macro level play a very important part in influencing whether and to what extent the six key principles are achievable in specific WD situations.

At its core, WD is a process or series of processes focused on finding mutual-gains solutions to water challenges through adaptive, collective management. At the groundwater governance level, it is the governance institutions that set the stage for these processes. In other words, at this level, it is the structure of institutions that govern groundwater in complex social-ecological systems (SESs) that determines the extent to which the principles needed for WD solutions are present. An SES is an ecological system connected to and impacted by one or more social systems; it is a subset of social systems in which interdependent human relationships are deeply intertwined with interactions involving biophysical and nonhuman biological units (Anderies et al. 2004).

Research pioneered by Elinor Ostrom (e.g., Ostrom 2009; Ostrom 2008; Ostrom 2000; Ostrom 1990) examined the processes that lead resource users to develop and sustain robust institutions to effectively manage CPRs in complex SESs. Groundwater is typically thought of as a CPR, meaning that it is a shared resource that is sufficiently large to make it difficult to exclude users, and each individual's use reduces benefits to other users who share the resource (Ostrom 2000). Ostrom (2008, 1990) identified certain conditions that seem to be present in robust CPR governance institutions. She also provided a road map for assessing the extent to which these conditions are present. Her research offers a way to connect the processes of CPR governance to the outcomes of adaptive, collective management. In the context of this research, the CPR is groundwater, and the desired outcome rests at the core of WD – sustainable groundwater governance. Thus, we argue that within the context of groundwater governance, WD can be considered a subset of Ostrom's work. In this chapter, we apply Ostrom's institutional design principles to U.S. state-level groundwater governance institutions in three states in the HPA region to assess the extent to which these states possess the conditions that lead to positive WD outcomes. In short, the HPA represents a substantial common-pool resource whose governance is largely conducted by individual states and their institutions, and this governance in turn affects the extent to which sustainable water resource outcomes are achievable.

Groundwater and Ostrom's eight design principles for governing sustainable common-pool resources

Like other CPRs, such as fisheries and forests, groundwater's lack of exclusivity and the subtractability of benefits to users who share it make governing its use difficult and overharvesting likely, leading to the eventual destruction of the aquifer from which the water is withdrawn (Ostrom 2008). Managing groundwater sustainably is further complicated by the long timescales of natural groundwater processes and the time lag of impacts associated with groundwater use, which are often inconsistent with policy time frames established for managing groundwater (Ross and Martinez-Santos 2010; Gleeson et al. 2012). Like Bleed and Babbitt (2015), we define a sustainable groundwater governance system as one that can sustain for current and future generations the benefits that society requires and desires from the water resource. In order to do this, the governance institution itself must be robust enough to be maintained, and the institution must be able to adapt to future challenges and condition in the resource (Bleed and Babbitt 2015).

There is no panacea for groundwater governance, and institutional arrangements are dependent on a number of factors including political processes, culture, historic paths to institutional change, and the scale of the resource system being governed (Meinzen-Dick 2007). While one approach may succeed in one place and fail in another, Ostrom (1990) identified eight institutional design principles that are frequently present in robust governance systems that sustainably manage CPRs. These principles are (Ostrom 1990; Ostrom 2000; Ostrom 2008; Bleed and Babbitt 2015):

1   Clearly defined boundaries: both the individuals who have rights to withdraw resources from the resource system and the boundaries of that resource system are clearly defined.
2   Proportional equivalence between benefits and costs: rules specifying the allocation of the resources are well-matched to local conditions (e.g., soils, climate, crops being grown, labor, etc.) and must be considered fair and legitimate by users.
3   Collective choice arrangements: most of the stakeholders affected by harvesting and protection rules are included in the group that can make and modify these rules.
4   Monitoring: Individuals and organizations that monitor biophysical conditions and user behavior are at least partially accountable to users and/or are users themselves.
5   Graduated sanctions: users who violate the rules are likely to receive sanctions that reflect the seriousness and context of the violation.
6   Conflict-resolution mechanisms: users and their officials have rapid access to low-cost, local avenues to resolve conflict between users and officials or among users.
7   Minimal recognition of rights to organize: users have the right to organize their own institutions and long-term tenure rights, and these rights are not challenged by external governmental authorities.
8   Nested enterprises: governance activities are organized in nested enterprises in which appropriation, monitoring, enforcement, conflict resolution, and other governance activities are organized in multiple layers.

Groundwater governance in the United States

Groundwater law in the United States is primarily created and administered at the state level, with some federal influence in situations in which groundwater use intersects with federal laws (e.g., the Endangered Species Act; Peck 2015). As a result, groundwater governance varies considerably between states. Groundwater governance is built around two fundamental principles: property rights and rules for allocation of groundwater.

Property rights structure determines who owns and controls the resource and influences what types of regulations can be used to manage it. A water right is a right to divert water for a period of time, from a specific source, to be used on a specific parcel of land (Peck 2015). States often consider the "public interest" or the "public welfare" when determining whether to grant a water right or to change existing water rights (Getches 2009; Peck 2015). Water rights as property rights differ by state; some states consider them to be private property and others do not. The distinction is important because the courts have provided strong protection for clearly defined private property rights under the Fifth Amendment of the U.S. Constitution, limiting options for regulatory management in some cases.

There are four general rules used in the U.S. to allocate groundwater to users. These rules govern who has a right to withdraw and use groundwater and how much and for what purposes they can withdraw, and they can clarify how to resolve disputes when there is a shortage. They are (Aiken and Supalla 1979; Kaiser and Skillern 2001; Peck 2007):

1   Rule of Capture (also known as Absolute Ownership): holds that landowners own all water that underlies their land and can pump water without limit, except where prohibitions on wasteful use exist.
2   Reasonable use doctrine: allows a landowner to use the groundwater underlying his or her land, provided that the water is used for a reasonable purpose and is used on the overlying tract of land.
3   Correlative rights: determines rights to groundwater based on ownership of land. Landowners overlying the same aquifer are limited to a reasonable share of the aquifer's total supply based on the amount of land owned by each.
4   Prior appropriation: follows the "first in time, first in right" principle, in which the rights to withdraw groundwater are allocated on a temporal basis rather than on a land ownership basis, and senior users have priority over more junior ones.

While it is sometimes useful to draw distinctions among the four rules that underlie groundwater rights and allocations, nearly all U.S. states have developed governance systems that have elements of more than one rule. Each of the three states included in this chapter uses its own combination of property rights and groundwater allocation rules. These form the foundations of each state's groundwater governance institutions as they relate to groundwater quantity, and each state has approached the design of these institutions differently. The remainder of this chapter describes these institutions and evaluates the extent to which they possess the conditions for positive WD outcomes by comparing them to Ostrom's eight principles.

Kansas

Irrigated agriculture is central to the economy of Kansas, and the state has approximately 1.2 million hectares of irrigated land, of which approximately 987,000 hectares are irrigated by water from the HPA (Table 8.1; Kenny and Juracek 2013). Agriculture accounts for 80% to 85% of all water diversions in Kansas, and about 96% of water used for irrigation comes from groundwater (Kenny and Juracek 2013; Kansas Water Office 2015). Almost all of the groundwater used for irrigation is withdrawn from the HPA, which underlies western Kansas and a portion of south-central Kansas (Figure 8.1; Kenny and Juracek 2013). Groundwater recharge rates in this part of the HPA are very low, with recharge equivalent to about 15% of current irrigation withdrawals (Steward et al. 2013). To date, approximately 30% of the groundwater has been pumped, and at current rates, an additional 39% will be depleted by 2060 (Steward et al. 2013). From predevelopment to 2015, the HPA region of Kansas saw an area-weighted average groundwater level decline of 8.0 meters (Table 8.1; McGuire 2017). There are considerable regional differences in HPA decline in the state. Since 1996, groundwater levels have fallen an average of 10.4 meters in southwest Kansas and just over 3 meters in west-central Kansas (Kenny and Juracek 2013; University of Kansas 2015). While groundwater levels continue to decline, recent trends have shown a slowing in the rate of decline (University of Kansas 2015).

The Kansas Water Appropriation Act of 1945 paved the way for statutory management of groundwater in the state. It asserted that water is dedicated to the use of the people of the state, charged the state with the duty to manage the resource, and adopted the prior appropriation allocation rule (*Kan. Stat. Ann. §§ 82a-701 to -733* 1989 & supp. 1994.). The Act was amended in 1957 to designate water rights as real property rights, subject to the principle of "beneficial use," and reasserted the state's duty to manage the resource under these conditions (Peck 2003). The designation of water rights as a real property right poses an obstacle to managing groundwater because it raises concerns regarding the authority of the state to restrict groundwater use without compensating water rights holders (Peck 2015).

In Kansas, groundwater is managed through the Kansas Division of Water Resources (KDWR) and through local institutions, called Groundwater Management Districts (GMDs). In 1972, the legislature enacted the Kansas Groundwater Management District Act (KGMDA) in response to extensive groundwater mining (Peck 2007; Hoffman and Zellmer 2013). The KGMDA allowed the establishment of GMDs, which have the authority to prepare district management plans, draft and recommend groundwater regulations to the KDWR, and tax, purchase, and sell property and water rights (Griggs 2014a). A locally elected board of directors governs each district (Peck 2006). There are five GMDs in Kansas, and all are located in areas that overlie the HPA (Figure 8.2). The GMDs are drawn along political boundaries (section, township, or county lines), and they roughly correspond

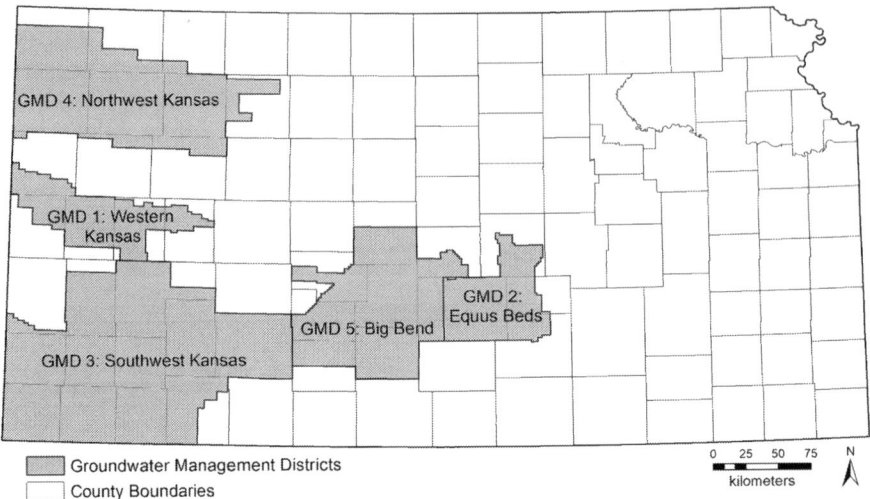

Figure 8.2 Groundwater management districts of Kansas.

Data sources: Kansas Department of Agriculture, Division of Water Resources, U.S. Census Bureau

to the subaquifers of the HPA. Collectively, they cover most of the ground overlying the HPA, however, there are areas of the aquifer that lie outside of GMD boundaries (Griggs 2014a; Kansas Department of Agriculture 2016).

In 1978, the KGMDA was amended to allow the establishment of Intensive Groundwater Use Control Areas (IGUCAs), which have additional authority to address groundwater use during times of water shortage (Griggs 2014a). IGUCAs can be established by the Chief Engineer of the KDWR, at the request of a GMD, or by petition of water rights owners in the area (Griggs 2014a). The IGUCA process allows the Chief Engineer to conduct hearings regarding areas of concern and approve *corrective control provisions* in areas determined to be critical, including limiting groundwater withdrawals (Griggs 2014b).

In 2012, the state legislature enacted a statute that allows for the creation of Local Enhanced Management Areas (LEMAs; *Kan Stat. Ann. §82a-1041* 2017). The LEMA legislation evolved because irrigators wanted to work collectively and locally to reduce groundwater use, but they viewed the IGUCA process as too unpredictable (Griggs 2014a). Under the LEMA process, either a GMD or 5% of the irrigators within a GMD can submit a management plan to the state for review (Griggs 2014b). The statute allows irrigators to self-organize to propose pumping restrictions on all users within a proposed boundary (Peck 2015). If the state approves the locally developed plan, it is responsible for enforcing the provisions but can only enforce what is in the plan (Peck 2015). The LEMA concept combines local control over the development of the plan with the central enforcement and administration of the IGUCA (Griggs 2014b).

Nebraska

Irrigated agriculture is also central to the economy of Nebraska. With 3.4 million hectares of irrigated crop and pastureland, of which approximately 2.9 million are irrigated with HPA water, Nebraska has more land under irrigation than any other state in the U.S. (Table 8.1; Bleed and Babbitt 2015). Nebraska accounts for almost half of all the irrigated area in the HPA region (Gollehon and Winston 2013). Similar to Kansas, irrigated agriculture is by far the largest user of groundwater in the state, accounting for 93% of total groundwater withdrawals in 2013 (Hoffman and Zellmer 2013). Groundwater recharge rates vary considerably across the state, with high positive net recharge rates in the Sand Hills and eastern parts of the state, where precipitation rates are highest, and negative net recharge rates in the southwestern part of the state (Szilagyi and Jozsa 2013). Between predevelopment and 2015, HPA levels have declined by an area-weighted average of 0.3 meters, or by only 0.5% to 1% from historical levels (Table 8.1; Stanton et al. 2011; Bleed and Babbitt 2015; McGuire 2017). Groundwater levels have declined more significantly in some parts of the state, most notably in the southwestern corner and in an isolated part of the western part of the state, but Nebraska has been able to slow or reverse these declines (Bleed and Babbitt 2015).

In Nebraska, all water is owned by the state "for the benefit of its citizens" (*Neb. Rev. Stat. §46–702 (Reissue 2010) 2007*). Under this ownership theory, water is subject to state control and never becomes private property. While there is no private property right of the water itself, landowners have a right to use groundwater, and this right automatically transfers with a change in land ownership (Aiken 1987; Bleed and Babbitt 2015). Groundwater is allocated through a hybrid of the reasonable use doctrine and modified correlative rights (Bleed and Babbitt 2015). The approach is unique among states and stipulates that landowners have the right to pump underlying groundwater, but those rights are subject to any current or future regulations, which provides the state with flexibility to legally change the right to withdraw and use groundwater (Aiken 1987).

Nebraska has a strong history of supporting local management of groundwater, dating back to the Groundwater Conservation Act of 1959 (Aiken 1980; Bleed and Babbitt 2015). Historically, natural resource management in the state was the domain of single-purpose districts organized along county boundaries (Aiken 1980). In 1972, the state reorganized more than 150 single-purpose districts into comprehensive Natural Resource Districts (NRDs; Aiken and Supalla 1979). There are 23 NRDs, which collectively cover the entire state (see Figure 8.3). The NRDs were developed largely along surface watershed boundaries "to provide effective coordination, planning, development, and general management of areas which have related resources problems" (*Neb. Rev. Stat. § 2–3203* 2007). A locally elected board of directors governs each NRD, and they are charged under state law with 12 areas of responsibility, including groundwater management and conservation (Nebraska Association of Resource Districts 2013). The NRDs have broad and flexible legislative

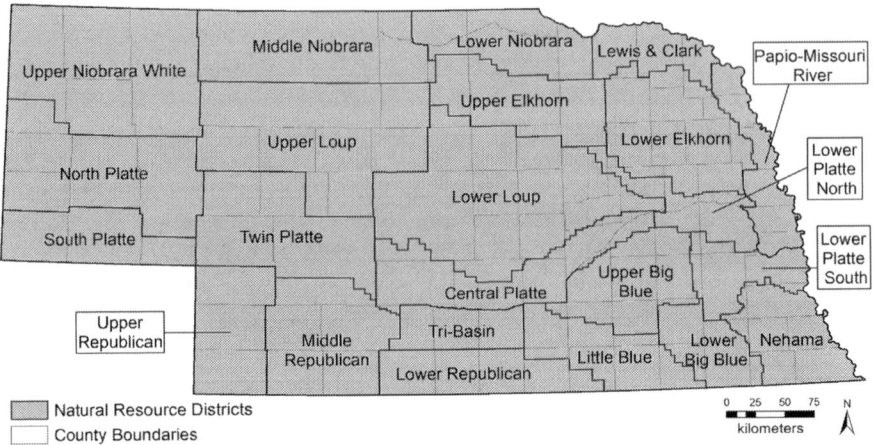

Figure 8.3 Natural resource districts of Nebraska.

Data sources: Nebraska Department of Natural Resources, U.S. Census Bureau

authority that includes the power to issue and enforce groundwater regulations (e.g., well spacing restrictions, irrigated acreage reductions, and moratoria on the drilling of new wells), levy taxes, and purchase, lease, and dispose of groundwater rights (Hoffman and Zellmer 2013).

While the NRD system facilitated local management of natural resources, by 2002, it had become apparent that the approach was not effectively addressing management of areas where surface water and groundwater are hydrologically connected (Hoffman and Zellmer 2013). In response, the state legislature passed LB 962 in 2004, which integrated surface and groundwater management planning by requiring the state Department of Natural Resources (DNR) and the local NRDs to work together to develop integrated water management plans for water-scarce basins (Hoffman and Zellmer 2013). Under this management structure, the DNR is responsible for all surface water–related issues, but the NRDs still have the sole legal authority to manage groundwater (Hoffman and Zellmer 2013).

Texas

In Texas, the HPA underlies the Northwest, or Panhandle, region of the state. While this region only accounts for 13% of the land area of Texas, it includes approximately 34% of the state's cropland and 1.4 million hectares of land irrigated with water from the HPA (Table 8.1; USDA–NASS 2012; McGuire 2017). Like Nebraska and Kansas, irrigated agriculture is the largest user of groundwater in the state. It accounts for 79% of total groundwater withdrawals, and 82% of the total groundwater used for irrigation is drawn from the HPA (Closas and Molle 2016). The HPA in Texas receives almost no recharge from precipitation, and extraction for irrigation far exceeds the recharge rate, resulting in a steady decline in groundwater levels (Kaiser and Skillern 2001). Between predevelopment and 2015, the HPA region of Texas saw an area-weighted average groundwater level decline of 12.5 meters (Table 8.1; McGuire 2017).

Over the course of more than a century of changes, local management and protection of private property rights have remained at the core of groundwater governance in Texas (Closas and Molle 2016; Maleki 2016). Groundwater is allocated based on the Rule of Capture, meaning landowners have virtually unrestricted use for any purpose. They can use as much water as they want, even if it harms their neighbor, unless interference with a neighbor's use is malicious or negligent (Aiken and Supalla 1979; Massey and Gordon 1984). Additionally, landowners expressly own the groundwater itself, and ground-water is not treated as a public resource (Massey and Gordon 1984). This governance structure significantly impacts the management of groundwater in Texas because regulations that restrict use without compensation may not be legal under the state constitution.

Litigation around groundwater rights and use has been common in Texas. The courts have repeatedly upheld the rights provided under the Rule of Capture and have ruled that if landowners come under regulation by public agencies, they need to be compensated and can seek damages for a violation of their real property rights (*Houston & Texas Central Railroad Company v. East* 1904; *Edwards Aquifer Authority v. Bragg Pecan Farm* 2013; Kaiser and Skillern 2001; Closas and Molle 2016). This legal precedent exposes Texas regulatory agencies to costly lawsuits and has made them hesitant to deny groundwater withdrawal permits or enforce regulations (Closas and Molle 2016).

Groundwater in Texas is regulated through a statewide system of local institutions called Groundwater Conservation Districts (GCDs). The Groundwater Conservation District Act of 1949 allowed for the establishment of GCDs as a political compromise, providing some form of groundwater regulation while giving local control and recognizing that landowners own the groundwater (Massey and Gordon 1984; Closas and Molle 2016). The GCDs are responsible for conserving, preserving, and protecting groundwater and are required to develop groundwater management plans that must be approved by the state (Sophocleous 2011; Closas and Molle 2016). They are the only institutions in the state with the legal authority to regulate groundwater, and they have the power to implement and enforce rules regarding licensing of new wells, well spacing, production of wells, and cross-county or cross-basin water transfers (Closas and Molle 2016). A board of directors governs each GCD, and members may be locally elected or appointed (Texas Water Development Board 2017a). There are currently 98 GCDs, and collectively they cover 70% of the state (Texas Water Development Board 2017a). The GCDs range in size from 259 to 31,079 km^2, and each district can encompass a partial county, a single county, or multiple counties (see Figure 8.4). There are eight separate GCDs that overlie the HPA in Texas.

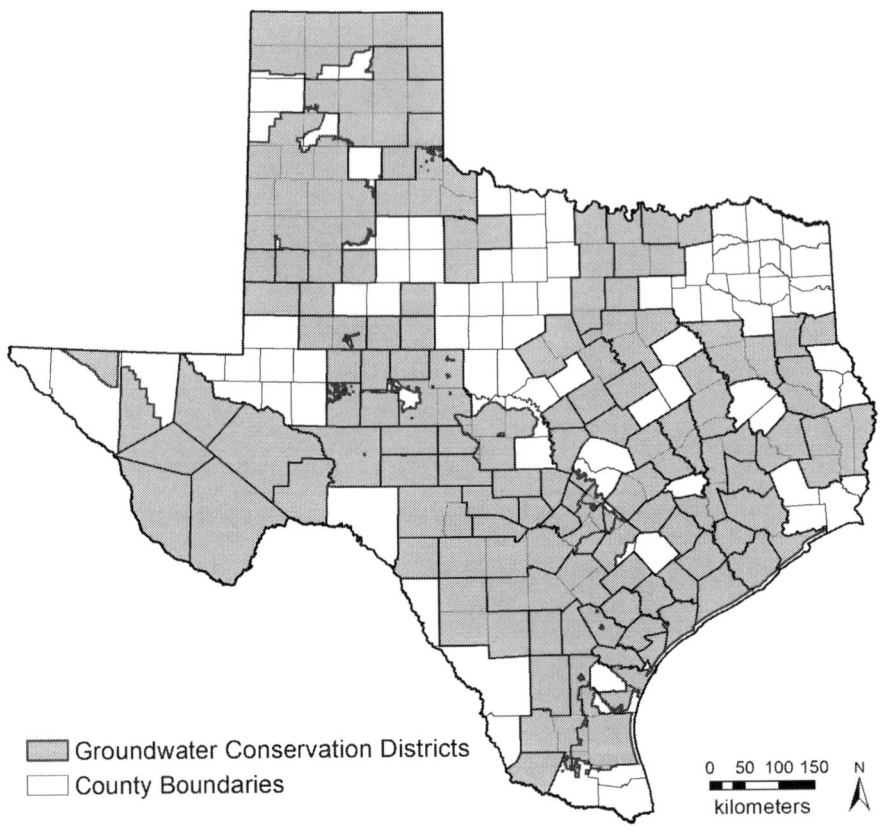

Figure 8.4 Groundwater conservation districts of Texas.
Data sources: Texas Commission on Environmental Quality, U.S. Census Bureau

After several decades of limited action in establishing GCDs, the Texas Legislature passed additional laws in 1985, 1997, and 2001 to encourage the formation of more districts (Texas Water 2014). These laws focused on critical
areas to address declining water tables in parts of the state and to mandate
establishment of a GCD under certain conditions (Kaiser and Skillern 2001).
The 2001 legislation also gave the Texas Water Development Board
(TWDB) the responsibility of creating Groundwater Management Areas
(GMAs; Sophocleous 2011). The GMAs are delineated along aquifer boundaries to encourage coordination of groundwater management and planning
across GCDs located in the same aquifer. There are 16 GMAs, and collectively
they cover the entire state (Texas Water Development Board 2017b). As of
2005, the GCDs within a GMA are required to engage in a joint planning

process to determine the desired future conditions for the management area (Sophocleous 2011; Closas and Molle 2016). These future conditions must be quantified, such as water levels or volumes, and physically possible, and they must be submitted and approved by the TWDB (Sophocleous 2011).

Analysis

In this section, we evaluate the extent to which the groundwater governance institutions in each state resemble the principles of WD. To assess this, we rely on a review of peer-reviewed literature, reports, and state government documents[2] and assess them within the context of Ostrom's eight design principles for sustainably governing CPRs. We begin each section describing our evaluative criteria for each of the eight principles then follow with an analysis for each state on the extent to which each state meets the principles. The results of this analysis are summarized in Table 8.2.

Overall, the state of Nebraska most fully meets the design principles, followed by Texas and finally Kansas. A detailed explanation for these rankings is provided in the following sections. In our concluding discussion, we assess these results in the context of WD and larger governance implications.

Clearly defined boundaries

Criterion: The boundaries of the resource system being governed and the individuals with rights to withdraw the resource must be clearly defined (Ostrom 1990; Ostrom 2008). Clearly defined boundaries allow for the exclusion of outsiders and ensure that conservation activities made by those with rights to withdraw the resource receive expected returns on investment (Ostrom 1990; Ostrom 2009; Bleed and Babbitt 2015).

Table 8.2 Assessment of the Extent to Which Each State in This Study Meets Ostrom's Eight Design Principles for Sustainably Governing CPRs

Sustainable Governance Design Principle	Kansas	Nebraska	Texas
Clearly defined boundaries	Fully meets	Fully meets	Partially meets
Proportional equivalence between benefits and costs	Partially meets	Fully meets	Partially meets
Collective-choice arrangements	Does not meet	Fully meets	Partially meets
Monitoring	Does not meet	Fully meets	Partially meets
Graduated sanctions	Fully meets	Fully meets	Fully meets
Conflict-resolution mechanisms	Does not meet	Partially meets	Does not meet
Minimal recognition of rights to organize	Partially meets	Fully meets	Fully meets
Nested enterprises	Does not meet	Fully meets	Partially meets

All three of the states' governance systems clearly define the boundaries of the groundwater resource and the individuals who have rights to withdraw the resource (Wagner and Kreuter 2004; Bleed and Babbitt 2015; *Kan Stat. Ann. §82a-1033* 2017). The key difference between the states is in how they address the interconnected nature of surface water and groundwater in some basins, where groundwater acts as a source of water for surface flows and where excessive withdrawal of groundwater can negatively impact surface water. Recognition of this hydrological connectivity goes beyond the common mindset of groundwater and surface water as separate, bounded entities and reflects an increasing understanding of how hydrological systems function holistically.

Both Kansas and Nebraska have established mechanisms to address hydrologically connected water resources. In Kansas, state law allows the Chief Engineer to establish IGUCAs and implement control provisions to protect surface flows from impairment from excessive groundwater pumping (Peck 2004; Griggs 2014b). Nebraska has addressed this challenge through its requirement that the DNR and NRDs collectively develop integrated water management plans in interconnected basins where water scarcity is a problem. Unlike Kansas and Nebraska, Texas regulates groundwater and surface water separately (Maleki 2016).

We argue that in situations in which surface water and groundwater are hydraulically connected, the boundaries of the governance system should be extended to include this interconnectivity; otherwise, the resource cannot be effectively governed. Both Kansas and Nebraska have established measures that do this, while Texas has not. Thus, while all three of the states clearly define the groundwater resource boundaries, technically meeting this design principle, we argue that in hydrological reality, both Kansas and Nebraska fully meet this design principle, while Texas only partially meets it.

Proportional equivalence between benefits and costs

Criterion: Rules specifying the quantity of the resource allocated to a user are related to local conditions and to rules requiring labor, materials, and/or financial inputs (Ostrom 2008). For these rules to achieve long-term sustainability and for them to be considered fair by users, the rules must be tailored to fit the local problem such that the costs of compliance with the rules do not outweigh the benefits (Ostrom 1990; Anderies and Janssen 2013). The first aspect of this principle requires that the approach to water governance reflects the conditions of a given locale. Each of the three states has some way in which the rules specifying the amount of groundwater allocated to users and the regulation of use are related to local conditions.

Of the three states, Nebraska's NRD system most closely aligns with this design principle. Under Nebraska law, all NRDs are charged with the same general responsibilities, but the law provides flexibility for each district to develop its own rules, determine which control measures to use, and take action based on local conditions (Edson 2005; Bleed and Babbitt 2015). Additionally, there are procedures that allow the NRDs to establish different rules within subareas of a single NRD based on data from a network of monitoring wells across the state, which further enhances their ability to address and tailor control measures to location-specific conditions (Nebraska Association of Resource Districts 2013; Bleed and Babbitt 2015; Nebraska Association of Resource Districts 2016).

In Kansas, all groundwater users must hold a water right issued by the KDWR, and all water rights are limited to an annual quantity that cannot be increased (Peck 1995; Griggs 2014b). However, allocation of these rights is determined through the prior appropriation doctrine rather than being based on local conditions, and there are more water rights granted than there is groundwater to supply the full authorized quantity (Griggs 2014b). Like Nebraska, Kansas has procedures, through the IGUCA process, to establish different rules in subareas of a district to tailor control measures to location-specific conditions. However, while the rules officially exist to address local shortages through IGUCAs, there hasn't been the political will to do so. Griggs (2017) points out that IGUCAs have only been established in areas where the groundwater supplies are renewable and connected to surface water bodies, where they have actually restored some degree of hydrological balance. However, he goes on to note that neither the GMDs nor the Chief Engineer have sought to establish IGUCAS in any of the areas above the HPA with the most severe depletion problems. IGUCAs were supposed to enable local irrigators to take the lead in reducing groundwater withdrawal by encouraging the Chief Engineer to reduce pumping, even if it conflicted with prior appropriation, but they have proven too powerful a tool to use (Griggs 2017).

In Texas, all GCDs are tasked with the same general responsibilities, and each district develops its own local management plan, goals, and rules to regulate groundwater based on local conditions. Texas has developed limits on total groundwater pumping that are set for a specific period of time and that can fluctuate based on local climate conditions – they can be lifted in wet years when recharge is higher and increased under drought conditions (Kelly 2010). This approach was incorporated into legislation applicable to all GCDs and requires GCDs in a GMA to agree to desired future conditions for the aquifer (Kelly 2010). It is designed to create limits on how much water can be extracted from an aquifer over a 50-year planning period and gives groundwater rights holders the flexibility to sell water to third parties willing to pay for their water rights (Wagner and Kreuter 2004; Kelly 2010). On paper, Texas's GCD system allows for adaptability to local conditions and for the flexibility of users to comply in ways that benefit them. However, efforts by the GCDs to enforce rules limiting groundwater use are hindered by the legal limitations that the Rule of Capture places on them. The fact is that, aside from a small number of case studies of how GCDs operate, there is little information beyond what the legislation prescribes about how these institutions make decisions and with what effect.

The second aspect of this design principle requires congruence between the rules that assign benefits and the rules that assign costs. Both Kansas and Nebraska have funding streams that allow them to collect taxes to fund regulatory and incentive-based control measures that can be tailored to local conditions and provide benefits to water users. In Texas, the state provides GCDs little financial or technical assistance, and some are restricted from imposing *ad valorem* property taxes, making them dependent on pumping fees, which can actually create a perverse incentive for more pumping in order to collect fees (Kelly 2010).

In summary, only Nebraska fully meets this design principle. Kansas and Texas partially meet this design principle because while they have the laws officially on the books to meet it, prior appropriation allocation and limited IGUCA establishment in Kansas, and the Rule of Capture and the reliance on pumping fees in some GCDs in Texas, prevent them from meeting the principle in practice.

Collective-choice arrangements

Criteria: When multiple users are dependent on the same resource for economic activity, the well-being of individuals is tightly interconnected to the actions of others in the same resource system (Bleed and Babbitt 2015). The basic problem they face is in organizing a governance system to avoid allowing individual water users to act independently of each other and in creating a system in which they adopt coordinated strategies to achieve higher collective benefits and/or reduce collective harm (Ostrom 1990; Bleed and Babbitt 2015). In other words, the collective-choice criterion is thought to represent the key to reducing the likelihood that water governance will lead to a "tragedy of the commons." To achieve collective choice, most individuals affected by governance rules should be included in the group who can modify them (Ostrom 1990). This does not mean that legally authorized decision makers must yield authority to the collaborating group; rather, stakeholders must have an active role in joint problem solving, formulating alternatives, and ranking preferred solutions (Bruns 2003; Bleed and Babbitt 2015).

This design principle appears to be strongest in Nebraska because the institutional structure of the NRDs encourages collective-choice arrangements, and the NRDs have broad regulatory and enforcement authority. Any citizen of an NRD may run for director or for the board, and the members, who make the rules, are part of the local community and must follow the rules themselves (Bleed and Babbitt 2015; personal interviews). In effect, those governed by the rules are able to participate directly in developing and modifying the rules by serving on the board and indirectly through the public meeting process of the NRD boards, where stakeholders can provide input as rules are created or modified.

While the Kansas GMDs were created with local management and stakeholder participation in mind, in practice this participation is narrowly drawn

(Peck 2006). Peck (2006) points out that the GMD boards of directors are made up primarily of irrigators, landowners, and other water users and that few positions on the boards are reserved for other stakeholders, such as municipal interests, environmental interests, or businesses that do not use large quantities of water or hold water rights, or for the public at large. The boards are locally elected, but the KGMDA defines an eligible voter narrowly as an adult person or corporation, municipality, or any other legal or commercial entity that: (1) owns 16.2 ha (40 acres) of land within a GMD and not within the corporate limits of any municipality, and (2) that withdraws at least 1,233 m^3 (1 acre-foot) of groundwater from within the GMD annually (*Kan Stat. Ann. §82a-1021* 2017). This definition of an eligible voter excludes the participation of all stakeholders who are affected by the rules and is problematic for this design principle. Additionally, while the GMDs are local institutions, they have limited authority to make and modify the rules regarding groundwater allocation and regulation because the state has primary authority. Thus, we argue that Kansas does not meet this design principle.

Local control is central to the GCD system in Texas, but unlike Nebraska and Kansas, not all GCD bylaws require that their boards of directors be locally elected (Lesikar et al. 2002). As in Nebraska, the GCD system allows for rules to be made and enforced at the local level, which meets conditions for the collective-choice design principle. However, Texas only partially meets this design principle for two key reasons. First, because in some GCDs board members can be appointed rather than elected, individuals affected by the rules in these districts do not have the opportunity to be included in the group that modifies them. Second, the way in which the GCD boundaries are drawn, almost totally by those individual people who form and propose creation of a GCD and primarily along county lines, negatively impacts collective-choice arrangements, albeit somewhat tangentially. The county-based GCD system means that, in some cases, the larger aquifer is carved up into multiple management jurisdictions. These smaller jurisdictions are more easily influenced by interests in one part of the aquifer. This can dilute the representation of stakeholders in the larger resource system and gives disproportionate weight to certain local constituencies, often leading to management decisions that may be counter to the interests of the stakeholders in the resource system as a whole (Dupnik 2012). If this is accurate, we would expect the consequence to be an inability to affect groundwater depletion. Indeed, although significant collaboration seems to have taken place among researchers and stakeholders in the two most prominent GCDs – the Panhandle Groundwater Conservation District and the North Plains Groundwater Conservation District – this collaboration does not seem to have altered the patterns of water consumption or depletion there (Johnson et al. 2011). This fragmented nature of groundwater governance is not unlike Nebraska, where some groundwater basins are governed by multiple NRDs. But Nebraska has largely overcome this through the required coordination between the NRDs and DNR in water-scarce hydrologically connected basins and by the legal precedent that NRDs must regulate groundwater in their districts to protect users in another NRD if the districts are hydrologically connected (*Upper Big Blue NRD v. State DNR* 2008; Hoffman and Zellmer 2013; Bleed and Babbitt 2015).

Monitoring

Criterion: Monitors who actively audit the resource and user behavior are at least partially accountable to users and/or are users themselves (Ostrom 1990; Ostrom 2008). With communication and neutral monitoring, no user can expect to overextract without other users learning of the noncompliance (Ostrom 1990; Bleed and Babbitt 2015).

Only Kansas and Nebraska actively monitor groundwater conditions and user behavior for all districts and make those data publicly available. In Kansas, every GMD requires meters on almost all irrigation wells, and every groundwater right owner must report annually to the Chief Engineer data such as quantity pumped, type of use, pump rate, and place of use (Sophocleous 2012; Griggs 2014b). At the state level, the Kansas Geological Survey maintains a network of about 1,400 monitoring wells that are tested annually in January, when irrigation activity is at a minimum (Sophocleous 2012).

In Nebraska, both the state and NRDs have widespread monitoring systems, but the extent and quality of monitoring varies by NRD. The Nebraska DNR has extensive, publicly available data on every registered well across the state (see https://dnr.nebraska.gov/data). Most NRDs require certification of groundwater–irrigated acres, permits for new wells, installation of meters on high–capacity wells, reporting on groundwater use, and have instituted moratoria on drilling new wells or on adding new irrigated acres without water use offsets (Nebraska Association of Resource Districts 2016). Depending on the NRD and the monitoring measures, these regulations may be implemented district-wide or in sub-areas of the district where water quantity is a concern (Bleed and Babbitt 2015). Only 2 of the 23 NRDs require no metering or water use reports, and one of those is located along the Missouri River, giving it access to a steady surface water source (Nebraska Association of Resource Districts 2016).

Only 86% of Texas GCDs actively monitor aquifer storage levels (Closas and Molle 2016). At the state level, the TWDB monitors groundwater and provides models for most aquifers in the state to facilitate the process of defining a sustainable yield for a given area (Wagner and Kreuter 2004; Kelly 2010). This program has been successful in providing accurate, publicly available models for regional water planning and in raising stakeholder awareness of the importance of groundwater management (Kelley et al. 2008; Sophocleous 2010). State law requires all wells to be registered and for GCDs to establish permitting programs for drilling new wells and for substantial alterations of existing wells (Lesikar et al. 2002). However, GCDs are authorized to exempt any and all wells from permitting requirements if those exemptions are documented in the district's management plan and if rules are in place allowing these exemptions (Lesikar et al. 2002). Finally, most GCDs do not require meters on wells, and as a result, data for monitoring withdrawal levels is not very accurate (Lesikar et al. 2002; Closas and Molle 2016).

While some form of monitoring is in place in all three states, only Nebraska fully meets and Texas partially meets this design principle. Ostrom (1990) indicates that the individuals and organizations who monitor the resource conditions and user behavior must be at least partially accountable to the users of that resource and/or should be users themselves. Kansas does not meet this requirement because the Chief Engineer is not a user, and because he is appointed rather than elected, he is not accountable to groundwater users. Texas has the lowest rate of active monitoring of the three states, and most GCDs do not require meters on wells, but where monitoring and metering are in place, the data are submitted to the GCDs. Since the GCD boards are composed of groundwater users, where monitoring is in place, Texas meets this design principle.

Graduated sanctions

Criterion: Groundwater users who violate the rules are likely to be assessed graduated sanctions, dependent on the seriousness and context of the offense, by other users and/or by officials accountable to these users (Ostrom 1990).

All three of the states include some form of graduated sanctions in their governance institutions. Kansas law allows for sanctions that vary based on the severity of the violation, including increasing civil penalties per violation, varying degrees of misdemeanor charges, and in extreme cases, jail sentences (*Kan. Stat. Ann. §82a-1214* 2017; *Kan. Stat. Ann. §82a-1216* 2017; Kansas 2017). Further, groundwater users can have their water rights revoked for nonuse or for failure to follow conditions imposed by the KDWR (Peck 1995). Bleed and Babbitt (2015) indicate that Nebraska's NRD system employs sanctions dependent on the seriousness and context of the offence. They note that the NRDs have shown flexibility in sanctioning, in some cases granting variances to their rules to allow a violator who is acting in good faith time to comply without incurring additional penalties. The Texas Water Code allows GCDs to issue civil penalties in the form of fines that are reflective of the severity of the violation (*Tex. Water Code §13.102* 2017).

Conflict–resolution mechanisms

Criterion: Groundwater users and their officials have rapid access to low–cost, local avenues to resolve conflict between users and officials or among users (Ostrom 1990). Users who possess a legal water right should be able to initiate action to enforce compliance without having to rely on a higher-level entity, such as costly and time-consuming lawsuits, to resolve noncompliance (Bleed and Babbitt 2015). Absent these mechanisms, users and their officials can feel powerless and ineffective in their efforts to manage the resource (Ostrom 1990; Ostrom 2009; Bleed and Babbitt 2015).

While none of the states fully meet this principle, Nebraska does provide a mechanism that meets these criteria for certain types of conflicts. In Nebraska, disputes among groundwater users can be resolved through a formal complaint process through the NRDs, which offers local, low-cost arenas to resolve conflicts among groundwater users (Bleed and Babbitt 2015). Nebraska state law also allows for disputes over hydrologically connected surface and groundwater, either between NRDs or between an NRD and the DNR, to be taken to an *ad hoc* five-member board appointed by the governor, the Interrelated Water Management Board (*Neb. Rev. Stat. §46–717–71* 2005; Bleed and Babbitt 2015). However, Nebraska state law has not established formal institutional alternatives to lawsuits to resolve disputes between individual surface and groundwater users or between groundwater users or for any other entity that is not the DNR or an NRD who has a dispute with water officials (Bleed and Babbitt 2015).

In Kansas, disputes over groundwater are resolved either by the Chief Engineer or in state general court (Joshi 2005; Griggs 2014b). While dispute resolution through the Chief Engineer offers a low-cost solution, we do not consider this a local arena because he is a central representative of the State. In Texas, disputes over groundwater are settled in the courts (Closas and Molle 2016). The reliance on courts as the only mechanism to resolve disputes is troublesome because it is neither a local nor low-cost arena (Bleed and Babbitt 2015).

In summary, we find that none of the states fully meet this design principle. We argue that Nebraska partially meets the principle because while it has no state law establishing alternatives to lawsuits, it has mechanisms for certain situations that can be resolved by individual NRDs and through the Interrelated Water Management Board. We argue that neither Kansas nor Texas meets this design principle due to the lack of a local avenue for conflict resolution or a reliance on the courts.

Minimal recognition of rights to organize

Criterion: The rights of groundwater users to devise their own institutions are not challenged by external governmental authorities, and users have long-term tenure rights to the resource (Ostrom 1990; Ostrom 2008).

Recognition of rights to organize is present in each of the three states. In each state, users have long-term, transferrable rights to groundwater. In Kansas, users can organize by petitioning the Chief Engineer to establish a GMD, IGUCA, or LEMA. However, because these institutions require approval of the nonlocal Chief Engineer and have limited authority to make rules, this principle is only partially met. In Nebraska, users organize through their local NRD, and the rights of the NRDs to devise their own rules to regulate groundwater are clearly recognized (Bleed and Babbitt 2015). In Texas, users can organize by petitioning the Texas Commission on Environmental Quality to create a GCD, and GCDs have clearly recognized rights to develop their own rules to regulate groundwater (Wagner and Kreuter 2004). This principle is fully met in both Nebraska and Texas.

Nested enterprises

Criterion: Governance activities are organized in nested enterprises in which appropriation, monitoring, enforcement, conflict resolution, and other governance activities are organized in multiple layers (Ostrom 1990; Ostrom 2008; Bleed and Babbitt 2015). Nested enterprises help ensure that management of the resource across different scales does not create harmful impacts to others without mitigation or compensation, and support from higher-level institutions helps overcome pressure to reduce use restriction from local users on local institutions (Peterson et al. 1993; Wiek and Larson 2012; Bleed and Babbitt 2015)

We argue that only the institutional structures of Nebraska and Texas qualify as nested enterprises with regard to groundwater governance, though to differing degrees. The exclusion of Kansas is due to the fact that the primary authority to regulate and control groundwater rests with the central authority of the Chief Engineer of the KDWR. In practice, the GMDs play only an advisory role, making recommendations, while the power to implement regulations remains at the state level in the hands of the Chief Engineer (Peck 2006; Hoffman and Zellmer 2013; Griggs 2014b).

In contrast, Nebraska's governance structure fully meets this design principle. NRDs are part of a nested hierarchy, with significant power to act at the local level and with coordination between NRDs when a groundwater basin underlies multiple districts and between NRDs and the DNR in areas with hydrologically connected surface water and groundwater (Hoffman and Zellmer 2013; Bleed and Babbitt 2015). Bleed and Babbitt (2015) point out that the Nebraska state government has very limited authority in the hierarchy, which impacts its ability to engage at multiple scales and across NRD boundaries. For example, the state Department of Environmental Quality (DEQ) can set standards for groundwater quality, but the NRDs have extensive flexibility in how to achieve those standards. DEQ has ultimate authority in verifying that these standards are met but does not necessarily have the resources for enforcement across the state (personal interviews). This does not mean that the NRDs are not part of a nested system, they are simply part of a nested system in which the state has granted them broad authority.

In addition to being nested within the state regulatory agencies, NRDs are nested above counties and municipalities. They encompass multiple counties or parts of counties and municipalities, and in turn, some counties may be part of more than one NRD. Some NRDs have been able to address some of the complexity that comes with these jurisdictional overlaps by establishing bridging organizations that allow for stakeholders to collaborate and find solutions at the appropriate scale (Bleed and Babbitt 2015).

The institutional structure of groundwater governance in Texas clearly represents a nested hierarchy in which the state and local communities interact to regulate and manage groundwater (Closas and Molle 2016). GCDs were created by the Texas legislature to act as local supervisory authorities, under which counties and municipalities are nested – though there is no legal connection between GCDs, county governments, and municipal governments, the connections are largely informal and, in some cases, symbiotic. However, groundwater in Texas is, in practice, governed by two opposing forces, the GCDs and the Rule of Capture (Maleki 2016). The Texas Supreme Court has reaffirmed the Rule of Capture as the law of the land and has repeatedly deferred to the legislature to develop rules to manage groundwater. The Court has asserted that the legislature has the constitutional authority to abolish, modify, or change the rule and replace it with state or local regulations (Kaiser and Skillern 2001). At the same time, the Court has also made it clear that landowners should be entitled to compensation if they are harmed by the regulations. so while the legislature is constitutionally allowed to regulate groundwater use, there are legal barriers to doing so, and it is not clear at what point restrictions on groundwater withdrawal would constitute a taking. This effectively hamstrings the actual authority of GCDs to govern groundwater, which then brings into question whether, *in practice*, Texas represents a nested hierarchy. For this reason, we argue that Texas only partially meets this design principle.

Discussion and conclusions

Earlier in this chapter, we argued that at its current nascent stage as a discipline, WD largely represents a subset of Ostrom's work in the context of water governance. As such, this WD analysis occurred through the lens of Ostrom's institutional design principles. It is important to note there is nothing in the Ostrom design principles that guarantees any particular result. It's entirely possible that a water governance system might have all the characteristics identified by Ostrom and still produce outcomes unfavorable to some stakeholders (i.e., aquifer depletion). Of course, the argument is that when these characteristics are present, the conditions necessary for WD are more likely. And when the conditions necessary for WD are more likely, interpersonal interactions will take place in a way that improves the likelihood of better outcomes (i.e., agreements will be reached to yield less water depletion).

Our analysis suggests that multiple variables impact the extent to which a groundwater governance institution exhibits the characteristics of WD and is thus more likely to achieve the goal of less groundwater depletion. Kansas demonstrates the extent to which the lack of an empowered local institutional setup can negatively impact achieving the design principles for sustainably governing CPRs. The state's reliance on the central, state-level authority of the Chief Engineer was responsible for it only partially meeting two of the principles and prevented it from fully meeting four of them.

Texas demonstrates that groundwater allocation rules can negatively impact achieving the eight design principles and can severely limit the options for managing the resource. Despite having legislatively created a local groundwater governance system that at least partially meets collective-choice principles, the Rule of Capture has created conditions where GCDs are wary of imposing limits on pumping for fear that they will be drawn into costly lawsuits over whether the restrictions constitute a taking of property rights. The Texas Supreme Court recently declined to hear a case that could have added much-needed clarity, which sets the stage for years of costly litigation over groundwater (Malewitz 2015). Meanwhile, groundwater levels continue to drop.

Like Texas, Kansas has designated groundwater as a property right. This poses a potential challenge to the state's ability to regulate groundwater use more sustainably. We noted in the section on proportional equivalence that there have been more water rights granted in Kansas than there is sustainable supply of groundwater. Currently, most groundwater users comply with their annual pumping limits, but this compliance does not address the problem of depletion because the resource itself is overappropriated (Griggs 2014b). If the state attempts to increase limits on pumping, it could find itself in a similar situation to Texas, where these restrictions are considered a taking of property by the government.

Nebraska no doubt faces important challenges with continuing to manage its supply of groundwater sustainably. There are parts of the state in which groundwater levels continue to decline, and the nature of groundwater means that there is a lag between management actions and their impacts (Bleed and Babbitt 2015). Because of this, it could be too soon to fully judge the success of Nebraska's current governance action (Bleed and Babbitt 2015). However, Nebraska is one of the most intensely irrigated and productive agricultural regions of the world, and it has achieved this without widespread groundwater depletion and has managed to reverse or slow depletion across much of the state (Bleed and Babbitt 2015). This is in stark contrast to the HPA regions of Kansas and Texas, where governance institutions are limited by allocation and property rights choices and where decades of groundwater governance have not abated the problem. This depletion has already made vast stretches of land in these states unable to support irrigated agriculture (Wines 2013). For an example of the scale of the problem, one need only look at the estimate that the HPA in Texas is expected to lose 52% of its volume by 2060 (Galbraith 2010).

It is important to note that Nebraska is fortunate to overlie the largest share of the HPA and has more favorable hydrogeology for recharge than do Kansas and Texas. However, the state's relative success in managing groundwater sustainably is attributable to more than fortunate geography. Nebraska has demonstrated the ability to adapt to conditions as monitoring data indicates control measures are necessary. The NRD system deserves a great deal of credit for helping to reverse or stabilize groundwater levels in areas of the state, even in areas with low recharge rates. Nebraska demonstrates that a groundwater governance system built around Ostrom's eight principles can create adaptive, collaboratively managed governance institutions for the sustainable management of this CPR. The NRD system appears to be gaining recognition for its achievement. Experts on the NRD system, interviewed by one of the authors for ongoing research, indicated that NRD representatives have been consulted by the state of California as it has developed its Sustainable Groundwater Management Program and that they have been invited to present on their governance system at national and international conferences.

This analysis provides insight on how to improve governance of groundwater through the lens of WD. The field of WD is new and evolving, and its application to the sustainable governance of groundwater is virtually absent in the existing literature. In this chapter, we extended WD principles to the macro, governance level by viewing it as a subset of Ostrom's work on CPR management. Further research is needed on how best to develop a WD framework and set of analysis tools that are specific to the discipline and that are focused on water. At its core, the WD Framework developed by Islam and Susskind (2013) has a good deal of synergy with Ostrom's eight principles. Both call for adequate stakeholder participation, and the Framework's joint fact-finding component that seeks to develop a shared understanding of the resource and the key variables that affect it is similar to Ostrom's monitoring principle. Nested enterprises relate to the Framework's fourth principle of linking informal problem solving to formal decision-making processes at the higher governance level. Nested enterprises and collective-choice arrangements are related to adaptive collective management and enable governance institutions to respond to new conditions and information over time (Ostrom 2000). They are also related to the Framework's sixth principle of capacity building and societal learning among individuals, organizations, and networks.

Further research is needed on expanding the WD Framework for specific application to governance institutions. Ostrom's principles are necessary conditions but not necessarily sufficient, and variables specific to water or other variables such as political will could be potential additions to a WD Governance Framework. Recent work by Bleed and Babbitt (2015) represents a useful step down the path to this process. Their assessment of Nebraska's NRDs expanded upon Ostrom's principles and incorporated additional ones for a total of 14 criteria. While not intended as a study on WD, their work was referenced extensively in this chapter and contributes to the discussion in the literature on the development of a WD Governance Framework. We hope that this chapter represents an important contribution on refining this evolving field.

Notes

1 Note that the terms "High Plains Aquifer" and "Ogallala Aquifer" are often used interchangeably in the literature (Verchick 1999; USGS 2014). The Ogallala Formation is the principle geologic unit of the High Plains Aquifer, which has led to the popular use of the term "Ogallala Aquifer" as a synonym (Sophocleous 2011). We use the technical term "High Plains Aquifer" throughout this chapter.

2 We have also included some references to personal interviews that were conducted as part of other research on Nebraska's NRDs. These are referenced as "personal interview" in the text.

References

Aiken, J.D., 1980. Nebraska ground water law and administration. *Nebraska Law Review*, 59, 917–1000.

Aiken, J.D., 1987. New directions in Nebraska water policy. *Nebraska Law Review*, 66, 8–75.

Aiken, J.D., and Supalla, R.J., 1979. Ground water mining and western water rights law: The Nebraska experience. *South Dakota Law Review*, 24, 607–648.

Anderies, J.M., and Janssen, M.A., 2013. Robustness of social-ecological systems: Implications for public policy. *Policy Studies Journal*, 41 (3), 513–536.

Anderies, J.M., Janssen, M.A., and Ostrom, E., 2004. A framework to analyze the robustness of social-ecological systems from an institutional perspective. *Ecology & Society*, 9 (1), 18.

Berndtsson, R., Madani, K., Aggestam, K., and Andersson, D.-E., 2017. The grand Ethiopian Renaissance Dam: Conflict and Water Diplomacy in the Nile Basin. *In*: S. Islam and K. Madani, eds. *Water Diplomacy in Action: Contingent Approaches to Managing Complex Water Problems*. London and New York: Anthem Press, 253–264.

Bleed, A., and Babbitt, C.H., 2015. *Nebraska's Natural Resources Districts: An Assessment of a Large-scale Locally Controlled Water Governance Framework*. Robert B. Daugherty Water for Food Institute, No. 1.

Brady, M., Li, T., and Yoder, J., 2015. The Columbia river treaty renegotiation from the perspective of contract theory. *Journal of Contemporary Water Research & Education*, 155, 53–62.

Bruns, B., 2003. *Water Tenure Reform: Developing an Extended Ladder of Participation*. Politics of the Commons: Articulating Development and Strengthening Local Practices. RCSD Conference, July 11–14, 2003, Chiang Mai, Thailand.

Choudhury, E., 2017. The nature of enabling conditions of transboundary water management: Learning from the negotiation of the Indus and Jordan Basin treaties. *In*: S. Islam and K. Madani, eds. *Water Diplomacy in Action: Contingent Approaches to Managing Complex Water Problems*. London and New York: Anthem Press, 181–202.

Closas, A., and Molle, G., 2016. *Groundwater Governance in America*. International Water Management Institute, No. 5.

Dennehy, K.F., Litke, D.W., and McMahon, P.B., 2002. The high plains aquifer, USA: Groundwater development and sustainability. *In*: K.M. Hiscock, M.O. Rivett, and R.M. Davison, eds. *Sustainable Groundwater Development*. Geological Society. London: Special Publications, 193, 99–119.

Dupnik, J.T., 2012. *A Policy Proposal for Regional Aquifer-Scale Management of Groundwater in Texas*. The University of Texas at Austin.

Edson, D.E., 2005. *A Unique System of Resource Governance Nebraska's Natural Resources Districts*. USCID Third International Conference, San Diego, CA.

Edwards Aquifer Authority v. Bragg Pecan Farm, 2013.

Galbraith, K., 2010. How bad is the Ogallala Aquifer's decline in Texas? *The Texas Tribune*. Available from: www.texastribune.org/2010/06/17/how-bad-is-the-ogallala-aquifers-decline-in-texas/

Getches, D.H., 2009. *Water Law in a Nutshell*. St. Paul, MN: Thomson West.

Gleeson, T., Alley, W.M., Allen, D.M., Sophocleous, M.A., Zhou, Y., Taniguchi, M., and Vandersteen, J., 2012. Towards sustainable groundwater use: Setting long-term goals, backcasting, and managing adaptively. *Ground Water*, 50 (1), 19–26.

Gollehon, N., and Winston, B., 2013. *Groundwater Irrigation and Water Withdrawals: The Ogallala Aquifer Initiative*. Available from: https://www.nrcs.usda.gov/Internet/FSE_DOCUMENTS/stelprdb1186440.pdf

Griggs, B.W., 2014a. Beyond drought: Water rights in the age of permanent depletion. *Kansas Journal of Law and Public Policy*, 62, 1263–1324.

Griggs, B.W., 2014b. *Lessons from Kansas: A More Sustainable Groundwater Management Approach*. Water in the West – Stanford University. Available from: http://waterinthewest.stanford.edu/news-events/news-press-releases/lessons-kansas-more-sustainable-groundwater-management-approach

Griggs, B.W., 2017. The political cultures of irrigation and the proxy battles of interstate water litigation. *Natural Resources Journal*, 57 (1), 1–73.

Gutentag, E.D., Heimes, F.J., Krothe, N.C., Luckey, R.R., and Weeks, J.B., 1984. *Geohydrology of the High Plains Aquifer in Parts of Colorado, Kansas, Nebraska, New Mexico, Oklahoma, South Dakota, Texas, and Wyoming*. U.S. Geological Survey Professional Paper.

Hoffman, C., and Zellmer, S., 2013. Assessing institutional ability to support adaptive, integrated water resources management. *Nebraska Law Review*, 91, 805–865.

Houston & Texas Central Railroad Company v. East, 1904.

Huntjens, P., 2017. Mediation in the Israeli-Palestinian water conflict: A practitioner's view. *In*: S. Islam and K. Madani, eds. *Water Diplomacy in Action: Contingent Approaches to Managing Complex Water Problems*. London and New York: Anthem Press, 203–228.

Islam, S., and Madani, K., eds., 2017. *Water Diplomacy in Action: Contingent Approaches to Managing Complex Water Problems*. London and New York: Anthem Press.

Islam, S., and Susskind, L., 2013. *Water Diplomacy: A Negotiated Approach to Managing Complex Water Networks*. New York: RFF Press.

Johnson, J.W., Johnson, P.N., Guerrero, B., Weinheimer, J., Amosson, S., Almas, L., Golden, B., and Wheeler-Cook, E., 2011. Groundwater policy research: Collaboration with groundwater conservation districts in Texas. *Journal of Agricultural and Applied Economics*, 43 (3), 345–356.

Joshi, S.R., 2005. *Comparison of Groundwater Rights in the United States: Lessons for Texas*. Texas Tech University.

Kaiser, R., and Skillern, F.F., 2001. Deep trouble: Options for managing the hidden threat of Aquifer depletion in Texas. *Texas Tech Law Review*, 32, 249–304.

Kansas, 2017. Findings and order expanding the boundaries of the Equus Beds Groundwater Management District No. 2. *2017 Session Laws of Kansas*, 1 (23).

Kansas Department of Agriculture, 2016. *Ogallala-High Plains Aquifer*. Available from: www.agriculture.ks.gov/divisions-programs/dwr/managing-kansas-water-resources/information-about-kansas-water-resources/ogallala-high-plains-aquifer/lists/regions-of-the-ogallala-high-plains-aquifer/equus-beds-(gmd2)

Kansas Water Office, 2015. *Vision for the Future of Water Supply in Kansas: A Long-Term Vision for the Future of Water Supply in Kansas*. Available from: https://agriculture.ks.gov/docs/default-source/attachment-j/kansas-water-vision.pdf?sfvrsn=add2b7c1_4

Kan. Stat. Ann. §§ 82a-701 to -733, 1989 & supp. 1994.

Kan. Stat. Ann. §82a-1214, 2017.

Kan. Stat. Ann. §82a-1216, 2017.

Kan Stat. Ann. §82a-1021, 2017.

Kan Stat. Ann. §82a-1033, 2017.

Kan Stat. Ann. §82a-1041, 2017.

Kelley, V., Mace, R., and Deeds, N., 2008. Groundwater availability modeling: The Texas experience. *The Water Report*, 54.

Kelly, M., 2010. *Nebraska's Evolving Water Law: Overview of Challenges & Opportunities.* Platte Institute for Economic Research. Available from: https://platteinstitute.org/Library/docLib/20100927_Kelly_Paper_-_FINAL.pdf

Kenny, B.J.F., and Juracek, K.E., 2013. *Irrigation Trends in Kansas, 1991–2011.* U.S. Geological Survey.

Koebele, E.A., 2015. Assessing outputs, outcomes, and barriers in collaborative water governance: A case study. *Journal of Contemporary Water Research & Education*, 155 (1), 63–72.

Lesikar, B., Kaiser, R., and Silvy, V., 2002. *Questions About Texas Groundwater Conservation Districts.* Texas Cooperative Extension, Texas A&M University.

Maleki, S., 2016. *An Insight into Groundwater Management and Policy in Texas.* Texas State University.

Malewitz, J., 2015. State supreme court punts on major water case. *The Texas Tribune.* Available from: www.texastribune.org/2015/05/01/supreme-court-punts-major-water-case/

Massey, D.T., and Gordon, R., 1984. Groundwater in the Ogallala Aquifer for irrigation. *Oklahoma City University Law Review*, 9 (3), 379–410.

McGuire, V.L., 2017. *Water-Level and Recoverable Water in Storage Changes, High Plains Aquifer, Predevelopment to 2015 and 2013–15.* U.S. Geological Survey Scientific Investigations Report 2017–5040.

Meinzen-Dick, R., 2007. Beyond panaceas in water institutions. *Proceedings of the National Academy of Sciences of the United States of America*, 104 (39), 15200–15205.

Moazezi, M.R., Madani, K., and Hipel, K.W., 2017. Strategic insights for California's delta conflict. *In*: S. Islam and K. Madani, eds. *Water Diplomacy in Action: Contingent Approaches to Managing Complex Water Problems.* London and New York: Anthem Press, 289–310.

Nebraska Association of Resource Districts, 2013. *NARD Summary on NRD Activities, August 2013.*

Nebraska Association of Resource Districts, 2016. *Nebraska's Natural Resource Districts: Water.* Available from: www.nrdnet.org/programs/water

Neb. Rev. Stat. § 2–3203, 2007.

Neb. Rev. Stat. §46–702 (Reissue 2010), 2007.

Neb. Rev. Stat. §46–717–71, 2005.

Ostrom, E., 1990. *Governing the Commons: The Evolution of Institutions for Collective Action.* New York: Cambridge University Press.

Ostrom, E., 2000. Reformulating the commons. *Swiss Political Science Review*, 6 (1), 29–52.

Ostrom, E., 2008. The challenge of common-pool resources. *Environment: Science and Policy for Sustainable Development*, 50 (4), 8–21.

Ostrom, E., 2009, July. A general framework for analyzing sustainability of social-ecological systems. *Science*, 325 (5939), 419–422.

Peck, J.C., 1995. The Kansas water appropriation act: A fifty-year perspective. *Kansas Law Review*, 43, 735–756.

Peck, J.C., 2003. Property rights in groundwater- some lessons from the Kansas experience. *Kansas Journal of Law & Public Policy*, 12 (3), 493–520.

Peck, J.C., 2004. Protecting the Ogallala Aquifer in Kansas from depletion: The teaching perspective. *Journal of Land, Resources, and Environmental Law*, 24 (1965), 349–354.

Peck, J.C., 2006. Groundwater management in Kansas: A brief history and assessment. *Journal of Law and Public Policy*, 15, 441–465.

Peck, J.C., 2007. Groundwater management in the high plains aquifer in the USA: Legal problems and innovations. *In*: M. Giordano and K.G. Villholth, eds. *The Agricultural*

Groundwater Revolution: Opportunities and Threats to Development. Oxfordshire and Cambridge, MA: CAB International, 296–319.

Peck, J.C., 2015. Legal challenges in government imposition of water conservation: The Kansas example. *Agronomy Journal,* 107 (4), 1561.

Peterson, E.W.F., Aiken, J.D., and Johnson, B.B., 1993. Property rights and groundwater in Nebraska. *Agriculture and Human Values,* 10 (4), 41–49.

Ross, A., and Martinez-Santos, P., 2010. The challenge of groundwater governance: Case studies from Spain and Australia. *Regional Environmental Change,* 10 (4), 299–310.

Sophocleous, M., 2010. Review: Groundwater management practices, challenges, and innovations in the high plains aquifer, USA: Lessons and recommended actions. *Hydrogeology Journal,* 18 (3), 559–575.

Sophocleous, M., 2011. Groundwater legal framework and management practices in the high plains aquifer, USA. *In:* A.N. Findikakis and K. Sato, eds. *Groundwater Management Practices.* Leiden, The Netherlands: CRC Press.

Sophocleous, M., 2012. The evolution of groundwater management paradigms in Kansas and possible new steps towards water sustainability. *Journal of Hydrology,* 414–415, 550–559.

Stanton, J.S., Qi, S.L., Ryter, D.W., Falk, S.E., Houston, N.A., Peterson, S.M., Westenbroek, S.M., and Christenson, S.C., 2011. *Selected Approaches to Estimate Water-Budget Components of the High Plains, 1940 Through 1949 and 2000 Through 2009.* U.S. Geological Survey Scientific Investigations Report 2011–5183.

Steward, D.R., Bruss, P.J., Yang, X., Staggenborg, S.A., Welch, S.M., and Apley, M.D., 2013. Tapping unsustainable groundwater stores for agricultural production in the high plains aquifer of Kansas, projections to 2110. *Proceedings of the National Academy of Sciences of the United States of America,* 110 (37), E3477– E3486.

Susskind, L., and Islam, S., 2012. Water diplomacy: Creating value and building trust in transboundary water negotiations. *Science & Diplomacy,* 1 (3).

Szilagyi, J., and Jozsa, J., 2013. MODIS-aided statewide net groundwater-recharge estimation in Nebraska. *Groundwater,* 51 (5).

Texas Water, 2014. *Groundwater Conservation Districts.* Texas A&M University. Available from: https://texaswater.tamu.edu/groundwater/groundwater-conservation-districts.html

Texas Water Development Board, 2017a. *Groundwater Conservation Districts.* Available from: www.twdb.texas.gov/groundwater/conservation_districts/facts.asp

Texas Water Development Board, 2017b. *Groundwater Management Areas.* Available from: www.twdb.texas.gov/groundwater/management_areas/index.asp

Tex. Water Code §13.102, 2017.

University of Kansas, 2015. *Groundwater Levels Decline in Western, Central Kansas.* Available from: http://news.ku.edu/2015/02/10/groundwater-levels-decline-western-central-kansas

Upper Big Blue NRD v. State, DNR, 2008.

USDA-NASS, 2012. *Census of Agriculture.* Available from: www.agcensus.usda.gov/Publications/2012/

USDA-NRCS, 2013. *Ogallala Aquifer Initiative 2013 Progress Report.* Available from: https://serc.carleton.edu/details/files/79740.html

USDA-NRCS, 2016. *Ogallala Aquifer Initiative 2016 Progress Report.* Available from: https://www.nrcs.usda.gov/wps/PA_NRCSConsumption/download?cid=nrcseprd1329465&ext=pdf

USGS, 2014. *USGS High Plains Aquifer Water-level Monitoring Study.* Available from: https://ne.water.usgs.gov/ogw/hpwlms/

van Rees, C., and Reed, J.M., 2015. Water diplomacy from a Duck's Perspective: Wildlife as stakeholders in water management. *Journal of Contemporary Water Research & Education*, 155 (1), 28–42.

Verchick, R.R.M., 1999. Dust bowl blues: Saving and sharing the Ogallala Aquifer. *Journal of Environmental Law and Litigation*, 14, 13–23.

Wagner, M.W., and Kreuter, U.P., 2004. Groundwater supply in Texas: Private land considerations in a rule-of-capture state. *Society & Natural Resources*, 17 (4), 349–357.

Wiek, A., and Larson, K.L., 2012. Water, people, and sustainability: A systems framework for analyzing and assessing water governance regimes. *Water Resources Management*, 26, 3153–3171.

Wines, M., 2013. Wells dry, fertile plains turn to dust. *The New York Times*, 19 May.

9 Creating flexibility in freshwater availability for the Eastern Nile Basin

Agustín Botteron

Introduction

Concerns have arisen in the international community and among local stakeholders regarding the availability of water resources and over forecasts of potential water crises in the Eastern Nile Basin (ENB), North Africa. Such concerns are primarily based on population increases, development plans, prevailing political situations, and climate change uncertainties. The total population in the ENB countries – Egypt, Ethiopia, Sudan, and South Sudan – increased by 44% between 2000 and 2015. The annual population growth rate, at 2.4%, is twice the world average. On top of demographic pressures, ENB countries are either low- or lower-middle-income countries, experience serious development constraints, and are subject to political instability (e.g., World Bank 2018; United Nations 2018; Fund For Peace 2018).

A recent study on climate change indicates that although water stress is not an issue at the regional scale within the Nile Basin, there are high risks with regard to seasonal water variability, flood occurrence, and drought severity. In particular, Ethiopia and Sudan are at high risk in terms of seasonal water variability and flood occurrence, while Egypt is projected to suffer from severe drought (Gassert et al. 2013). Issues of equity and sustainability in terms of water use are pressing. Egypt withdraws more water than the country's estimated renewable water resources (FAO 2015); South Sudan, the newest country in the world, is demanding revisions to the water agreements with Sudan in order to promote its agriculture (Salman 2011); and Ethiopia is building the Grand Ethiopian Renaissance Dam on the Blue Nile River, which will significantly affect the hydropolitical dynamics of the Nile Basin (Abdelhady et al. 2015). In summary, "water is clearly a major factor in socio–economic recovery and development in Africa. The continent appears to be blessed with substantial rainfall and water resources. Yet it has severe and complex natural and man-made problems that constrain the exploitation and proper development of its water resources potential" (United Nations 2003, p. 28).

The Eastern Nile Basin

Located in the northeast corner of the African continent, the entire Nile River basin encompasses eleven countries, covers approximately 3.2 million km²,

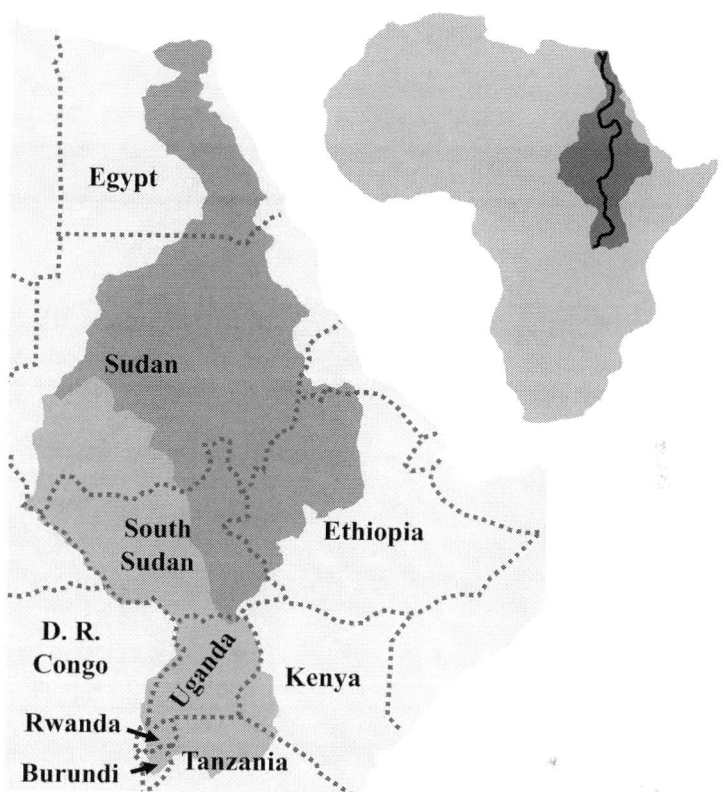

Figure 9.1 Study area. Left: Detail of Nile Basin and Eastern Nile Basin with political boundaries drawn as dotted lines and major basins shown as shaded regions. The Eastern Nile basin is shown in the darkest shade. Right: The Nile Basin's location within the African continent.

and hosts one of the world's longest rivers (see Figure 9.1). The Nile River traverses 6,695 km from its origin in Central Africa to its delta on the Mediterranean Sea (Nile Basin Initiative 2012). The Nile Basin can be divided into eight major subbasins based on catchment delineation, subbasin characteristics, and the location of river gauging points (Conway and Hulme 1993). The Blue Nile River, formed by the residual of seasonal rainfall on the Ethiopian Highlands, provides about 60% of the flow of the Main Nile River. Its contribution, however, is highly seasonal between July and October (Conway and Hulme 1993; Sutcliffe 2009).

The Eastern Nile Basin (ENB) covers 2 million km^2 (60% of the Nile River Basin), and it is of regional relevance for three primary reasons: (i) within three to four months of the year, more than 80% of the Main Nile streamflow originates in this area and is characterized by seasonal and interannual variability; (ii) the Ethiopian Highlands offer hydropower generation and water-saving

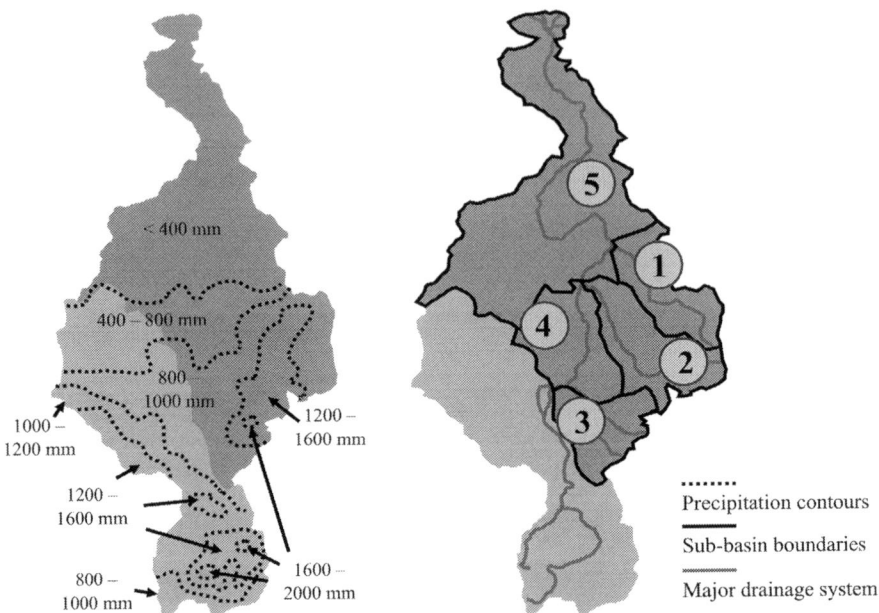

Figure 9.2 Study area. Left: Annual precipitation distribution drawn with data from Awulachew et al. (2007). Right: Major drainage system with subbasins: (1) Tekeze-Atbara, (2) Blue Nile; (3) Baro-Akobo-Sobat, (4) Lower White Nile; (5) Main Nile.

potential; (iii) the geographical layout of the countries allows for interconnecting infrastructure such as power lines, roads, and canals (Nile Basin Initiative 2012).

Freshwater in the ENB is available from rainfall, rivers, and aquifers, but not all basin countries benefit from these sources to the same degree (see Figure 9.2). Precipitation is insignificant in Egypt, where 98% of water withdrawals come from outside its national borders (Abdin and Gaafar 2009). The Nile River is Egypt's main source of water with an annual allocation of 55.5 billion cubic meters (BCM) under the 1959's Nile Waters Agreement with Sudan (Egypt and Sudan 1959; FAO 2015). Unlike Egypt, Ethiopia is rich in renewable water resources, mainly because of the country's western highlands, where annual average rainfall ranges from 1,600 mm to 2,122 mm (Awulachew et al. 2007). For this study, the precipitation volume in Ethiopia's Nile Basin was estimated at 784 BCM, using data found in Conway and Hulme (1993), Awulachew et al. (2007), and Awulachew (2012). Total runoff volumes range from 5% to 21% of precipitation across the subbasins, thus most of the precipitated water either evaporates or infiltrates. Groundwater is not significantly tapped in the region (FAO 2015), even though Egypt and Sudan lie partially over the Nubian Sandstone Aquifer System, a fossil

Table 9.1 Water Balance in Eastern Nile Basin. Own computations based on Water Accounting+ (2016). Values in BCM/year for period 2005–2010. Utilized flow: "part of available water that is depleted for uses"; Landscape ET: "Water from rainfall that evaporates locally from leaves, litter, soil, and via plants that extract moisture from the unsaturated zone."

Inflow		Outflow	
Precipitation	841	Consumed water	828
		Landscape ET	*784*
		Utilized flow	*44*
Surface water and groundwater inflow	54	Losses to groundwater	18
		Outflow to Mediterranean Sea	24
		Non-reported outflows	25
Total inflow	895	Total outflow	895

aquifer believed to be the largest reserve of freshwater in the world (Thorweihe and Heinl 2002).

Present and future water availability are controversial. Some have claimed that North Africa ran out of renewable freshwater to meet its food requirements decades ago (Qadir et al. 2007). Additionally, climate change projections indicate the region might become hotter and drier (Conway 2005; Islam and Susskind 2015), and variability in precipitation and streamflow will increase, causing more extreme droughts and heavier floods (Siam and Eltahir 2017). Current debates in the region are about the management of 84 BCM/year, a volume agreed upon more than half a century ago on a treaty between Egypt and Sudan only. However, the estimated total water inflow for the Eastern Nile Basin is 895 BCM/year, much greater than the debated volume (Table 9.1).

Research aims and methodology

Can water be a flexible resource? Is there a way to increase water availability in the region? These questions are addressed in this chapter through the lens of the Water Diplomacy Framework (WDF) developed by Islam and Susskind (2013). On the water–demand side, the focus of this research is on agriculture in Egypt and Sudan. This was done because agriculture is the largest water-demanding sector in the basin, and these two countries dominate in terms of cultivated and irrigated land, as well as irrigation water withdrawals, according to FAO (2015). This work also only considered major cereal crops – namely wheat, maize, sorghum, and sugarcane – as they represent 48% of the harvested area (11.3 million ha) and 45% of the irrigated area (3.6 million ha). On the water-supply side, strategies related to irrigation efficiency, rainwater collection, crop yields, and cropping patterns were evaluated and discussed with a motivation to reframe water as a flexible resource.

The research methodology had three major steps: a comprehensive literature review, a definition of context-based assumptions, and a formulation and evaluation of strategies. The literature review showed knowledge gaps about the study area. To overcome this, values and assumptions were taken from a collection of sources and documented accordingly. Brouwer and Heibloem (1986) was used for crop water requirements, and Steduto et al. (2012) provided values of yield and water productivity. Additionally, the data repositories FAOSTAT (FAO 2016) and AQUASTAT (FAO 2015) were used extensively for this research. Information from the existing literature and these databases was used to establish a baseline upon which the proposed strategies are evaluated. Such strategies respond to either actions being taken already within the basin, successful experiences in other parts of the world that might be translated into the study area, or international standards for agricultural best practices. For example, existing initiatives in the region are reported by Simons et al. (2012) and Karrou et al. (2012); actions potentially transferable into the region are found in Basán Nickisch et al. (2016); and best practices in the agriculture sector are found in Kirda (2002) and Steduto et al. (2012). More detail on the methodology is presented within the discussion of each strategy.

Exploring flexible strategies

The agricultural sector

Agriculture is critical for the Nile Basin countries (see Table 9.2). On average, it generates 75% of the basin's employment and accounts for 33% of its gross domestic product (GDP; OECD 2006; cited in Karimi et al. 2012, p. 133). Sudan and Ethiopia dominate in terms of cultivated land, whereas Egypt leads in irrigated land. Together, Egypt and Sudan are responsible for 95% of agricultural water withdrawals, and 90% of the irrigation in these countries is conducted using surface techniques, such as flood and furrow (FAO 2015), widely known as highly inefficient. Outside Egypt, rainfed agriculture dominates water consumption in the Nile Basin (Johnston 2012).

Despite water constraints and a desert climate, all of Egypt's potential arable land is under production and irrigated. The most important crops are wheat and maize, accounting for 41% of the total harvested area (FAO 2015). Wheat is a staple food, and even though Egypt produces large amounts domestically at very high yields, the country has ranked as the world's leading importer for decades to satisfy its domestic demand. Egypt does not implement agricultural water tariffs, and most of the sectoral cost falls in the government's hands (Barnes 2014a). Agriculture in Egypt is largely inefficient, full of subsidies, and characterized by postharvest losses and water misuse (Barnes 2014a).

Agriculture in Sudan is highly relevant for both the economy and the country's social development. Sudan features the largest cultivated area in the Nile Basin. The most important crop is sorghum, which covers 43% and 49% of the irrigated and rainfed harvested area, respectively (FAO 2015). The arable land is divided into traditional rainfed cultivation, rainfed agriculture supported

Table 9.2 Harvested Area and Irrigation Requirements for the Most Relevant Crops in Egypt and Sudan. Relevance is based on harvested area and water demand. Harvested area was retrieved from FAO (2015), water demand estimates are based on Simons et al. (2012), Brouwer and Heibloem (1986), and Keith et al. (1998).

Crop production				Estimated annual water demand (ETa)	
Crop	Yield [tons/ha]	Harvested area			
		[1000 ha]	[% of total]	Total [BCM]	% of total
Egypt					
Wheat	6.7	1418.7	22	5.6	14
Maize	7.7	1030.3	16	5.4	13
Sugarcane	114.1	138.2	2	2.7	7
Total irrigated area (all crops)		6333.0		Total agriculture ETa 41.0	
				Total agriculture withdrawal 59.0	
Irrigated Sudan					
Sorghum	1.5 (Gezira 2.2)	678.7	43	3.7	22
Wheat	2.1	254.6	16	1.3	8
Sugarcane	92.7	70.7	5	1.4	8
Total irrigated area (all crops)		1563.0		Total agriculture withdrawal 25.9	
Rainfed Sudan					
Sorghum	0.7	7698.9	49.0	34.65	43.0
Total non-irrigated area (fodder area N/A; all crops)		15720.6		Total agriculture consumption in rainfed areas 80.64	

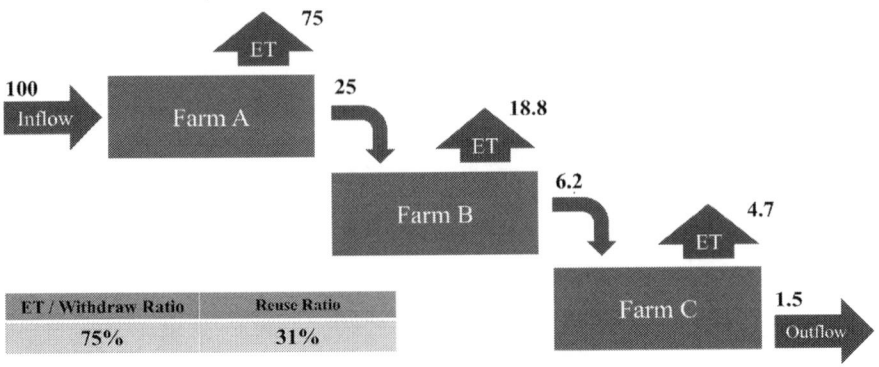

Figure 9.3 An example of how agricultural water consumers "cascade" in Egypt's irrigation
system based on Simons et al. (2015). The numbers are made up for illustration
purposes. The left-hand arrow represents the initial water withdrawal by Farmer
A. The smaller curved arrow represents the sum of runoff and percolation
captured by the drainage system. The drainage system works as a water source
for downstream users. The arrow on the far right of the diagram stands for
the outflow of the drainage system into the Mediterranean Sea. In this model,
the farm-level ratio between evapotranspiration and withdrawal is 75%. Water
losses to infiltration and runoff are entirely reused downstream. As result, the
total irrigation withdrawal is 131, meaning that 31% reuse is achieved.

by mechanization and irrigated agriculture, with the coverage percentage of each
farming technique split 58%, 33%, and 9%, respectively (Mahgoub 2014). Irrigation is crucial for Sudan's economy. Most cash crops are produced in irrigation
schemes (FAO 1997), sugar being the most relevant. The Gezira Scheme is the
oldest and largest gravity irrigation system in the country (870,000 ha). Despite
its relevance, irrigation is only developed at 40% of its potential (FAO 2015).

Irrigation efficiency

About 88% of Egypt's irrigation is implemented through inefficient techniques
(FAO 2015). Egypt's irrigation practice, however, is an example of a multiple-
use-cycle system with low farm-level efficiency but high region/nation-level efficiency (Keller and Keller 1995; Allen 2000). Through a rotation system, farmers
apply large amounts of water in a relatively short period of time to account for
days when the system will be shut down. Part of this water is used up by crop
evapotranspiration, and the rest is lost to percolation and runoff (Figure 9.3).
Water, however, is not necessarily lost for overall use, as it recharges an underlying unconfined aquifer (Postel 1992, 1993; cited in Maliva and Missimer 2012,
p. 686) flows into the well-developed drainage system, both of which act as a
water source for downstream farmers (Barnes 2014b).

Irrigation efficiency improvements aim to reduce water losses to soil evaporation, runoff, and deep percolation without affecting transpiration losses

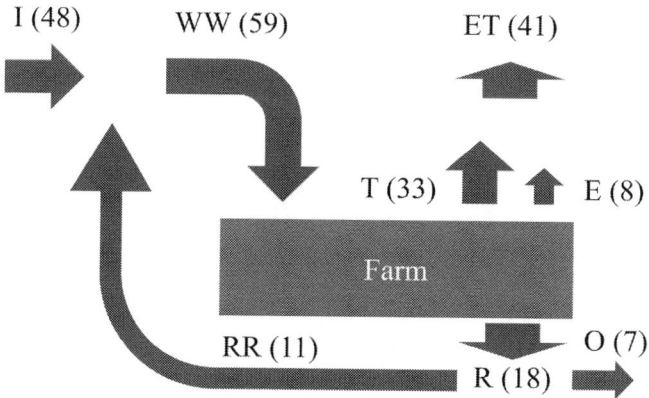

Figure 9.4 Conceptual model for Egypt's irrigation system, considered as a "big farm." I: inflow; WW: total water withdrawal; ET: evapotranspiration; E: evaporation; T: transpiration; R: return flow; RR: return flow reuse; O: outflow (nonreused reflow).

from plants. In this research, a future scenario of efficient irrigation at farm level was envisioned for Egypt and Sudan. Thus, a new "on-farm" application efficiency of 80% was assumed, comparable to the center pivot sprinkler irrigation efficiencies found in Howell (2003). The implementation in Sudan is straightforward, as the literature indicates irrigation is primarily conducted through surface application and no drainage water reuse is in place (Mahgoub 2014; FAO 2015). It demands, however, further considerations in Egypt. Following Simons et al. (2015), Egypt can be considered altogether as a "big farm," as the one shown in Figure 9.4. Water withdrawal and return flow are currently reported at 59 BCM/year and 18 BCM/year, respectively, meaning the overall evapotranspiration is 41 BCM/year (FAO 2015). Evapotranspiration was assumed as 20% soil evaporation and 80% plant transpiration, based on values found in Simons et al. (2012) for wheat and maize in the Nile Delta. Using Eq. 9.1 and Eq. 9.2, the current local (on-farm) and overall application efficiency were computed at 56% and 68%, respectively.

$$\text{Local efficiency} = \frac{\text{Transpiration}}{\text{Agriculture withdrawal}} \times 100 \qquad \text{Eq.9.1}$$

$$\text{Overall efficiency} = \frac{\text{Transpiration}}{\text{Agriculture withdrawal} - \text{Total reuse}} \times 100$$

$$\text{Eq.9.2}$$

For the future scenario, crop transpiration was kept fixed, so food production is not at stake. Based on findings in Ding et al. (2013), a rearrangement of 10% to 90% for evaporation and transpiration, respectively, was assumed. The return flow reuse was considered at the same proportion as present, i.e., 61% of the return flow (formed by both groundwater and surface water). The overall efficiency for the improved performance scenario was estimated at 86%. Efficiency values were used along with the crop's actual evapotranspiration (ETa) and the efficient irrigation coverage as indicated in equation Eq. 9.3. Water savings were computed comparing water demand for both current and improved performance scenarios.

$$\text{Water Demand} = \text{ETa} \times \left[\frac{\text{coverage}}{\text{effic}_{ip}} + \frac{(1 - \text{coverage})}{\text{effic}_{cp}} \right] \qquad \text{Eq.9.3}$$

Where:

- "coverage" is the percentage of croplands with efficient irrigation (sprinkler or better)
- "effic" is the overall efficiency
- "cp" is the current performance level
- "ip" is the improved performance level

Rainfall exploitation

According to Falkenmark and Rockström (2005), while rainfall is the only actual natural input of water to a watershed, it is often overlooked as a resource, as hydrologists and planners tend to focus on secondary water supplies such as rivers and aquifers. In the ENB, precipitation occurs primarily in upstream savannahs, forests, and rainfed agriculture lands. A small amount is utilized by humans, while the bulk of water evaporates at low or no beneficial use. More than 90% of rainfall is lost through landscape evapotranspiration (ET) in areas of limited human activities, such as modified land use (MLU), utilized land use (ULU), and protected land use (PLU)[1] (own computations based on Water Accounting+ 2016).

The average consumed water for the period 2005–2010 amounts to 828 BCM/year (Table 9.1). Only 5% is managed, and the rest is lost as landscape ET (44 BCM/year and 784 BCM/year, respectively). About 76% of this landscape ET occurs in three subbasins, namely Baro-Akobo-Sobat, Blue Nile, and Lower White Nile, at rates of 233 BCM/year, 224 BCM/year, and 142 BCM/year, respectively. Two facts deserve further attention. Firstly, MLU and ULU classified land use types are responsible for the largest share of evapotranspiration, with the shrubland type being particularly problematic. Secondly, nonbeneficial evaporation[2] accounts for 36% (220 BCM) of total ET. Therefore, a significant amount of (green) water might be gained by implementing rainwater harvesting (RWH) and aquifer recharge throughout the shrublands of these subbasins.

This strategy aims at implementing RWH over the shrublands in the subbasins Baro-Akobo-Sobat, Blue Nile, and White Nile. Water collection takes place in schemes sized between 100,000 ha and 200,000 ha, an extension

Table 9.3 Characteristics of Subbasins Selected for RWH Implementation. Precipitation values are estimated from maps in Awulachew (2012), Nile Basin Initiative (2012). ET values are from Water Accounting+ (2016). RWH ratio calculated based on assumed based on Herweg and Ludi (1999), Mekuria et al. (2015), Basán Nickisch et al. (2016).

Subbasin	Area [1000 Ha]	Evapotranspiration (ET)				RWH schemes characteristics		
		All land use	Land use "Shrubland"			Country	Precip. [mm]	RWH ratio
			Total	T share	E share			
Blue Nile	30820	224	76	46	30	Sudan	300	40%
						Ethiopia	800	60%
						Ethiopia	1000	60%
Baro-Akobo-Sobat	24778	233	127	70	57	South Sudan	1000	60%
						South Sudan	700	60%
Lower White Nile	22094	142	82	32	50	Sudan	500	50%

comparable to the five largest Sudanese irrigation schemes (except Gezira; FAO 2015). Schemes were placed in Sudan, South Sudan, and Ethiopia, under precipitation regimes ranging between 300 mm and 1000 mm. Schemes were located as follows: three in the Blue Nile, two in the Baro-Akobo-Sobat, and one in the Lower White Nile.

Since some rainfall is always lost to local infiltration, a RWH ratio was introduced (Table 9.3). This ratio is the relationship between precipitation and water captured for exploitation. The consulted literature shows the infiltration depth tends to be fixed, thus the higher the typical precipitation the higher the RWH ratio. Values of the RWH ratio were set according to total annual precipitation (TAP) as follows: for TAP $\geq$ 700mm, RWHR = 60%; for TAP $\leq$ 300mm, RWHR = 40%; and a RWHR = 50% for TAP in between.

Changing cropping patterns

One of the main concerns in the Nile region is food production under water-scarce scenarios. To cope with this, water productivity should be improved in existing irrigated lands (Viala 2008). In addition, Fedoroff et al. (2010), cited in Maliva and Missimer (2012, p. 687), claim that "meeting future water demand involves a radical rethinking of agriculture [...] both land and water productivity will have to be substantially improved."

Any "radical thinking" regarding changes in cropping patterns should involve at least three different but complementary scenarios: (i) using better varieties of the crops currently cultivated (high–yield, low water-demand); (ii) completely replacing current crops with those that offer better ratios of water demand to nutritional quality (fruits, vegetables, oatmeal); and (iii) trading virtual water. This strategy aims to increase water availability by both introducing more water into the system, namely virtual water, and improving production in rainfed areas where water consumption is not currently accounted for. The rationale is reducing agriculture water demand by changing the production matrix and international trade relations in Egypt and Sudan. Virtual water trade is the approach recommended for Egypt, whereas crop productivity is the target in Sudan.

Actions in Egypt

Embedded within its imports and exports, Egypt currently buys and sells virtual water. In this research, a rearrangement of virtual water flows was proposed that fulfills two requirements: (i) satisfying domestic demand for crops and produce and (ii) not jeopardizing Egypt's trade balance. The water footprint was used as a metric to assess water flows. Water footprint is linked to water demand, thus, the lower the value the better. Further, water cost, computed here as the ratio between merchandise trade and total water volume used for a given commodity, accounts for the price paid for virtual water. Higher prices are better for exporters who want to conserve water. Under these concepts, crop candidates for production augmentation are potatoes, onions, grapes, and vegetables (Table 9.4). For instance, Egypt pays $0.20 USD per cubic meter of embedded water in imported wheat, whereas it receives $5.33 USD per same unit of water in the form of exported frozen vegetables.

This crop change strategy proposes reducing domestic production (and increasing importation) of wheat and maize. At the same time, it foresees cropland reallocation toward more profitable agricultural products in terms of

Table 9.4 Egypt's Most Relevant Agricultural Products, Imports, and Exports for Virtual Water Gains. Trade value from FAOSTAT database, water footprint from Chapagain and Hoekstra (2003). Water used/traded and water cost are based on author's own computations with data from FAO (2015).

Product	Trade value [1000 USD]	Water		
		Footprint [m3/ton]	Used/Traded [BCM]	Cost [USD/m^3]
Products				
Wheat	*	930	8.80	*
Maize	*	1031	8.20	*
Imports				
Wheat	2715936	930	13.7	0.20
Maize	1984982	1031	5.2	0.38
Exports				
Potatoes	205901	308	0.13	1.58
Onions, dry	202553	258	0.09	2.25
Grapes	183357	537	0.05	3.67
Vegetables, frozen	159957	322	0.03	5.33

Table 9.5 Harvested Area for Different Crops in Egypt and Sudan and Water Saving/Gains for Scenarios in Cropping Pattern Strategy

Crop/Region	Harvested area [1000 ha]			Water saving/gain [BCM]	
	Present	*Scenario 1*	*Scenario 2*	*Scenario 1*	*Scenario 2*
Egypt					
Wheat	1419	709	709	5.2	
Maize	1030	515	515	(both scenarios)	
Potatoes	160	144	144		
Onions, dry	53	44	44		
Grapes	64	95	104		
Vegetables	127	322	274		
Totals	2854	1829	1791		
Sudan's sorghum					
Gezira	522	0	0	4.4	5.7
Irrigated (non–Gezira)	156	156	0		
Rainfed	7700	7350	7280		
Rainfed improved	0	350	420		
Totals	8378	7856	7700		

water demand. Alternatives of cropland reallocation and the corresponding impacts on water demand were found through linear optimization, using a Simplex linear programming algorithm in Solver, Microsoft Excel. Yields were maintained along the optimization; hence, a production increase for a given crop is subject to additional land. In turn, the agriculture area was limited to the total harvested area for these six crops, thus more land for one product means less land for another. Domestic demands for all products were kept fixed as well. Water demand using the water footprint approach differs from the values in the baseline. Per the water footprint approach (Table 9.5), demand for wheat and maize was estimated at 17 BCM; per baseline (Table 9.2), the demand is at 11 BCM. This mismatch between Chapagain and Hoekstra (2003) and FAO (2015) was amended by introducing a factor equal to 0.647 (11 BCM / 17 BCM) that allows comparisons with the other strategies.

Three optimization scenarios were assessed for Egypt. Scenario zero is the least restrictive. It sets no limits on the reduction of wheat and maize harvested area and no limit on the increase in vegetable production. Scenarios one and two are more restrictive; they introduce limits for cereal cropland as well as for vegetable production and importation based on international values used as benchmarks (Table 9.7).

Actions in Sudan

Sorghum is extensively produced in both rainfed and irrigated areas; irrigated sorghum is responsible for the largest agricultural water withdrawals. Unlike the approach in Egypt, the strategy in Sudanese territory aims at supporting rainfed sorghum. The implementation of agricultural best practices, e.g., better seeds, land leveling, correct use of fertilizers, reduction in soil compaction, and RWH, is expected in this study to increase current rainfed sorghum yields from 0.7 tons/ha up to 4 tons/ha based on Steduto et al. (2012) and international values (Table 9.7). Rising yields in rainfed areas frees up some schemes (land and water) for more-profitable uses.

Two scenarios were considered. Scenario 1 addresses eliminating irrigated sorghum in Gezira, and scenario 2 considers the total elimination of irrigated sorghum in Sudan. To eliminate sorghum irrigation in Gezira, yield has to be improved in 350,000 ha of currently rainfed croplands. To totally eliminate irrigated sorghum in Sudan, improvements must be applied to 420,000 ha of currently rainfed croplands.

Results and discussion

Table 9.6 summarizes the results of the three strategies developed. The savings and gains identified in *impact on water* refers to water kept from use and water introduced into the managed system, respectively. The *impact on crops* indicates whether water savings and gains imply changes in the current crop pattern. Under *actions* is a summary of the strategy rationale, which is supported by the references stated right beside. The *feasibility* column indicates what the criteria were to compute feasible minimum and maximum values of *water increase*. The three strategies were defined upon findings in published literature and well-known databases. Additionally, local and regional experiences exist for most of the strategies developed. Although scientifically possible and technically feasible, some of these strategies may face implementation challenges. Some of these challenges are addressed below.

Irrigation efficiency

Irrigation is key to produce more food and cash crops. It renders higher yields (Schoengold and Zilberman 2007) when water is used efficiently and wisely. Efficiency, however, needs to be evaluated at both local and regional scales to avoid negative outcomes. Upgrading on-farm application techniques minimizes water losses to nonproductive use. This will allow more water to remain in the canals instead of percolating to the aquifer and running off to the drainage ditches. The irrigation network will have to be redesigned for the exceeding water to be available for downstream users. Nonetheless, if water is not properly priced, efficiency might lead to an increase in total irrigated land and water withdrawals (colloquially known as the efficiency paradox).

The improvement potential for irrigation efficiency in Egypt and Sudan was evaluated through a coverage factor. This factor was determined by considering the percentage of efficient irrigation and the total efficient irrigation area in

Table 9.6 Summary of Strategies and Their Impact on Water Availability

Strategy		Impacts	Countries of Implementation				Water Increase [BCM]	
#	Name		Egypt	Ethiopia	Sudan	South Sudan	min	max
1	Irrigation efficiency	on water …	saving	–	saving	–	1.2	2.9
		on crops …	keep	–	keep	–		
2	Rainfall exploitation	on water …	–	gain	gain	gain	1.2	2.9
		on crops …	–	none	none	none		
3	Major changes in cropping patterns	on water …	gain	–	saving	–	9.6	10.9
		on crops …	change	–	change	–		

Strategy		Crops (region)	Actions	References	Feasibility
#	Name				
1	Irrigation efficiency	wheat, maize, sugarcane, sorghum	Upgrade from flood and furrow to sprinkler irrigation. Overall field application efficiency from 68% to 86% in Egypt and 65% to 80% in Sudan	Brouwer et al. (1989), Keller and Keller (1995), Simons et al. (2015), own computations	Implementation coverage between 28% and 56% of irrigated land, based on efficient irrigation coverage in Morocco and India
2	Rainfall exploitation	Subbasins: Blue Nile, Baro–Akobo–Sobat, Lower White Nile	Rainwater harvesting schemes similar to the White Nile pumping schemes in Sudan, found in FAO 2015	Mekuria et al. (2015), Basán Nickisch et al. (2016)	6 schemes proposed with area between 100 and 200 thousand hectares. Possible implementation in stages.
3	Major changes in cropping patterns	wheat, maize (vegetables, grapes, onions, potatoes), sorghum	In Egypt, crop replacement based on virtual water trade. In Sudan, yield improvement in mechanized rainfed sorghum areas	Chapagain and Hoekstra (2003), FAO (2016), own computations	Limits in vegetables production and frozen vegetables exports in Egypt. Limits in yield improvement in Sudan. Both limits based on comparable countries and global benchmarks.

comparable countries. Comparability was defined so as to account for economic (i.e., per-capita GDP and agricultural contribution to GDP) and climate (i.e., annual precipitation) factors (Table 9.7). A lower limit of 28% was assumed based on findings for Morocco. The upper limit of 56% is the quotient between the efficient irrigation area in India and the total irrigated area in Egypt. It was therefore assumed that Egypt might be able to manage a similar extension of efficient irrigation practices.

Harnessing rainfall

Rainfall is an untapped resource in the Eastern Nile Basin that can bridge the gap between water availability and allocation. The strategy of rainfall exploitation aims to bring more water into the managed system by converting lands of limited use into rainwater collection schemes. The water gained through this strategy is currently not accounted for, as it evaporates or is transpired by plants locally to no productive use.

Irrigation schemes are ancient institutions rooted in the study area. Therefore, implementing RWH schemes should be familiar to local communities. The RHW scheme's size was defined based on the New Haifa irrigation scheme, the second largest in the country. No additional considerations were factored into the recommended scale of application for this strategy. Further refinement of this strategy should consider other factors such as local and regional economics, agronomic characteristics, and precipitation regimes. Runoff tends to be greater when soil is saturated; thus rainwater harvesting should take place in regions of intense precipitation. For instance, the Ethiopian Highlands is widely known for its prevailing monsoon regime.

As RWH would take place in regions with limited land use, its implementation, operation and maintenance might be difficult. The strategy shows water gains at about 5 BMC/year, but gains might well be larger if more land is allotted and additional technology deployed. In regions where access to conventional water resources is limited, options like this are alluring. However, governmental intervention, transnational agreements, as well as the enactment of water markets are needed for this approach to scale.

Rethinking cropping patterns

The strategy of changing existing cropping patterns offers huge potential water gains. In contrast to the previous strategies, it has little to do with water infrastructure, and its realization requires both high-level policy making and profound farm-level decisions. Scenario 1 considers a full balance in the harvested area, whereas scenarios 2 and 3 are less restrictive and allow a reduction in the total harvested land. The analysis suggests that Egypt should substitute all domestic production of wheat and maize with imports. In addition, land currently utilized for these crops should be reallocated for more profitable and less water-demanding crops such as vegetables and grapes.

Table 9.7 Feasibility Metrics for Modern Irrigation Implementation and Sorghum Yield Improvements

Country	India	Morocco	Egypt	Sudan	Tanzania	Uzbekistan	Moldova
Per-capita GDP [2010 US$][1]	1699	3143	2654	1703	813	1749	1980
Average precipitation [mm/year][1]	N/R	N/R	51	250	1071	206	450
Value added by agriculture [% of GDP][1]	17	13	11	29	31	19	15
Irrigated area [1000 ha][2]	62286 (2008)	1341 (2011)	3422 (2002)	993 (2011)	184 (2002)	N/R	N/R
Efficient irrigation area [1000 ha][2]	2024 (2008)	414 (2012)	393 (2002)	N/A	0	N/R	N/R
% of efficient irrigation[2]	3 (2004)	28 (2011)	12 (2002)	N/A	0	N/R	N/R
Sorghum yield [Tons/Ha] (2013)[3]	N/R	N/R	5.4	0.64	3.7	12.7	3

Sources:
[1]World Bank (2018),
[2]AQUASTAT (FAO 2015),
[3]FAOSTAT (FAO 2016) database.
N/A: not available. N/R: not relevant.

From the optimization results here, it was found that vegetables, in the form of frozen vegetables, are the most rewarding agricultural products. The combination of good price and low water footprint is a key factor in the optimization. Grapes, conversely, are linked to a good water export cost but play a secondary role in the optimization. Finally, all optimization scenarios render improvements in the trade balance, which in the current situation is negative.

Feasibility in Egypt was assessed by limiting both the total harvested area available for vegetables and the required levels of vegetable export to international values found in the FAOSTAT database (FAO 2016). The total area of vegetable production was limited to 10% of the arable land (similar to China, Vietnam, Nepal, Philippines, and Laos) and the maximum level of vegetable export was limited to Belgium's level at about 1.2 million tons/year (highest value in database). Some of the limitations of this strategy are a lack of consideration of: (i) the full cost structure (e.g., net revenues or marginal costs of each crop/good exchanged) and (ii) any economies of scale that might impact crop production.

In the Sudanese case, the strategy aims at vacating sorghum's irrigated land and using it for more profitable economic activities. Yield improvements should start in the mechanized rainfed areas, where technology, training, and institutional organization exist already. An improved sorghum yield was assumed at 4 ton/ha, a value between Moldova's (similar GDP and precipitation), Tanzania's (large sorghum area), and Egypt's (same basin).

Concluding remarks

There is a growing concern that the Eastern Nile Basin suffers from water scarcity and insecurity. Regional stakeholders and international experts claim North Africa, and the Eastern Nile Basin in particular, is running out of water. Population pressure, urbanization, changes in food and nutrition, and climate change will exacerbate the situation. A closer examination of the water balance and hydrologic processes over the region, however, shows water resources are abundant. The apparent scarcity arises from the regional and temporal scattering of water resources as well as a lack of efficiency in harnessing and managing those resources. The total utilized water represents 5% of the total water input to the region. Thus, water availability may not be a binding constraint for sustainable and equitable development throughout the basin.

Through a focus on the agriculture sector and following the Water Diplomacy Framework's premise of water as a flexible resource, this work aims at reframing the management of available water resources so as to create flexibility and increase water productivity in the basin. Three strategies were evaluated, and findings suggest that there is a potential to increase the annual water availability by up to 11 BCM, a value that represents 13% of the Nile Waters Agreement flow.

The proposed strategies present challenges and opportunities that need further examination prior to effective implementation. The barriers to application include water pricing and wheat-related subsidies in Egypt, smallholding subsistence farming in irrigated areas of Egypt and Sudan, and socioeconomic conditions in Sudan, Ethiopia, and South Sudan as well as fears about food security and sovereignty in all ENB countries. Along with efficient irrigation, sound water allocation policies need to be in place before encouraging techno-logical intervention and adoption of more profitable crops. On the other hand, rainfed areas such as the mechanized sector in Sudan offer opportunities to increase yields, because some technology and agricultural best practices are already in place. In addition to increase water availability, some strategies offer considerable potential to save land. This might eradicate the need for rec-lamation of desert lands and enable the allocation of fertile land for more prof-itable uses.

The present work is a movement to reframe the current conversation – where water allocation between countries and sectoral water use and manage-ment within countries view water as a fixed and limited resource – to more collaborative and inventive efforts toward finding sustainable solutions by thinking about and managing water as a flexible resource.

Notes

1 MLU: "land where vegetation is replaced with the intention to increase the utilization of land resources. Examples are plantation forests, pastures and rainfed crops, among others." ULU: "land use classes with a low to moderate utilization of natural resources, such as savannah, woodland and mixed pastures." PLU: "environmentally sensitive land uses and natural ecosystem that cannot be modified due to protective measure" (Water Accounting+ 2016).
2 Nonbeneficial evaporation: evaporation for nonintended use, for instance from soil and leaves.

References

Abdelhady, D., Aggestam, K., Andersson, D., Beckman, O., Berndtsson, R., Palmgren, K.B., Madani, K., Ozkirimli, U., Persson, K.M., and Pilesjö, P., 2015. The Nile and the Grand Ethiopian Renaissance dam: Is there a meeting point between nationalism and hydrosolidarity? *Journal of Contemporary Water Research and Education*, 155 (1), 73–82.
Abdin, A., and Gaafar, I., 2009. Rational water use in Egypt. *In:* M. El Moujabber, L. Mandi, G. Trisorio Liuzzi, I. Martin, A. Rabi, and R. Rodriguez, eds. *Technological Per-spectives for Rational Use of Water Resources in the Mediterranean Region*. Bari, Italy: Medi-terranean Agronomic Institute of Bari, 11–28.
Allen, T., 2000. *The Middle East Water Question: Hydropolitics and the Global Economy*. New York: I.B. Tauris.
Awulachew, S.B., 2012. *The Nile River Basin: Water, Agriculture, Governance and Livelihoods*. Abingdon: Routledge.
Awulachew, S.B., Yilma, A.D., Loulseged, M., Loiskandl, W., Ayana, M., and Alamirew, T., 2007. *Water Resources and Irrigation Development in Ethiopia*. Columbo, Sri Lanka: International Water Management Institute.
Barnes, J., 2014a. *Cultivating the Nile: The Everyday Politics of Water in Egypt*. Durham: Duke University Press.

Barnes, J., 2014b. Mixing waters: The reuse of agricultural drainage water in Egypt. *Geoforum*, 57 (1), 181–191.

Basán Nickisch, M., Lahitte, A., Sosa, D., Sánchez, L., and Tosolini, R., 2016. *Aguadas para Ganadería Bovina en Los Bajos Submeridionales y Áreas de Influencia*. Fave: Sección Ciencias Agrarias. Available from: https://bibliotecavirtual.unl.edu.ar/publicaciones/index.php/FAVEAgrarias/article/view/6747

Brouwer, C., and Heibloem, M., 1986. *Irrigation Water Management: Irrigation Water Needs, Training Manual 3*. Rome, Italy: FAO.

Brouwer, C., Prins, K., and Heibloem, M., 1989. *Irrigation Water Management: Irrigation Scheduling, Training Manual 4*. Rome, Italy: FAO.

Chapagain, A.K., and Hoekstra, A., 2003. *Virtual Water Flows Between Nations in Relation to Trade in Livestock and Livestock Products*. Delft: UNESCO-IHE.

Conway, D., 2005. From headwater tributaries to international river: Observing and adapting to climate variability and change in the Nile Basin. *Global Environmental Change*, 15 (1), 99–114.

Conway, D., and Hulme, M., 1993. Recent fluctuations in precipitation and runoff over the Nile Sub-basins and their impact on main Nile discharge. *Climatic Change*, 25 (1), 127–151.

Ding, R., Kang, S., Zhang, Y., Hao, X., Tong, L., and Du, T., 2013. Partitioning evapotranspiration into soil evaporation and transpiration using a modified dual crop coefficient model in irrigated maize field with ground-mulching. *Agricultural Water Management*, 127 (1), 85–96.

Egypt and Sudan, 1959. *United Arab Republic and Sudan Agreement (with Annexes) For the Full Utilization of the Nile Waters*. Available from: https://www.internationalwaterlaw.org/documents/regionaldocs/uar_sudan.html

Falkenmark, M., and Rockström, J., 2005. *Rain: The Neglected Resource*. Stockholm: Stockholm International Water Institute.

FAO, 1997. *Irrigation in the Near East Region in Figures*. Available from: www.fao.org/docrep/w4356e/w4356e0s.htm#sudan

FAO, 2015. *Aquastat: Countries, Regions, Transboundary River Basins*. Available from: www.fao.org/nr/water/aquastat/countries_regions/

FAO, 2016. *FAOSTAT: Food and Agriculture Data*. Available from: www.fao.org/faostat/

Fedoroff, N.V., Battisti, D.S., Beachy, R.N., Cooper, P.J.M., Fischhoff, D.A., Hodges, C.N., Knauf, V.C., Lobell, D., Mazur, B.J., Molden, D., Reynolds, M.P., Ronald, P.C., Rosegrant, M.W., Sanchez, P.A., Vonshak, A., and Zhu, J.-K., 2010. Radically rethinking agriculture for the 21st century. *Science*, 327 (5967), 833–834.

Fund For Peace, 2018. *Fragile States Index 2018*. Available from: https://fundforpeace.org/2018/04/24/fragile-states-index-2018-annual-report/

Gassert, F., Reig, P., Luo, T., and Maddocks, A., 2013. *Aqueduct Country and River Basin Rankings: A Weighted Aggregation of Spatially Distinct Hydrological Indicators*. World Resources Institute. Available from: http://wri.org/publication/aqueduct-country-river-basin-rankings

Herweg, K., and Ludi, E., 1999. The performance of selected soil and water conservation measures: Case studies from Ethiopia and Eritrea. *Catena*, 36 (1–2), 99–114.

Howell, T.A., 2003. Irrigation efficiency. *In*: B.A. Stewart and T.A. Howell, eds. *Encyclopedia of Water Science*. New York: Marcel Dekker, 467–472.

Islam, S., and Susskind, L.E., 2013. *Water Diplomacy: A Negotiated Approach to Managing Complex Water Networks*. New York: RFF Press.

Islam, S., and Susskind, L.E., 2015. Understanding the water crisis in Africa and The Middle East: How can science inform policy and practice? *Bulletin of the Atomic Scientists*, 71 (2), 39–49.

Johnston, R., 2012. Availability of water for agriculture in the Nile Basin. *In*: S.B. Awulachew, V. Smakhtin, D. Molden, and D. Peden, eds. *The Nile Basin: Water, Agriculture, Governance and Livelihoods*. Abingdon: Routledge, 61–83.

Karimi, P., Molden, D., Notenbaert, A., and Peden, D., 2012. Nile Basin farming systems and productivity. *In*: S.B. Awulachew, V. Smakhtin, D. Molden, and D. Peden, eds. *The Nile Basin: Water, Agriculture, Governance and Livelihoods*. Abingdon: Routledge, 133–153.

Karrou, M., Oweis, T., El Enein, R.A., and Sherif, M., 2012. Yield and water productivity of maize and wheat under deficit and raised bed irrigation practices in Egypt. *African Journal of Agricultural Research*, 7 (11), 1755–1760.

Keith, J., Hussein, S., and Mahdy, E. S., 1998. *Egypt's Sugarcane Policy and Strategy for Water Management*. Available from: http://pdf.usaid.gov/pdf_docs/pnach359.pdf

Keller, A.A., and Keller, J., 1995. *Effective Efficiency: A Water Use Efficiency Concept for Allocating Freshwater Resources*. Columbo, Sri Lanka: International Water Management Institute.

Kirda, C., 2002. Deficit irrigation scheduling based on plant growth stages showing water stress tolerance. *FAO Water Reports*, 22 (1), 102.

Mahgoub, F., 2014. *Current Status of Agriculture and Future Challenges in Sudan*. Upsala: The Nordic Africa Institute.

Maliva, R.G., and Missimer, T.M., 2012. *Arid Lands Water Evaluation and Management*. New York: Springer.

Mekuria, W., Chanie, D., Admassu, S., Akal, A. T., Guzman, C.D., Zegeye, A.D., Tebebu, T.Y., Steenhuis, T., and Ayana, E.K., 2015. *Sustaining The Benefits of Soil and Water Conservation in the Highlands of Ethiopia*. Available from: https://cgspace.cgiar.org/rest/bitstreams/64401/retrieve

Nile Basin Initiative, 2012. *The State of the Nile River Basin*. Kampala, Uganda: Graphic Systems Ltd.

OECD, 2006. *Promoting Pro-poor Growth – Agriculture*. Paris: Organisation for Economic Co-operation and Development.

Postel, S., 1992. *The Last Oasis: Facing Water Scarcity*. London: Earthscan.

Postel, S., 1993. Water and Agriculture. *In*: P. H. Gleick, ed. *Water In Crisis: A Guide to the World's Fresh Water Resources*. Oxford: Oxford University Press.

Qadir, M., Sharma, B.R., Bruggeman, A., Choukr-allah, R., and Karajeh, F., 2007. Non-conventional water resources and opportunities for water augmentation to achieve food security in water scarce countries. *Agricultural Water Management*, 87 (1), 2–22.

Salman, S.M.A., 2011. The new state of South Sudan and the hydro-politics of the Nile Basin. *Water International*, 36 (2), 154–166.

Schoengold, K., and Zilberman, D., 2007. The economics of water, irrigation, and development. *Handbook of Agricultural Economics*, 3 (1), 2933–2977.

Siam, M.S., and Eltahir, E.A.B., 2017. Climate change enhances interannual variability of the Nile river flow. *Nature Climate Change*, 7 (1), 350–354.

Simons, G., Bastiaanssen, W., and Immerzeel, W., 2015. Water reuse in river basins with multiple users: A literature review. *Journal of Hydrology*, 522 (1), 558–571.

Simons, G., Terink, W., Badawy, H., Van Den Eertwegh, G., and Bastiaanssen, W., 2012. *Egypt: Assessing the Effects of Farm-level Irrigation Modernization on Water Availability and Crop Yields*. Available from: www.futurewater.nl/wp-content/uploads/2013/10/final_report_egypt_wb_jan_2013.pdf

Steduto, P., Hsiao, T.C., Fereres, E., and Raes, D., 2012. *Crop Yield Response to Water*. Rome, Italy: FAO.

Sutcliffe, J.V., 2009. The Hydrology of the Nile Basin. *In*: H. Dumont, ed. *The Nile*. Dordrecht, The Netherlands: Springer, 335–364.

Thorweihe, U., and Heinl, M., 2002. *Groundwater Resources of the Nubian Aquifer System, NE Africa*. Paris: Observatoire Du Sahara Et Du Sahel.

United Nations, 2003. *Africa Water Vision For 2025: Equitable And Sustainable Use of Water For Socioeconomic Development*. Available from: http://repository.uneca.org/handle/10855/5488

United Nations, 2018. *Multidimensional Poverty Index*. Available from: http://hdr.undp.org/en/2018-MPI

Viala, E., 2008. Water for food, water for life: A comprehensive assessment of water management in agriculture. *Irrigation And Drainage Systems*, 22 (1), 127–129.

Water Accounting+, 2016. *Water Accounting: Independent Estimates of Water Flows, Fluxes, Stocks, Consumption, and Services*. Available from: http://wateraccounting.org

World Bank, 2018. *DataBank*. Available from: https://data.worldbank.org/

10 Confronting the natural domain

Strategies for addressing ecology and conservation in complex water management challenges

Charles B. van Rees, Gabriela Marie Garcia, and Jessica Rozek Cañizares

Introduction

The multifarious interdependencies of hydrological and ecological systems make biodiversity conservation an unavoidable and potentially complicating factor in water management (Johnson et al. 2001; Falkenmark and Rockström 2006). The influence of freshwater availability, distribution, and condition on ecological systems is undisputed and well-studied; indeed, water is called "the bloodstream of the biosphere" (Ripl 2003). Consequently, a major characteristic of modern water resources management is the consideration of the potential impacts of water management decisions on ecological factors (Schoeman et al. 2014; Arthington et al. 2018); for example, the European Water Framework Directive, a major international agreement on water management, explicitly includes ecological objectives as a major priority (European Commission 2000). Ecosystem effects on freshwater resources are less well understood, but appreciation for their ability to influence almost all aspects of the hydrological cycle is growing (Farley et al. 2005; Le Maitre et al. 2007), and they are becoming an important component of water security and sustainable development strategies (Harrison et al. 2018; Vörösmarty et al. 2018). Indeed, nature-based solutions are gaining increasing prominence as tools for addressing water scarcity (WWAP/UN-Water 2018).

Increased recognition of the complex and manifold interactions between biodiversity and water resources has generated increasing overlap in the interests of global water development and wildlife conservation (van Rees et al. 2019; Darwall et al. 2018). It is common to find language concerning ecological systems in the mission statements and declarations of major water development initiatives, for example "environmental need for water" (The Melbourne Declaration; Pigram 2000), "the ecosystems approach" (Convention on the Protection and Use of Transboundary International Lakes; United Nations Economic Commission for Europe 2013), and "protection and preservation of ecosystems" (Convention on the Law of the Non-navigational Uses of International Watercourses; United Nations 1997). Similarly, many conservation organizations have designated branches for addressing freshwater issues (e.g., The International Union for the Conservation of Nature Water Program, The Nature Conservancy's Water for People and Nature, World Wildlife Fund Water Stewardship). The Convention on Biological Diversity's Aichi Targets include specific mention of protecting inland waters (Target 11) and hydrological ecosystem services (Target 14; Hagerman and Pelai 2016).

The growing confluence of water management and ecology did not arise solely out of academic interest but from an increasing sense of urgency among practitioners in both fields, who see firsthand the consequences of management that ignores complex linkages between water governance and ecology. Freshwater scarcity experienced by nations of diverse socioeconomic status is leading to severe impacts on human well-being, economic viability, and societies as a whole (Johnson et al. 2001; Vörösmarty et al. 2010). The loss of biodiversity and ecosystem functions as a result of water resource mismanagement has often exacerbated effects on people by reducing important ecosystem services (Millennium Ecosystem Assessment Panel 2005). This is especially true in the case of low-income households, which are often directly and disproportionately reliant on biodiversity and ecosystem services for their livelihoods. Examples of such services include food, fuel, and fresh water (Kumar and Yashiro 2014). At the same time, freshwater ecosystems are more threatened and degraded than both terrestrial and marine systems, with alarming rates of decline and extinction among freshwater biotas (Johnson et al. 2001; Dudgeon et al. 2006; WWF 2016). The extinction rates of aquatic animal species in North America are estimated to be five times greater than terrestrial ones (Ricciardi and Rasmussen 1999). It has been repeatedly said without exaggeration that freshwater ecosystems are in a state of "biodiversity crisis" (Abell 2002; Strayer and Dudgeon 2010; Harrison et al. 2018).

Interestingly, researchers in the societal domain of water management look to the natural domain for solutions (Vörösmarty et al. 2018; WWAP/UN-Water 2018), while ecologists and conservation biologists, working in the natural domain, emphasize the need for governance solutions to halt declines in freshwater habitats and species populations (van Rees et al. 2019; Baynham-Herd et al. 2018; Irvine 2018). The overlap of these domains and strategies for successfully managing their dynamic interactions are thus of great importance, but they occupy particularly troublesome ground. The interconnectedness, vulnerability, and basin–specific spatial structure of freshwater biodiversity and the ubiquity of human impacts and water demand in the world's watersheds make the conservation of freshwater ecosystems difficult (Dudgeon et al. 2006). Meanwhile, as many water managers have discovered, ecological factors (van Rees and Reed 2015) increase complexity in water systems through a variety of feedbacks, time lags, and interrelationships. Ecological dynamics can frustrate well-meaning management schemes if not recognized and addressed but can also provide opportunities for creative water solutions.

We briefly review common forms of complexity in ecological systems and the challenges they pose to effective water management and the frameworks that facilitate practical water decision making by providing a conceptual structure for the role of ecological variables in water resources management. Lastly, we present a deeper analysis of three case studies, demonstrating the diversity of forms that ecological complexity can take in water management systems and also how this complexity can lead to unexpected problems and unforeseen solutions.

Complexity and ecological systems

Understanding ecological systems is essential to optimize water resource decisions and protect ecosystem services and biodiversity, but these systems add complexity and uncertainty to water resources management. As with cultural and societal systems (Majumder 2017), ecosystems and their components have dynamic behavior requiring an adaptive approach that accounts for inherent uncertainty (Holling 2001; Islam 2017). Here, we discuss five complexity characteristics of ecological factors (interdependence, multiple scales, feedbacks, emergence, and nonlinearity; Maguire et al. 2006) and address how these concepts manifest in ecological and social-ecological systems.

Interconnectedness and Interdependence

Complex systems are made up of many connected, dynamic components (Nicolis and Prigogine 1989). Interconnectedness and interdependence in ecological systems span spatial and temporal scales, manifesting within and between ecosystems and between ecosystems and society. Probably the most well-recognized interdependencies in the natural world are food webs, which indicate trophic relationships (i.e., "who eats whom") with directional arrows from consumer to consumed. Misunderstandings of food webs have led to unpleasant surprises in water resources management. For example, attempts to boost numbers of the economically important kokanee salmon (*Oncorhynchus nerka*) in the Flathead River–Lake ecosystem (Montana, USA) by introducing supplemental prey to their lake habitats instead led to declines in those fish (salmon in MacDonald Creek dropped from the tens of thousands to just 330) and local bird species (bald eagle, *Haliaeetus leucocephalus*, numbers declined from a high of 639 to a low of 25; Spencer et al. 1991, Figure 10.1). The negative effects on the food web extended beyond the natural domain to the societal domain, with dramatic losses in economic and recreational opportunities (Figure 10.1).

As evidenced by the Flathead River–Lake example, social-ecological systems are highly interdependent. Human societies are ultimately dependent on ecosystem services, which they exploit for a variety of functions and resources (MEA 2005). This exploitation, however, can lead to loss of services at certain levels of intensity, so a balance is required to maintain both the preservation and use of an ecological resource (Daily 1999). For example, overutilization of marine communities by global fisheries can lead to decreased

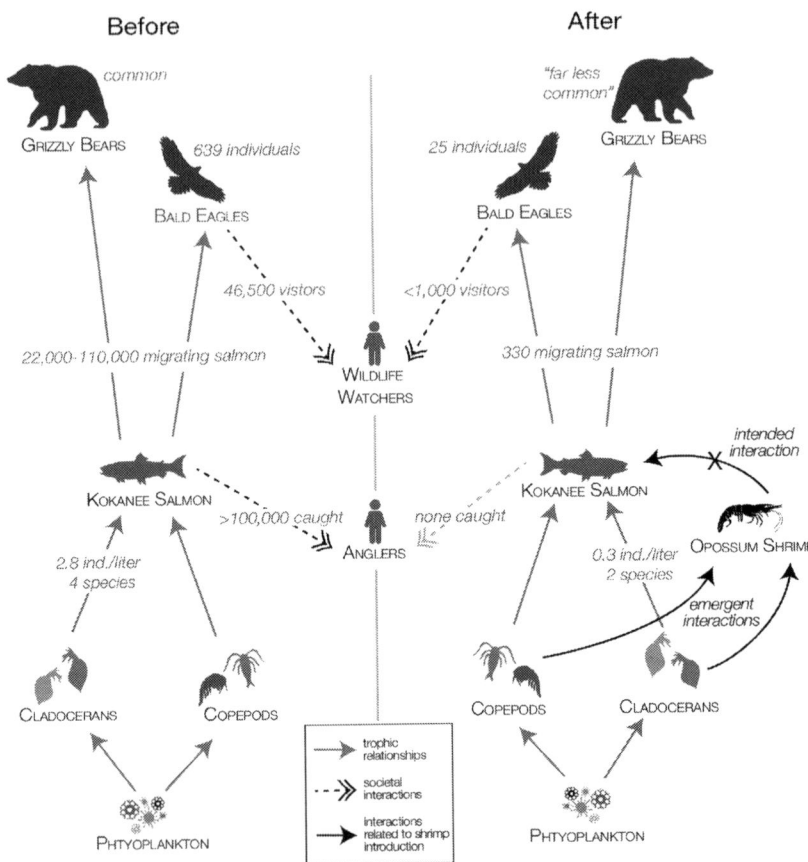

Figure 10.1 Trophic webs capture the interconnectedness and interdependence typical of
ecosystems by illustrating the relationship between species and functional
groups. Water management decisions can create perturbations in trophic
interactions which can result in unpredictable responses. The left panel of the
figure depicts the trophic interactions (gray arrows) and societal benefits
(dashed arrows) of the Flathead-River Lake Ecosystem. After the introduction
of opossum shrimp (right panel), intended to supplement the diet, and
therefore increase abundance, of the kokanee salmon, the trophic interactions
changed. The salmon did not heavily feed upon the shrimp; instead the
shrimp established themselves as a major competitor for the salmon's preferred
food, aquatic invertebrates. The decrease of available prey for the salmon
resulted in cascading effects through the trophic web and decreased economic
and recreational benefits in the societal domain.

biodiversity and higher rates of fishery collapse. In contrast, an equilibrium
between fisheries and marine reserves that conserves biodiversity allows for
higher average catches and increased fishery resilience on local and regional
scales (Worm et al. 2006).

Multiple spatial and temporal scales

Cross-scale ecological processes are a specific type of interconnectedness and interdependence worth special mention for their prominence in ecological systems. When components of a system interact and respond at multiple spatial and temporal scales, they increase uncertainty in system behavior. Without scale-relevant monitoring, important interactions may be overlooked or misunderstood (Ramalingam et al. 2008). Interactions in ecological systems span space and time and often cross geographic and political boundaries, compounding their complexity.

Migratory species such as waterbirds and fish pose significant challenges to conservation because they move across hydrological and governance boundaries (López-Hoffman et al. 2017). While waterbirds can exploit ephemeral resources, they are dependent on a specific set of spatial and temporal factors throughout their life cycle and may suffer delayed consequences if conditions are suboptimal (Myers et al. 1987). For example, the date at which rice fields are drawn down during the agricultural cycle affects wetland plant succession in subsequent months (Meeks 1969) and, consequently, how migrating waterfowl use the fields later in the year (Kadlec 1962). Thus, water management decisions made even months prior have a profound effect on the reproduction and resources of species with large geographic ranges and highly specific spatiotemporal needs.

Another form of temporal cross-scale phenomenon is extinction debt, when populations show decline or extinction that is temporally separated from environmental perturbation (Kuussaari et al. 2009). In such cases, water management does not appear to have significant impacts at first but has strong effects later, sometimes long after modification or reversal of a decision is possible. If monitoring is applied only at the spatial scale at which effects are expected, then the ecological response of hydrological management will be entirely misunderstood. Extinction debts have been demonstrated in riparian fish and freshwater mussels after the construction of upstream dams (Olden et al. 2010; Vaughn 2012). Extinction debts are a challenge for conservation and environmental planning, as they can be difficult to predict and detect (Kuussaari et al. 2009).

Feedback

Feedback mediates the behavior of complex systems by amplifying or dampening changes in a system (positive and negative feedbacks, respectively; Ramalingam et al. 2008). Feedback processes or loops can be complicated and difficult to predict; in complex systems, multiple feedback processes can cause simultaneous stabilizing and destabilizing effects in different parts of the system (Ramalingam et al. 2008).

Complex interactions like feedbacks are a ubiquitous source of unpredictable behavior in ecological systems. For example, willow (*Salix* spp.) recruitment along the banks of rivers in the American West results in the accumulation of alluvial soils, improving conditions for further willow recruitment (Rood et al. 2010). A well-known example of feedbacks in freshwater ecology is eutrophication in freshwater lakes and ponds. Nutrient inputs to these systems result in oxygen depletion by respiring bacteria, which decrease the clarity of the water, reduce photosynthesis, and lead to further depletion of oxygen, causing turbid and anoxic conditions, unsuitable for fishing and other forms of recreation (Brothers et al. 2014).

Intensified agriculture and unrestricted irrigation practices in arid or semi-arid climates can lead to desertification, an extreme and semipermanent ecological shift (Bathurst et al. 2003). The drivers of the desertification process (abstraction of freshwater from above and below ground, destruction of native vegetation) are also its symptoms, leading to rapid and potentially irreversible losses of natural resources (see Tervonen et al. 2015).

Emergence

Unexpected behaviors or properties that result from component interactions are referred to as emergent. Emergence is a product of other aspects of complexity, including interconnectedness and feedback relationships (Maguire et al. 2006). A chief difficulty of emergent properties is the unpredictability as to when and on what scale they will arise. In order to maximize the adaptiveness of a system, space for emergence must exist; attempting to control or restrict complex systems can result in management failures (Ramalingam et al. 2008).

For example, fallen trees entering a river channel can initiate wood jams that have a dramatic effect on the surrounding ecosystem and landscape. The jams facilitate sedimentation and create elevated patches of soil that are resistant to erosion, altering streamflow, and influencing broader floodplain behavior, thus contributing to local and downstream conditions through geological, hydrological, and ecological mechanisms. The conditions for the creation of log jams and their subsequent effects are unpredictable, which can be problematic for managers that assume stationarity in riparian systems (Collins et al. 2012).

Nonlinearity

Feedback loops, emergent dynamics, and adaptive behaviors can all lead to non-linearity; unpredictable system behavior in which changes in outcome are disproportionate to changes in input. Nonlinear relationships between streamflow and ecological response are a common and often problematic part of managing water in river systems (see Wohl et al. 2015). Nonlinearity requires a holistic perspective of the system in which managers must recognize and critically examine common assumptions like "input equals output," as changing system dynamics may generate unanticipated results (Ramalingam et al. 2008).

Regime shifts are a common example of nonlinearity in ecological systems. These changes can happen over a very short period of time with little observed change in system inputs. Instead of a gradual linear change in response to external forces, a dramatic switch occurs when the perturbations from external forces pass a threshold (Scheffer et al. 2001). Examples of regime shifts include rapid eutrophication of lakes (Wang et al. 2012), fisheries collapse due to over-harvest (Vasilakopoulos and Marshall 2015), and rapid reduction in Amazonian stream fish diversity after watershed deforestation (Brejao et al. 2018). A reliable method for predicting regime shifts does not exist; instead, research has focused on defining early warning signals of such transitions (e.g., Veraart et al. 2012) with few applications in the field (but see Rozek et al. 2017).

Frameworks for addressing ecosystems in water management

As with sociocultural systems, when the behavior and complexity of ecological systems is not accounted for in water management decisions, unforeseen consequences may result. The difficulties of simultaneously addressing societal, ecological, and hydrological variables in water resources management necessitates approaches that increase the connection between research fields, for example highlighting the connection between ecological variables and economic or social ones. The last three decades have seen a proliferation of socio-ecological frameworks that bridge this gap by putting ecological phenomena in the context of human needs or management actions. We touch briefly upon several of these here but emphasize that many such frameworks exist, both explicitly and implicitly addressing water issues.

Integrating the social and natural domains with interdisciplinary approaches

Interdisciplinary approaches are crucial for solving complex socioecological problems in water management (Schoeman et al. 2014), and many frameworks have been developed to guide decision making in such contexts. Ecosystem service valuation (Costanza et al. 1997) is perhaps the most widely recognized framework to integrate human and natural resource requirements. Ecological factors are viewed through the lens of the goods and services that ecosystems offer humanity, including sustained hydrological function, water quality, and flood regulation (MEA 2005). The ecosystem services approach has been widely applied to address ecological complexity in water management, from restoring river flows and alleviating poverty in South Africa (Turpie et al. 2008) to understanding the relationships between wetlands and human health (Horwitz and Finlayson 2011).

Frameworks like ecohydrology (Rodriguez-Iturbe 2000), natural and environmental flows or e-flows (Poff et al. 2016; Arthington 2012; Arthington et al. 2018), and blue and green water (Falkenmark and Rockström 2006) explicitly consider the interaction of hydrological and ecological systems, permitting a quantitative approach to characterizing the potential impacts of hydrological change on local ecology. The ELOHA framework, for example, provides a conceptual scaffold around which models of flow–ecology relationships can be built for managed river systems, comparing ecological responses to flow variability and alterations to conditions under baseline or unaltered hydrological regimes (Poff et al. 2010). Ecological information in the form of flow–response curves can then be used to inform management models and actions taken by decision makers or stakeholders.

Stakeholder-based frameworks such as Integrated Water Resources Management (Agarwal et al. 2000) and Water Diplomacy (Islam and Susskind 2013; Susskind and Islam 2012) are envisioned to include ecological information in effective decision making, but there are few formal guidelines for how – and what type of – ecological information should be included in the decision-making process. van Rees and Reed (2015) outlined a practical approach to bridge the gap between ecological information like flow–response curves and stakeholder-based decision making. The ecological stakeholder analog framework (ESA) treats ecological factors as stakeholder analogs in the decision–making process (van Rees and Reed 2015; van Rees et al. 2019). As demonstrated in the previous section, the relationship between cause and effect in ecological systems can often be nonlinear, context dependent, and unpredictable; a unique advantage of the ESA concept is that it distinguishes ecological mechanisms from outward responses and explicitly accounts for the dynamic causes underlying an ecological effect. Through prioritization of context-driven and actionable ecological research, the ESA concept promotes effective characterization of ecological complexity in stakeholder-based water management and avoids reduction of complex ecological phenomena to simple metrics or artificially hard rules that may constrain available solutions.

As in the ecosystem services framework, the ESA approach allows ecological factors to be considered as opportunities for added value rather than simply as strict constraints to economic and social needs. This non–zero–sum perspective opens the door to mutual-gains solutions that satisfy the interests of multiple stakeholders by shifting attention away from the superficial ecological response toward the underlying mechanisms. Van Rees and colleagues (2019) provide conceptual elaboration and exploration of the concept, demonstrating how well-established tools from stakeholder theory can readily be used to identify, classify, and prioritize ecological stakeholder analogs.

As highlighted in the case studies that follow, interdisciplinary approaches are instrumental in the effective management of interrelated hydrological and ecological systems that span the social, natural, and political domains. Since no single approach is universally appropriate for complex water management, the most effective interdisciplinary framework will depend on the management context. We outline below three cases featuring ecological factors of different types (an ecosystem, a suite of ecosystem functions, and an endangered species), the degrees of complexity which they contribute to larger contexts of water resources management, and how interdisciplinary approaches open the door for creative solutions.

Case studies of ecological factors in water management

Case study I: the changing hydrology of the Caroni Swamp: complexity and uncertainty

Caroni Swamp, Trinidad's second-largest wetland, encompasses a variety of habitats including mangrove forests, herbaceous marshes and estuarine channels. Designated a National Park in 1979 and a wetland of international importance under the Ramsar Convention in 2005, the swamp is a mixed-use area of high ecological, economic, and cultural importance. More than 150 species of bird are found in the swamp and adjacent marshes (ffrench 1977) including the scarlet ibis (*Eudocimus ruber*), the national bird of Trinidad. It is also a site of economic importance, supporting the largest oyster fishery in the country (Ramdial 1975) and a major source of fish, crabs, rice, and charcoal (Food and Agriculture Organization 2007; Ghermandi 2015). Since the mid-1970s, ecotours in the Caroni Swamp highlighting mass roosts of the ibis have produced more than $1 million (TTD) annually (Bacon 1987).

The swamp's unique hydrology and biodiversity, its large size, and its social-ecological importance lend it a complexity that has resulted in unintended consequences to otherwise well-meaning management decisions made throughout the 20th century. In 1922, Caroni's complex hydrology was altered by the construction of a network of drainage canals, waterways, and sluice gates to create 900 ha of agricultural lands for rice farming (Bacon 1970; Gerald 1985). Unfortunately, rice cultivation never thrived, so the project was officially abandoned in 1954, and the related water management infrastructure fell into disrepair (Bildstein 1990). With insufficient awareness of the myriad ecosystem services provided by the swamp, land managers likely assumed that there was little cost to developing this portion of the wetland.

Additional lack of understanding of the hydrological dynamics of the wetland led to subsequent and increasingly extensive problems. By the mid-1960s, the northern sluice gate was damaged, allowing saltwater intrusion from the adjacent Gulf of Paria (Bildstein 1990), leading to saline conditions in all of the canals and former freshwater marsh area (Bildstein 1990). The shift in salinity has had cascading effects on the rest of the Caroni system, promoting conditions in which mangroves outcompete salt-intolerant plants: between 1957 and 1986 more than 1000 ha of mangrove were established in the former marsh area, an expansion that has continued until the last land cover measurements in 2007 (Juman and Ramsewak 2013).

Although formal studies have not been conducted, this has no doubt led to dramatic shifts in the ecological communities inhabiting the wetland (Tilman 1997), including more than 70 breeding bird species (ffrench 1977). It has been suggested that the loss of the marsh area had a major effect on the prime ecotourism draw, the scarlet ibis, resulting in their abandonment of the Caroni Swamp as a breeding area for nearly two decades (Bildstein 1990). In addition, the swamp's altered hydrology also appears to be driving an increased incidence of local severe flooding (Rozek, unpublished data). The underlying ecological and hydrological complexity of the Caroni Swamp system has resulted in far-reaching consequences to simple water management decisions, which might otherwise have been avoided under modern interdisciplinary water frameworks like ecosystem services, environmental flows, or ESA.

Another fascinating example of complexity in Caroni comes from the diverse spatial scales at which ecological systems operate and the unexpected results that arise from interconnectedness. It has been hypothesized that, due to the increasing salinity of the Caroni Swamp, the scarlet ibis has favored the river floodplains in northern South America in order to feed their young freshwater invertebrates (Bildstein 1990). Interestingly, a study conducted between 1997 and 1998 found that some patches of red mangroves (*Rhizophora mangle*) in Caroni Swamp had unusually high incidences of mutations, and the soil surrounding them was contaminated with mercury (Klekowski et al. 1999). They concluded that the ibis were feeding in contaminated waters near gold mines in Venezuela, and when they molted their feathers at their roosting sites in Caroni Swamp, they contaminated the local soil around the mangroves (Klekowski et al. 1999). Thus, declines in the bird responsible for Caroni's ecotourism value were due not only to changes in local hydrology but also to economic decisions made in a different country, separated from the roost by geographical and political boundaries. Once again, these impacts, which span ecological and economic factors, could not be understood, much less anticipated, without integrating detailed knowledge of the ecological systems and their components that make up a large part of this complex hydrological system.

Case study II: coffee landscapes in Santa Maria, Costa Rica

Ecological concerns in water management systems are not limited to natural landscapes. Water management across agroecosystems faces new sustainability challenges in the context of global change (Molden et al. 2007). Compared to irrigated crops, rainfed agroecosystems are especially vulnerable to climate changes (Rockström et al. 2010) and account for the vast majority of agriculture in developing countries (International Water Management Institute 2010). Coffee (*Coffea arabica*), a tropical rainfed crop, is the developing world's most economically important agricultural export commodity (Donald 2004). As greenhouse gas concentration increases, major droughts and floods are predicted to become more frequent on inter- and intra-annual scales (Sala et al. 2015), with negative implication for coffee sustainability and farmer livelihoods. To exacerbate the matter, the majority of coffee worldwide is grown in rural areas on small-scale farms where adaptive capacity is limited by technical, cultural, and socioeconomic obstacles (Quiroga et al. 2015). Given the lack of access to irrigation, farmer resilience to precipitation changes depends largely on their ability to maximize on-farm water use efficiency.

On-farm water management has implications not only for the farmer but for tropical watershed function. Soil erosion from agricultural lands is a major determinant of water quality worldwide (Ramos-Scharrón and Figueroa-Sánchez 2017), and coffee farms in Latin America are often situated at the montane headwaters where heavy rainfall and steep slopes increase soil erosion and chemical runoff into streams passing through the farms (de Jesús-Crespo et al. 2016). Water quantity at the watershed scale is also negatively affected by coffee farm intensification, which can involve dense planting and the removal of shade trees (Silvano et al. 2005; de Jesús-Crespo et al. 2016). Poor on-farm management therefore poses a significant threat not only to farmer livelihoods but to downstream water quality and availability for human consumption, hydropower generation, and aquatic ecosystems (Chomitz et al. 1999; Gómez-Delgado et al. 2011).

The primary challenges to effective water management in coffee landscapes are (1) risk of nonsource pollution by sediment erosion and chemical runoff, (2) limited farmer access to water resources and inefficient use of precipitation, and (3) limited adaptive capacity. These may appear disparate issues if considered individually, but analysis through the ecosystem services lens reveals an emergent ecological link: soil health, or the capacity of the soil to support plant production and provide other essential ecosystem services (Karlen et al. 1997). Soil health is the key determinant of erosion and chemical runoff, as well as on-farm water infiltration. Its proper management provides an opportunity to benefit both on-farm and downstream water use.

Farm management practices largely determine whether soil health is progressively built or deteriorated through a positive feedback loop. Coffee cultivation under full sun without shade trees degrades the soil with negative downstream effects such as increased sedimentation and turbidity (de Jesús-Crespo et al. 2016). As nutrients and stability are lost through erosion, the soil becomes more susceptible to further degradation and has decreased water-retention capacity. The downward spiral has negative implications for farmer adaptive capacity, as well as on-farm and downstream water use. Incorporating abundant and diverse shade trees, however, especially those with larger root systems (i.e., agroforestry), provides numerous ecosystem services, including soil restoration, stability, and enhanced water infiltration and retention, ultimately maximizing on-farm use of precipitation and building adaptive capacity (Donald 2004; de Jesús-Crespo et al. 2016; Lin 2007). As shade trees flourish, their leaf litter fertilizes the plants and further builds soil health, allowing for reduced chemical inputs and lower risk of runoff (Vaast et al. 2005).

Certification schemes such as The Rainforest Alliance (RA) have applied an ecosystem services approach to watershed management in coffee landscapes to incentivize agroforestry (Rainforest Alliance 2016). Certifying organizations offer a higher price for coffee farmed under agroforestry management because the shade trees are a source of provisional (timber and fruit products), supporting (nutrient cycling; biodiversity maintenance) and regulatory (watershed protection; erosion control) ecosystem services (Perfecto et al. 2005; de Jesús-Crespo et al. 2016; Rainforest Alliance 2016). Agroforestry boosts water efficiency and minimizes erosion and runoff, thereby building adaptive capacity. There are still improvements to be made, however, before certifications are adopted at the watershed level. In 2016, only 1 out of 9 farmers associated with the Santa Maria cooperative, CoopeDota, were RA certified (Garcia, *unpublished*).

Several interviewees claimed the monetary payments offered by RA were disproportionate to the challenge of meeting the lengthy and rapidly changing criteria, and similar attitudes have been recorded among farmers in other coffee-producing regions (Ibanez and Blackman 2016). Further complexity is introduced through the interaction between shade trees and coffee bushes. Farmer knowledge and empirical research have also indicated potential trade-offs of agroforestry, such as reduced yields and greater susceptibility to pests (Vaast et al. 2006; Jha et al. 2014). Under the current scenario, farmers are weighing the benefits of certification against those of coffee intensification with major implications for tropical watersheds. Further interdisciplinary research efforts should be directed towards generating actionable knowledge through the integration of farmer knowledge and experience, watershed function, coffee plant physiology, and livelihood stability.

Although an exhaustive exploration of interactions between hydrological and agroecological systems is beyond the scope of the chapter, the highlighted case study demonstrates the advantages of interdisciplinary frameworks for water management in agricultural landscapes. Analysis through the ecosystem services framework may allow for a mutual-gains watershed management solution (agroforestry) that otherwise has been underexplored. The approach applied in the coffee system can be adapted to other agroecosystems for their effective integration in complex management decisions.

Case study III: waterbirds and reef health on Oʻahu, Hawaiʻi

The island of Oʻahu, Hawaiʻi has undergone dramatic changes in land and water management since European colonization in the late 18th century. Rapid urbanization to accommodate burgeoning populations has greatly increased the impermeable land cover on the island, reducing groundwater recharge and increasing runoff into the ocean (Ridgley and Giambelluca 1991; Giambelluca 1996). Losses in groundwater recharge are concerning given the island's dependence on aquifers for freshwater and signs that these aquifers are being overdrawn in some areas (Ridgley and Giambelluca 1991). Urbanization has also increased flooding during storm events and generated massive inputs of sediment, nutrients, and heavy metals into nearshore marine environments (Banner 1974; Lapointe and Bedford 2011). These allochthonous inputs induce physiological stress and mortality in corals (Fabricius 2005) and contaminate fisheries (Hunter et al. 1995; McCarthy et al. 2008). Recent work on Oʻahu shows that most residents are concerned with runoff impacts on nearshore ecosystems, including coral reef health, the safety of waters for recreation, and the contamination of coastal fisheries (Hawaii Conservation Alliance 2015; Peng and Oleson 2017).

O'ahu's landscape development also drove extensive wetland loss (van Rees and Reed 2014), which likely exacerbated the island's hydrological problems, given wetlands' ability to remove nutrients and sediment from runoff (Tournebize et al. 2013). Wetland loss also caused population declines in a suite of endemic wetland birds, including the 'Alae 'ula (Hawaiian gallinule, *Gallinula galeata sandvicensis*). Although the establishment of protected refuges has helped 'Alae 'ula populations begin recovering, remnant subpopulations are spatially isolated, and the connectivity among remaining populations is a conservation concern (U.S. Fish and Wildlife Service 2011; van Rees et al. 2018a).

Approaching O'ahu's water management issues from the lens of ecological stakeholder analogs (van Rees and Reed 2015), it may be possible to align the behavior of the island's vulnerable ecosystems with the priorities of major stakeholders (stormwater control and nearshore marine water quality). Coral reefs and associated fish communities are a major ecological factor in this system, with the operating position (van Rees and Reed 2015) that status quo (no management action) will continue to decrease reef health and the safety of fish catch. These ecosystems have high economic and cultural value to the Hawaiian Islands (Friedlander 2004; Hawaii Conservation Alliance 2015), making current declines undesirable. In this way, coral reefs are salient ecological stakeholders (van Rees et al. 2019) in the Hawaiian urban water management system and are also aligned well with the interests of the general public.

By contrast, Hawaiian waterbirds cannot live without wetland habitats, and their current extinction risk is poorly understood (Reed et al. 2011). These ecological stakeholders have high legitimacy as native species and some power derived from their prominence in native Hawaiian legends (Westervelt 1910) but are not considered economically important. While the outward trends of 'Alae 'ula populations show some evidence of negative impacts (in this case, habitat fragmentation) due to current land use, the mechanisms (ecological interests) behind these impacts are less well understood. van Rees et al. (2018b) demonstrated that water features like canals, drainage swales, streams, and small wetlands increase gene flow in the 'Alae 'ula, suggesting that small lotic systems, previously thought to be unimportant for waterbird conservation, may contribute to their population connectivity, decreasing their extinction risk.

Water and natural resources managers on O'ahu are thus faced with three apparently unrelated problems: burgeoning populations beginning to overdraft groundwater supplies, deteriorating offshore conditions, and threatened, culturally important bird species with a potentially bleak conservation prognosis. Considering the inherent interdependencies of terrestrial and oceanic systems and the ecological function of wetlands, unexpected solutions become evident that are not intuitively obvious from analyzing each problem separately (emergence). van Rees and Reed (2015) highlighted the possibility that constructed wetlands and green infrastructure might simultaneously alleviate population fragmentation in Hawaiian waterbirds and future water supply concerns on

Oʻahu. Such wetland systems increase recharge, purify, and absorb runoff (Engelhardt and Ritchie 2001) and decrease floods (Magmedov et al. 1996) but could also act as stepping-stone habitats for waterbirds moving between fragmented habitats. New information on the importance of near-shore water quality to the general public on Oʻahu adds additional weight to this potential solution. Accordingly, the restoration and creation of coastal and low-elevation wetlands and green infrastructure could ameliorate a variety of Oʻahu's water resources–related issues while also satisfying the interests of both human and ecological stakeholders.

This case study illustrates how the complexities of ecological and hydrological systems do not always lead to greater difficulties in management but can present unanticipated, synergistic solutions to seemingly separate problems. Interdisciplinary approaches like ESA and the Water Diplomacy Framework permit the sort of holistic view that is necessary to capture this complexity and find such solutions.

Conclusions

Ecological phenomena have a diversity of strong interactions with hydrological systems. As biologists and water managers have repeatedly discovered, water management inevitably leads to ecological change, which means that environmental impacts are a consequence of every alteration of local hydrology. The inclusion of ecological variables often leads to substantial increases in the complexity of the management system and necessitates new and adaptive frameworks for management (e.g., Arthington 2012; van Rees and Reed 2015; Islam 2017). The gradual maturation of ecology's role in water management has seen it progress from a minor system input to a major source of system controls and constraints, with increasing understanding of its inherent complexity. Now that theoretical and applied understanding has begun to capture a more accurate image of the ecology–water nexus, better conceptual approaches are needed to link the equity of Habermasian stakeholder-based water management approaches with increasingly refined ecological data. The development of frameworks and approaches that can reconcile complex scientific information with bottom-up management strategies might also enable taking greater advantage of the currently under-appreciated benefits and synergies of healthy freshwater ecosystems and the internalization of ecological goals in water management.

References

Abell, R., 2002. Conservation biology for the biodiversity crisis: A freshwater follow-up. *Conservation Biology*, 1 (5), 1435–1437.

Agarwal, A., delos Angeles, M.S., Bhatia, R., Chéret, I., Davila-Poblete, S., Falkenmark, M., Villarreal, F.G., Jønch-Clausen, T., Kadi, M.A., Kindler, J., and Rees, J., 2000. *Stockholm Integrated Water Resources Management*. Stockholm: Global Water Partnership.

Arthington, A.H., 2012. *Environmental Flows: Saving Rivers in the Third Millennium*. Oakland: University of California Press.

Arthington, A.H., Kennen, J.G., Stein, E.D., and Webb, J.A., 2018. Recent advances in environmental flows science and water management: Innovation in the Anthropocene. *Freshwater Biology*, 63, 1022–1034.

Bacon, P.R., 1970. *The Ecology of the Caroni Swamp, Trinidad*. Central Statistical Office Printing Unit, Port of Spain, Trinidad.

Bacon, P.R., 1987. Use of wetlands for tourism in the insular Caribbean. *Annals of Tourism Research*, 14 (1), 104–117.

Banner, A.H., 1974. Kaneohe Bay, Hawaii; urban pollution and a coral reefs ecosystem. *Proceedings of the Second International Coral Reef Symposium*, 2, 685–702.

Bathurst, J.C., Sheffield, J., Leng, X., and Quaranta, G., 2003. Decision support system for desertification mitigation in the Agri basin, southern Italy. *Physics and Chemistry of the Earth*, Parts A/B/C, 28 (14–15), 579–587.

Baynham-Herd, Z., Amano, T., Sutherland, W.J., and Donald, P.F., 2018. Governance explains variation in national responses to the biodiversity crisis. *Environmental Conservation*, 45, 407–419.

Bildstein, K.L., 1990. Status, conservation and management of the scarlet ibis Eudocimus ruber in the Caroni Swamp, Trinidad, West Indies. *Biological Conservation*, 54 (1), 61–78.

Brejao, G.L., Hoeinghaus, D.J., Perez-Mayorga, M.A., Ferraz, S.F.B., and Casatti, L., 2018. Threshold responses of Amazonian stream fishes to timing and extent of deforestation. *Conservation Biology*, 32 (4), 860–871.

Brothers, S., Kohler, J., Attermeyer, K., Grossart, H.P., Mehner, T., Meyer, N., Scharnweber, K., and Hilt, S., 2014. A feedback loop links brownification and anoxia in a temperate, shallow lake. *Limnology and Oceanography*, 59 (4), 1388–1398.

Chomitz, K.M., Brenes, E., and Constantino, L., 1999. Financing environmental services: The Costa Rican experience and its implications. *Science of the Total Environment*, 240 (1–3), 157–169.

Collins, B.D., Montgomery, D.R., Fetherston, K.L., and Abbe, T.B., 2012. The floodplain large-wood cycle hypothesis: A mechanism for the physical and biotic structuring of temperate forested alluvial valleys in the North Pacific coastal ecoregion. *Geomorphology*, 139, 460–470.

Costanza, R., d'Arge, R., De Groot, R., Farber, S., Grasso, M., Hannon, B., Limburg, K., Naeem, S., O'neill, R.V., Paruelo, J., and Raskin, R.G., 1997. The value of the world's ecosystem services and natural capital. *Nature*, 387 (6630), 253.

Daily, G.C., 1999. Developing a scientific basis for managing Earth's life support systems. *Conservation Ecology*, 3 (2), 14.

Darwall, W., et al., 2018. The alliance for freshwater life: A global call to unite efforts for freshwater biodiversity science and conservation. *Aquatic Conservation: Marine and Freshwater Ecosystems*, 28 (4), 1015–1022.

de Jesús-Crespo, R., Newsom, D., King, E.G., and Pringle, C., 2016. Shade tree cover criteria for non-point source pollution control in the rainforest alliance coffee certification program: A snapshot assessment of Costa Rica's Tarrazú coffee region. *Ecological Indicators*, 66, 47–54.

Donald, P.F., 2004. Biodiversity impacts of some agricultural commodity production systems. *Conservation Biology*, 18 (1), 17–38.

Dudgeon, D., Arthington, A.H., Gessner, M.O., Kawabata, Z.I., Knowler, D.J., Lévêque, C., Naiman, R.J., Prieur-Richard, A.H., Soto, D., Stiassny, M.L., and Sullivan, C.A.,

2006. Freshwater biodiversity: Importance, threats, status and conservation challenges. *Biological Reviews*, 81 (2), 163–182.

Engelhardt, K.A.M., and Ritchie, M.E., 2001. Effects of macrophyte species richness on wetland ecosystem functioning and services. *Nature*, 411 (6838), 687.

European Commission, 2000. Directive 2000/60/EC of the European Parliament and of the council of 23 October 2000: Establishing a framework for community action in the field of water policy. *Official Journal of the European Communities*, 40 (327).

Fabricius, K.E., 2005. Effects of terrestrial runoff on the ecology of corals and coral reefs: Review and synthesis. *Marine Pollution Bulletin*, 50 (2), 125–146.

Falkenmark, M., and Rockström, J., 2006. The new blue and green water paradigm: Breaking new ground for water resources planning and management. *Journal of Water Resources Research Planning and Management*, 132 (3), 129–132.

Farley, K.A., Jobbágy, E.G., and Jackson, R.B., 2005. Effects of afforestation on water yield: A global synthesis with implications for policy. *Global Change Biology*, 11 (10), 1565–1576.

ffrench, R., 1977. Birds of the Caroni Swamp and marshes. *Living World*, 1977, 42–44.

Food and Agriculture Organization of the United Nations, 2007. *Mangroves of North and Central America 1980–2005: Country Reports – Forest Resources Assessment Working Paper 138.* Rome: United Nations.

Friedlander, A.M., et al., 2004. *Status of Hawai'i's Coastal Fisheries in the New Millennium, Revised.* Proceedings of the 2001 Fisheries Symposium sponsored by the American Fisheries Society, Hawai'i Chapter.

Gerald, L., 1985. The changing Caroni Swamp: 1922–1985. *The Naturalist*, 5, 14–17.

Ghermandi, A., 2015. *Mapping Ecosystem Service Values in Trinidad and Tobago.* St. Augustine, Trinidad, and Tobago: University of the West Indies.

Giambelluca, T.W., 1996. Water balance, climate change and land-use planning in the Pearl Harbor Basin, Hawai'i. *International Journal of Water Resources Development*, 12 (4), 515–530.

Gómez-Delgado, F., Roupsard, O., Maire, G.L., Taugourdeau, S., Pérez, A., Oijen, M.V., Vaast, P., Rapidel, B., Harmand, J.M., Voltz, M., and Bonnefond, J.M., 2011. Modelling the hydrological behaviour of a coffee agroforestry basin in Costa Rica. *Hydrology and Earth System Sciences*, 15 (1), 369–392.

Hagerman, S.M., and Pelai, R., 2016. As far as possible and as appropriate: Implementing the Aichi biodiversity targets. *Conservation Letters*, 9 (6), 469–478.

Harrison, I., Abell, R., Darwall, W., Thieme, M.L., Tickner, D., and Timboe, I., 2018. The freshwater biodiversity crisis. *Science*, 362 (6421), 1369.

Hawaii Conservation Alliance, 2015. *Effective Conservation Program Community Engagement Results.* Available from: www.hawaiiconservation.org/images/uploads/pages/HCA_Community_ECP_v2-2.pdf

Holling, C.S., 2001. Understanding the complexity of economic, ecological, and social systems. *Ecosystems*, 4 (5), 390–405.

Horwitz, P., and Finlayson, C.M., 2011. Wetlands as settings for human health: Incorporating ecosystem services and health impact assessment into water resource management. *BioScience*, 61 (9), 678–688.

Hunter, C.L., Stephenson, M.D., Tjeerdema, R.S., Crosby, D.G., Ichikawa, G.S., Goetzl, J.D., Paulson, K.S., Crane, D.B., Martin, M., and Newman, J.W., 1995. Contaminants in oysters in Kaneohe bay, Hawaii. *Marine Pollution Bulletin*, 30 (10), 646–654.

Ibanez, M., and Blackman, A., 2016. Is eco-certification a win–win for developing country agriculture? Organic coffee certification in Colombia. *World Development*, 82, 14–27.

International Water Management Institute, 2010. *Managing Water for Rainfed Agriculture.* Available from: www.iwmi.cgiar.org/Publications/Water_Issue_Briefs/PDF/Water_Issue_Brief_10.pdf

Irvine, K., 2018. Aquatic conservation in the age of the sustainable development goals. *Aquatic Conservation: Marine and Freshwater Ecosystems*, 28 (6), 264–1270.

Islam, S., 2017. Complexity and contingency: Understanding and managing complex water problems. *In*: S. Islam and K. Madani, eds. *Water Diplomacy in Action: Contingent Approaches to Managing Complex Water Problems*. London: Anthem Press, 3–18.

Islam, S., and Susskind, L.E., 2013. *Water Diplomacy: A Negotiated Approach to Managing Complex Water Networks*. New York: RFF Press.

Jha, S., Bacon, C.M., Philpott, S.M., Ernesto Méndez, V., Läderach, P., and Rice, R.A., 2014. Shade coffee: Update on a disappearing refuge for biodiversity. *BioScience*, 64 (5), 416–428.

Johnson, N., Revenga, C., and Echeverria, J., 2001. Managing water for people and nature. *Science*, 292 (5519), 1071–1072.

Juman, R.A., and Ramsewak, D., 2013. Land cover changes in the Caroni Swamp Ramsar Site, Trinidad (1942 and 2007): Implications for management. *Journal of Coastal Conservation*, 17 (1), 133–141.

Kadlec, J.A., 1962. Effects of a drawdown on a waterfowl impoundment. *Ecology*, 43 (2), 267–281.

Karlen, D.L., Mausbach, M.J., Doran, J.W., Cline, R.G., Harris, R.F., and Schuman, G.E., 1997. Soil quality: A concept, definition, and framework for evaluation (a guest editorial). *Soil Science Society of America Journal*, 61 (1), 4–10.

Klekowski, E.J., Temple, S.A., Siung-Chang, A.M., and Kumarsingh, K., 1999. An association of mangrove mutation, scarlet ibis, and mercury contamination in Trinidad, West Indies. *Environmental Pollution*, 105 (2), 185–189.

Kumar, P., and Yashiro, M., 2014. The marginal poor and their dependence on ecosystem services: Evidence from South Asia and Sub-Saharan Africa. *In*: J. von Braun and F. Gatzweiler, eds. *Marginality*. Dordrecht, The Netherlands: Springer, 169–180.

Kuussaari, M., Bommarco, R., Heikkinen, R.K., Helm, A., Krauss, J., Lindborg, R., Öckinger, E., Pärtel, M., Pino, J., Roda, F., and Stefanescu, C., 2009. Extinction debt: A challenge for biodiversity conservation. *Trends in Ecology & Evolution*, 24 (10), 564–571.

Lapointe, B.E., and Bedford, B.J., 2011. Stormwater nutrient inputs favor growth of non-native macroalgae (Rhodophyta) on O'ahu, Hawaiian Islands. *Harmful Algae*, 10 (3), 310–318.

Le Maitre, D.C., Milton, S.J., Jarmain, C., Colvin, C.A., Saayman, I., and Vlok, J.H.J., 2007. Linking ecosystem services and water resources: Landscape-scale hydrology of the Little Karoo. *Frontiers in Ecology and the Environment*, 5 (5), 261–270.

Lin, B.B., 2007. Agroforestry management as an adaptive strategy against potential microclimate extremes in coffee agriculture. *Agricultural and Forest Meteorology*, 144 (1–2), 85–94.

López-Hoffman, L., Chester, C.C., Semmens, D.J., Thogmartin, W.E., Rodríguez-McGoffin, M.S., Merideth, R., and Diffendorfer, J.E., 2017. Ecosystem services form transborder migratory species: Implications for conservation governance. *Annual Review of Environment and Resources*, 42, 509–539.

Magmedov, V.G., Zakharchenko, M.A., Yakovleva, L.I., and Ince, M.E., 1996. The use of constructed wetlands for the treatment of run-off and drainage waters: The UK and Ukraine experience. *Water Science and Technology*, 33 (4–5), 315–323.

Maguire, S., McKelvey, B., Mirabeau, L., and Öztas, N., 2006. Complexity science and organization studies. *In*: S. Clegg, C. Hardy, T. Lawrence, and W. Nord, eds. *The SAGE Handbook of Organization Studies*. London: SAGE Publications, 165–214.

Majumder, M., 2017. The blind men, the elephant and the well: A parable for complexity and contingency. *In*: S. Islam and K. Madani, eds. *Water Diplomacy in Action: Contingent Approaches to Managing Complex Water Problems*. London: Anthem Press, xiii–xvii.

McCarthy, S.G., Incardona, J.P., and Scholz, N.L., 2008. Coastal storms, toxic runoff, and the sustainable conservation of fish and fisheries. *In*: K.D. McLaughlin, ed. *Mitigating Impacts of Natural Hazards on Fishery Ecosystems*. Bethesda: American Fisheries Society, 7–27.

Meeks, R.L., 1969. The effect of drawdown date on wetland plant succession. *The Journal of Wildlife Management*, 33 (4), 817–821.

Millennium Ecosystem Assessment Panel, 2005. *Ecosystems and Human Wellbeing: Synthesis*. Washington, DC: Island Press.

Molden, D., et al., 2007. *Water for Food, Water for Life: A Comprehensive Assessment of Water Management in Agriculture*. Sterling: Earthscan.

Myers, J.P., Morrison, R.I.G., Antas, P.Z., Harrington, B.A., Lovejoy, T.E., Sallaberry, M., Senner, S.E., and Tarak, A., 1987. Conservation strategy for migrating species. *American Scientist*, 75 (1), 8–26.

Nicolis, G., and Prigogine, I., 1989. *Exploring Complexity*. New York: W.H. Freeman.

Olden, J.D., Kennard, M.J., Leprieur, F., Tedesco, P.A., Winemiller, K.O., and Garcia-Berthou, E., 2010. Conservation biogeography of freshwater fishes: Recent progress and future challenges. *Biodiversity Research*, 16 (3), 496–513.

Peng, M., and Oleson, K.L.L., 2017. Beach recreationalists' willingness to pay and economic implications of coastal water quality problems in Hawaii. *Ecological Economics*, 136 (C), 41–52.

Perfecto, I., Vandermeer, J., Mas, A., and Pinto, L.S., 2005. Biodiversity, yield, and shade coffee certification. *Ecological Economics*, 54 (4), 435–446.

Pigram, J.J., 2000. The Melbourne declaration: The challenge of sharing and caring for the world's water. *Water International*, 25 (2), 320.

Poff, N.L., Brown, C.M., Grantham, T.E., Matthews, J.H., Palmer, M.A., Spence, C.M., Wilby, R.L., Haasnoot, M., Mendoza, G.F., Dominique, K.C., and Baeza, A., 2016. Sustainable water management under future uncertainty with eco-engineering decision scaling. *Nature Climate Change*, 6 (1), 25.

Poff, N.L., Richter, B.D., Arthington, A.H., Bunn, S.E., Naiman, R.J., Kendy, E., Acreman, M., Apse, C., Bledsoe, B.P., Freeman, M.C. and Henriksen, J., 2010. The Ecological Limits of Hydrologic Alteration (ELOHA): A new framework for developing regional environmental flow standards. Freshwater Biology, 55, 147–170.

Quiroga, S., Suá Rez, C., and Solís, J.D., 2015. Exploring coffee farmers' awareness about climate change and water needs: Smallholders' perceptions of adaptive capacity. *Environmental Science and Policy*, 45, 53–66.

Rainforest Alliance, 2016. *Current and Upcoming Impact Evaluations*. Available from: www.rainforest-alliance.org/impact/upcoming-evaluations

Ramalingam, B., Jones, H., Reba, T., and Young, J., 2008. *Exploring the Science of Complexity: Ideas and Implications for Development and Humanitarian Efforts*. London: Overseas Development Institute.

Ramdial, B.S., 1975. *The Social and Economic Importance of the Caroni Swamp in Trinidad and Tobago*. Ph.D. thesis, The University of Michigan.

Ramos-Scharrón, C.E., and Figueroa-Sánchez, Y., 2017. Plot-, farm-, and watershed-scale effects of coffee cultivation in runoff and sediment production in western Puerto Rico. *Journal of Environmental Management*, 202, 126–136.

Reed, J.M., Elphick, C.S., Ieno, E.N., and Zuur, A.F., 2011. Long-term population trends of endangered Hawaiian waterbirds. *Population Ecology*, 53 (3), 473–481.

Ricciardi, A., and Rasmussen, J.B., 1999. Extinction rates of North American freshwater fauna. *Conservation Biology*, 13 (5), 1220–1222.

Ridgley, M.A., and Giambelluca, T.W., 1991. Drought, groundwater management and land use planning: The case of central Oahu, Hawaii. *Applied Geography*, 11 (4), 289–307.

Ripl, W., 2003. Water: The bloodstream of the biosphere. *Philosophical Transactions of the Royal Society of London B: Biological Sciences*, 358 (1440), 1921–1934.

Rockström, J., Karlberg, L., Wani, S.P., Barron, J., Hatibu, N., Oweis, T., Bruggeman, A., Farahani, J., and Qiang, Z., 2010. Managing water in rainfed agriculture: The need for a paradigm shift. *Agricultural Water Management*, 97 (4), 543–550.

Rodriguez-Iturbe, I., 2000. Ecohydrology: A hydrologic perspective of climate-soil-vegetation dynamics. *Water Resources Research*, 36 (1), 3–9.

Rood, S.B., Goater, L.A., Gill, K.M., and Braatne, J.H., 2010. Sand and sandbar willow: A feedback loop amplifies environmental sensitivity at the riparian interface. *Oecologia*, 165 (1), 31–40.

Rozek, J.C., Camp, R.J., and Reed, J.M., 2017. No evidence of critical slowing down in two endangered Hawaiian honeycreepers. *PLoS ONE*, 12 (11), e0187518.

Sala, O., Gherardi, L., and Peters, D.P.C., 2015. Enhanced precipitation variability effects on water losses and ecosystem functioning: Differential response of arid and mesic regions. *Climatic Change*, 131 (2), 213–227.

Scheffer, M., Carpenter, S., Foley, J.A., Folke, C., and Walker, B., 2001. Catastrophic shifts in ecosystems. *Nature*, 413 (6856), 591.

Schoeman, J., Allan, C., and Finlayson, C.M., 2014. A new paradigm for water? A comparative review of integrated, adaptive and ecosystem-based water management in the Anthropocene. *International Journal of Water Resources Development*, 30 (3), 377–390.

Silvano, R.A., Udvardy, S., Ceroni, M., and Farley, J., 2005. An ecological integrity assessment of a Brazilian Atlantic Forest watershed based on surveys of stream health and local farmers' perceptions: Implications for management. *Ecological Economics*, 53 (3), 369–385.

Spencer, C.N., McClelland, B.R., and Standford, J.A., 1991. Shrimp stocking, salmon collapse, and eagle displacement. *Bioscience*, 41 (1), 14–21.

Strayer, D.L., and Dudgeon, D., 2010. Freshwater biodiversity conservation: Recent progress and future challenges. *Journal of the North American Benthological Society*, 29 (1), 344–358.

Susskind, L., and Islam, S., 2012. Water diplomacy: Creating value and building trust in transboundary water negotiations. *Science & Diplomacy*, 1 (3), 1–7.

Tervonen, T., Sepehr, A., and Kadziński, M., 2015. A multi-criteria inference approach for anti-desertification management. *Journal of Environmental Management*, 162, 9–19.

Tilman, D., Knops, J., Wedin, D., Reich, P., Ritchie, M., and Siemann, E., 1997. The influence of functional diversity and composition on ecosystem processes. *Science*, 277 (5330), 1300–1302.

Tournebize, J., Passeport, E., Chaumont, C., Fesneau, C., Guenne, A., and Vincent, B., 2013. Pesticide de-contamination of surface waters as a wetland ecosystem service in agricultural landscapes. *Ecological Engineering*, 56, 51–59.

Turpie, J.K., Marais, C., and Blignaut, J.N., 2008. The working for water programme: Evolution of a payments for ecosystem services mechanism that addresses both poverty and ecosystem service delivery in South Africa. *Ecological Economics*, 65 (4), 788–798.

United Nations, 1997. *Convention on the Law of the Non-navigational Uses of International Watercourses.* Available from: http://legal.un.org/ilc/texts/instruments/english/conventions/8_3_1997.pdf

United Nations Economic Commission for Europe, 2013. *Convention on the Protection and Use of Transboundary Watercourses and International Lakes.* Available from: www.unece.org/fileadmin/DAM/env/documents/2013/wat/ECE_MP.WAT_41.pdf

U.S. Fish and Wildlife Service, 2011. *Recovery Plan for Hawaiian Waterbirds, Second Revision.* Portland: U.S. Fish and Wildlife Service.

Vaast, P., Beer, J., Harvey, C., and Harmand, J.M., 2005. *Environmental Services of Coffee Agroforestry Systems in Central America: A Promising Potential to Improve the Livelihoods of Coffee Farmers' Communities.* Integrated Management of Environmental Services in Human-Dominated Tropical Landscapes, CATIE, November 1–3, 2005, Turrialba, Costa Rica, pp. 35–39.

Vaast, P., Bertrand, B., Perriot, J.-J., Guyot, B., and Génard, M., 2006. Fruit thinning and shade improve bean characteristics and beverage quality of coffee (*Coffea arabica L.*) under optimal conditions. *Journal of the Science of Food and Agriculture*, 86 (2), 197–204.

van Rees, C.B., and Reed, J.M., 2014. Wetland loss in Hawai'i since human settlement. *Wetlands*, 34 (2), 335–350.

van Rees, C.B., and Reed, J.M., 2015. Water diplomacy from a duck's perspective: Wildlife as stakeholders in water management. *Journal of Contemporary Water Research & Education*, 155 (1), 28–42.

van Rees, C.B., Reed, J.M., Wilson, R.E., Underwood, J.G., and Sonsthagen, S.A., 2018a. Small-scale genetic structure in an endangered wetland specialist: Possible effects of landscape change and population recovery. *Conservation Genetics*, 19 (1), 129–142.

van Rees, C.B., Reed, J.M., Wilson, R.E., Underwood, J.G., and Sonsthagen, S.A., 2018b. Landscape genetics identifies streams and drainage infrastructure as dispersal corridors for an endangered wetland bird. *Ecology and Evolution*, 8 (16), 8328–8343.

van Rees, C.B., Cañizares, J.R., Garcia, G.M., and Reed, J.M., 2019. Ecological stakeholder analogs as intermediaries between freshwater biodiversity conservation and sustainable water management. *Environmental Policy and Governance*, 29 (4), 303–312.

Vasilakopoulos, P., and Marshall, C.T., 2015. Resilience and tipping points of an exploited fish population over six decades. *Global Change Biology*, 21 (5), 1834–1847.

Vaughn, C.C., 2012. Life history traits and abundance can predict local colonisation and extinction rates in freshwater mussels. *Freshwater Biology*, 57 (5), 982–992.

Veraart, A.J., Faassen, E.J., Dakos, V., van Nes, E.H., Lürling, M., and Scheffer, M., 2012. Recovery rates reflect distance to a tipping point in a living system. *Nature*, 481 (7381), 357–359.

Vörösmarty, C.J., McIntyre, P.B., Gessner, M.O., Dudgeon, D., Prusevich, A., Green, P., Glidden, S., Bunn, S.E., Sullivan, C.A., Liermann, C.R., and Davies, P.M., 2010. Global threats to human water security and river biodiversity. *Nature*, 467 (7315), 555.

Vörösmarty, C.J., et al., 2018. Ecosystem-based water security and the Sustainable Development Goals (SDGs). *Ecohydrology & Hydrobiology*, 18 (4), 317–333.

Wang, R., Dearing, J.A., Langdon, P.G., Zhang, E., Yang, X., Dakos, V., and Scheffer, M., 2012. Flickering gives early warning signals of a critical transition to a eutrophic lake state. *Nature*, 492 (7429), 419–422.

Westervelt, W.D., 1910. *Legends of Ma-ui: A Demi God of Polynesia, and of His Mother Hina.* Honolulu: Hawaiian Gazette.

Wohl, E., Lane, S.N., and Wilcox, A.C., 2015. The science and practice of river restoration. *Water Resources Research*, 51, 5974–5997.

Worm, B., Barbier, E.B., Beaumont, N., Duffy, J.E., Folke, C., Halpern, B.S., Jackson, J.B., Lotze, H.K., Micheli, F., Palumbi, S.R., and Sala, E., 2006. Impacts of biodiversity loss on ocean ecosystem services. *Science*, 314 (5800), 787–790.

WWAP/UN-Water, 2018. *The United Nations World Water Development Report 2018: Nature-Based Solutions for Water*. Paris: UNESCO.

WWF, 2016. *Living Planet Report 2016: Risk and Resilience in a New Era*. Gland, Switzerland: WWF International.

11 Access to safe drinking water across the Navajo Nation

Laura Corlin

Introduction

The Navajo Nation Tribal Land encompasses more than 27,000 square miles near the Four Corners region of the United States (US). This area is about as large as West Virginia and is home to over 173,600 individuals. It is divided among 110 political and cultural chapters in five geographically defined agencies with an average population density of 6.33 people per square mile – or about 2% of the average population density in the US (Navajo Division of Health and Navajo Epidemiology Center 2013). Due in part to the low population density and in part to other sociopolitical reasons, a variety of mechanisms exist to provision water across the Navajo Nation. The primary water utility is the Navajo Tribal Utility Authority (NTUA). The NTUA operates 90 water systems across eight districts in the Navajo Nation. These water systems collectively serve over 39,000 customers, or about 25% of the population residing in the Navajo Nation (Navajo Division of Health and Navajo Epidemiology Center 2013; Navajo Tribal Utility Authority 2015; Navajo Tribal Utility Authority 2017). In addition to the NTUA, a number of smaller public water systems are operated by public and private schools, coal and gas companies, hospitals, and governmental actors (Navajo EPA 2017). Nevertheless, the United States Environmental Protection Agency (US EPA) estimates that approximately 30% of residents of the Navajo Nation do not have piped water in their homes. These people must either haul water from safe watering points or use unregulated water sources (US EPA 2016a). Ensuring consistent access to safe drinking water for all residents of the Navajo Nation is a critical issue that the Water Diplomacy Framework can help address.

Ensuring consistent and safe access to drinking water in isolated areas of the Navajo Nation is a complex problem. It is transboundary in nature as the groundwater and surface water sources cross between US and sovereign Navajo territory. The transboundary issues are more than geographical; there are US laws and agreements that govern water rights on the Navajo Nation (Brougher 2011; Environmental Justice Atlas 2016; Stern 2017). Additionally, the uncertainties regarding the exposure science and epidemiology and the lack of consensus regarding responsibility for water provisioning on the Navajo Nation require complex decision making processes (Islam and Susskind 2013). The legacy of mistrust between the Navajo people and the US government complicates these processes and suggests that true stakeholder engagement will be necessary, though challenging, to achieve. As part of this stakeholder engagement, the strong interdependent relationship between the Navajo people and their natural resources should be considered (Blessing for the Tribal Environmental Health Summit by a Tribal Elder). Finally, effective steps to mitigate this problem and ensure that 54,000 people across the Navajo Nation have reliable access to safe drinking water will likely require stakeholders to work together to find creative, non–zero–sum options grounded in the principles of equity and sustainability (US EPA 2016a). These aspects of the problem–defining and problem–solving processes can be explored through the Water Diplomacy Framework.

Water provisioning in rural, isolated areas across the Navajo Nation: a complex problem

Legacy effects

Understanding the historic legacy of exploitation is critical to understanding the current water access issues on the Navajo Nation. Centuries of colonization and oppression led to power asymmetries and a lack of trust among stakeholders. But decisions made in the past 150 years also laid the foundation for many of the present opportunities to improve water security. For example, the Navajos' legal claim to adequate water resources was established and protected by a series of treaties and Supreme Court decisions. Water rights across much of the western US function through a prior appropriation system following the idea of "first in time, first in right." Essentially, in times where there is inadequate water to meet all users' demands, those with earlier, or senior, water rights are allocated the quantity of water they need first. Under the Winters Doctrine established through a 1908 US Supreme Court case, when Congress reserves land, it also reserves water resources to "fulfill the purpose of the reservation" (Brougher 2011). Since the Navajo Reservation was established in the 1868 treaty that ended the internment of the Navajo at Fort Sumner (Smithsonian National Museum of the American Indian 2016), the Navajo hold fairly senior water rights. Nevertheless, the fact that the Winters Doctrine did not specify the quantity of water allocated and since the "use it or lose it" principle does not apply to Navajo water rights, there have been numerous legal and practical challenges to their water rights over the past century. Additionally, while water rights are crucial to the well-being of residents of reservations in arid areas, the fact that they were given in part to change the lifestyle of indigenous groups to be more pastoral suggests power asymmetries and a disrespect for different cultures of various indigenous groups (US Supreme Court 1978).

Another part of the legacy of exploitation that is inextricably linked with the current water provisioning discussions is the mining history. The region has substantial natural resources including coal, uranium, oil, and natural gas. The issues regarding uranium mining are of particular importance because of the harm inflicted by external actors on the Navajo and because of ongoing concern about uranium contamination of drinking water. Uranium mining started as early as the 1920s (Chenoweth 1985), and from the 1940s through the 1980s, uranium mining was encouraged by the US government. From 1948 through 1971, the US government was the only buyer of uranium ore mined in the US (Pearson 1980; Brugge and Goble 2002). Although the employment opportunities were initially welcomed by the Navajo people, insufficient attention was paid to worker safety. Miners were not adequately warned about potential health hazards, and working conditions were underregulated. This resulted in harm to the workers, primarily through radiation exposure in the mines, and to the communities (National Research Council 1999; Brugge and Goble 2002; Voyles 2015). The Navajo have led efforts to recover damages from mining companies and from the US Department of Energy. In 1984, an Advisory Committee on Human Radiation Experiments concluded that the health risks of continued mining were not justified on grounds of national security. By 1990, the Radiation Exposure Compensation Act was passed, and in 2000, it was amended to address certain environmental justice concerns left by the legacy of mining on Navajo lands (Brugge and Goble 2002). By 2005, the Navajo Nation Council banned mining on their land, and more recently, the US government has spent more than $100 million on remediation efforts under two Five-Year Plans (2008–2012, 2014–2018) that were developed in collaboration with the Navajo (Brugge and Goble 2002; Kane 2013; US EPA and NNEPA 2014; US EPA 2017). Despite some recent recognition by the US government of their responsibility, hundreds of abandoned uranium mines have yet to be remediated (US EPA 2016b). Not only does this continued exposure lead to psychological and potentially physical harm (Markstrom and Charley 2003), it adversely affects the ability of stakeholders to create sustainable and equitable solutions to water access problems.

Sources of uncertainty

The uncertainty regarding potential for water contamination by uranium is one reason that the continuing presence of abandoned uranium mines affects water provisioning discussions. Uranium is present in some drinking water sources across the Navajo Nation but may be the result of either natural processes or human-related activities. Water quality is monitored within the public water systems as mandated by the Navajo Nation Primary Drinking Water Regulations (Navajo EPA 2015). Each year, the NTUA is responsible for publishing consumer confidence reports that document water quality details. Although each system does not report the same set of water contaminants, some report levels of uranium and arsenic, among other contaminants. In 2015, the quality of water provided by the NTUA was generally adequate, though there were a few reported violations of water quality standards for uranium ($n = 1$), arsenic ($n = 11$), and other contaminants (Navajo Tribal Utility Authority

2015). Moreover, within private water supplies, including livestock wells used for drinking water, there is less consistent monitoring of water quality, and violations may be relatively common. For example, a 2016 study examining the concentrations of uranium and arsenic within a convenience sample of wells used for drinking water found that 5% of wells had uranium concentrations in violation of the US EPA 30-µg/L standard, 22% of wells had arsenic concentrations in violation of the 10-µg/L standard, and 4% of wells had concentrations that exceeded both standards (Corlin et al. 2016). A 2017 study found similar results on a different subset of wells, with 15.1% of wells in exceedance of the arsenic standard, 12.5% in exceedance of the uranium standard, and 3.9% in exceedance of both (Hoover et al. 2017). These studies indicate that individuals may be exposed to uranium in drinking water, though questions about the regularity of exposure remain.

Given the potential for chronic exposures to uranium in drinking water, there is a natural question of whether exposure has health impacts. Despite a strong perception of risk (Panikkar and Brugge 2007; US Government Accountability Office 2014), there remains scientific uncertainty about the health effects of exposure through this pathway. A recent review indicated that uranium in drinking water may be associated with renal and reproductive health effects, along with DNA damage. The evidence is primarily from toxicology studies, however, and the epidemiologic evidence is still limited (Corlin et al. 2016). Nevertheless, some evidence suggests that the current water quality standards for uranium are not sufficiently protective (Raymond–Whish et al. 2007; Brugge and Buchner 2011). We might better understand the health risks of chronic, low-level exposures to uranium in drinking water if studies focused more on individual-level exposure assessment of cumulative exposures within the context of large prospective cohort studies.

Even if additional research on health impacts of uranium exposure would be useful in certain ways, questions remain about whether more research is the best use of resources. Conflicting views exist for how much certainty regarding health consequences is required before there is an ethical or legal burden to minimize exposure to uranium in drinking water (Rothman 1990; Kriebel et al. 2001; Navajo Nation Department of Water Resources 2011; US EPA 2016a). Taking action to minimize uranium exposure could have large co-benefits in reducing individuals' exposures to other known contaminants in drinking water, such as arsenic or bacteria. The Precautionary Principle, which advocates for "taking preventive action in the face of uncertainty" and "increasing public participation in decision making" (Kriebel et al. 2001), may be applied. Nevertheless, to apply this principle, there are questions of who is responsible to whom and for what.

Transboundary questions with multilevel interactions

Shared responsibility among federal, state, and tribal actors increases the complexity of water provisioning on the Navajo Nation. Per the Winters

Doctrine, the US federal government may legally be required to reserve sufficient quantities of water for the purposes of the reservation, but that does not necessarily mean the US federal government must pay for capital investments (Price and Weatherford 1976). Actual water provisioning is in the purview of the Navajo government. The ability and responsibility to do this is a component of tribal sovereignty (Price and Weatherford 1976; Tsosie 2012), yet US federal and state governments are partially responsible for water provisioning on the Navajo Nation based on specific settlements and agreements. For example, to partially compensate for damages caused by uranium mining, the Indian Health Service, the US EPA, and the US Department of Housing and Development jointly distributed $58 million so that people in more than 3,000 homes near abandoned uranium mines could access safe drinking water (US EPA 2016a). Additionally, to settle water access concerns in Utah, a bill was introduced to the US Congress in March of 2017 that, if passed, would protect the Navajos' right to 81,500 annual acre-feet of water and would contribute nearly $200 million towards Navajo water projects (US Senate 2017). While a diverse group of stakeholders seem generally supportive of this bill, it is not without its dissenters. Some who oppose the bill are concerned that the process did not include Navajo grassroots groups and that the bill may not adequately ensure that the Navajo receive the promised amount of water (Minard 2016). These serious concerns have a historical precedent. In 2012, a similar agreement failed after decades of negotiation partially due to the action of grassroots activists. This agreement would have affected water rights to the Little Colorado River in Arizona and would have guaranteed investments in water infrastructure projects (Minard 2012).

Interdependencies

There have been recent positive steps to rebuild trust among actors at different levels and to more explicitly acknowledge the interdependencies that exist. Some negotiations acknowledge the interdependencies of the Navajo way of life and their natural resources. Some also recognize the interdependencies of the Navajo and non-Navajo users of water. These acknowledgments often reference the fact that the Navajo have unquantified senior water rights, so more junior (non-Navajo) water rights holders cannot predict the amount of water they will have available, especially with increasing water scarcity due to climate change and population growth. Other interdependencies are also occasionally acknowledged, as in the recent debates about the future of the Navajo Generating Station. The Navajo Generating Station is one of the largest power-generating facilities in the western US. It is located within the Navajo Nation and is key in powering the movement of water towards cities such as Phoenix and Tucson (Lustgarten 2015). The plant contributes to 22% of the Navajo annual general budget and is a major employer in the region (Hurlbut et al. 2016; Electric Light and Power 2017). While many Navajo families rely on the plant for their livelihoods, the plant is also responsible for extensive environmental degradation (Clay 2014; Leake et al. 2016). Recent discussions have focused on whether to shut down the plant and, in exchange, to continue bringing electricity and water to rural communities (Frisch 2017) [Note: In March 2019, Navajo Nation Council delegates voted to not take over the plant. This will likely result in closure of the plant (Randazzo and Lyn Smith 2019)]. In this example, it seems that relevant interdependencies are at least recognized even if they may not be incorporated into the final decision-making process. In the more general case of provisioning water to rural, isolated residents of the Navajo Nation, however, critical interdependencies are sometimes missed.

In particular, while the Navajo understand and respect the coupling of the human and natural systems, non-Navajo actors may not fully understand the cultural context of these interdependencies. The Navajo have strong cultural and religious ties to the land where their ancestors lived. Through a rich oral tradition, the Navajo pass down an understanding of sacred places, such as four mountains that roughly mark the boundaries of the Navajo Nation and the lake in a mountain where humans are thought to originate (Jett 1992). The Navajo are physically and spiritually linked with their land from birth. Infants' umbilical cords are carefully placed by families in locations that are thought to be beneficial for the child and family's future (Schwarz 2013). The Navajo way of life is also practically tied with the land, as many Navajo's livelihoods depend on the natural resources present. More generally, the idea that humans are interrelated with their environment and thus should live in harmony with nature is integral to Navajo culture (Navajo Nation Council 2002; Kahn-John Diné and Koithan 2015). This is reflected in Navajo beliefs about environmental resource management. In one study designed to understand Navajo perspectives on energy development on the Navajo Nation, everyone interviewed discussed the use of natural resources for cultural purposes, and 90% of those interviewed discussed concerns about environmental protections. Other common concerns related to future generations, sovereignty, and maintaining the rural character of the area (Necefer et al. 2015). Therefore, when potential water provisioning solutions are suggested that do not account for the interrelationships among the Navajo and the environment, it demonstrates a lack of respect and understanding for critical narratives that need to be part of the conversation.

Moving forward: determining the process and values that could lead to mutual-gains solutions

Principled pragmatism

In the previous section, we explored factors that contribute to the complexity of the problem of water provisioning across the Navajo Nation. The next question is whether the Water Diplomacy Framework can also help stakeholders address this complex issue. Since actors with different goals and interests must work together to identify opportunities for creative, mutual-gains solutions, they must first find common ground. Common ground can start from an understanding of the shared values that all parties agree are fundamental to any proposed solution. Starting with values rather than interests, positions, or tools can help build social, political, and intellectual capital. It contributes to consensus building because there is a focus on both process and outcomes (Innes and Booher 1999). Particularly where there is a long history of disrespect and exploitation, an inclusive, collaborative process that builds trust is crucial to ensuring sustainable, equitable, and actionable solutions. This is the idea behind the Water Diplomacy concept of principled pragmatism – essentially, focus on practical real-world constraints, capacity, and context guided by fundamental principles that shape how progress and opportunities are assessed (Islam 2017).

The fundamental principles suggested within the Water Diplomacy Framework are equity and sustainability (Islam and Susskind 2013). Precisely what each term means in practice varies by situation. For example, in considering different aspects of equity, stakeholders might debate the relationship between health equity and economic equity. From a health equity standpoint, consistent access to safe drinking water is essential for life and is a right protected by the United Nations 2010 Resolution 64/292, among other agreements (UNDESA 2014). But even the idea that everyone should have access to sufficient quantities of safe drinking water is an ill-defined statement. There can be arguments regarding the definition of "safe" drinking water or about how much water is "sufficient." Assuming that a sufficient amount of water is available, there are economic equity questions. These questions might address the economically responsible parties or the appropriate response when a responsible party is unable to pay. Additionally, there may also be equity arguments about debts due to a history of exploitation or about cultural understandings of water as sacred versus economic understandings of water as a commodity (Chief et al. 2016). These and many other questions would have to be considered jointly by stakeholders. The goal would be actionable and contingent definitions rather than a single set of meanings for nuanced terms. Practically, it may be helpful to have facilitators who can help establish common language and build trust among stakeholders (Fisher and Ball 2003).

Stakeholder identification and engagement

Initiating the types of participatory processes that can lead to shared understanding and mutually beneficial solutions presents many practical challenges. Even identifying and bringing together all relevant stakeholders can be a fraught endeavor. First, there is a question of who has the power to identify and convene stakeholders. Actors within the US federal government and the Navajo government are logical choices because they hold formal power. As an example of how this formal power is expressed, the Navajo Nation holds primacy for regulating oil- and gas-related injection wells under the Safe Drinking Water Act (US EPA 2015). But the definition of stakeholders should not be limited to only those with formal power. A thorough and inclusive process should identify other stakeholders. Beyond having formal (or informal) power, stakeholders may be defined as those holding urgent or legitimate claims to action or at least claims to participation (Mitchell et al. 1997). For example, the voice of individuals residing in homes without ready access to safe drinking water must be represented. These individuals have urgent and legitimate claims. Understanding their perspectives is useful in coming up with creative, workable solutions because they may have local knowledge that other actors do not. Additionally, it would be worthwhile to consider how ecological factors can or should be represented by surrogate stakeholders (van Rees and Reed 2015). More environmentally equitable solutions may be prioritized if ecological interests are considered from the outset.

Identifying stakeholders is only one part of the process. Meaningful engagement of diverse stakeholders requires deliberate action at every stage from problem identification to joint fact-finding to solution generation to decision making to implementation and evaluation. Each stakeholder should be able to contribute in some way to each part of the process, even if there are varying degrees of engagement at different stages. Certain stages might require a higher level of engagement for true citizen control (Arnstein 1969). To achieve this, each stakeholder must have the power and capacity to actually contribute (Richards et al. 2007). Additionally, stakeholders must trust each other enough to be willing to learn from each other (Hurlbert and Gupta 2015). To build trust and address water security issues for the Navajo Nation, stakeholders should engage in culturally sensitive practices, stakeholders should explicitly recognize that there are different knowledge systems and types of knowledge, community members should determine the goals, and tribes should have at least some oversight of the process (Chief et al. 2016). Trust building is a long-term process that requires investment.

Efforts have already been made to support meaningful stakeholder engagement for issues related to water security on the Navajo Nation. For example, since 2015, a Community Outreach Network representing numerous Navajo and US federal actors has existed to support engagement around uranium contamination. The goals seem to be primarily focused on informing community members and perhaps understanding community members' perspectives with some acknowledgment that community members also should be able to contribute to decision making (US EPA 2016c). Additionally, actors with formal power are investing in long-term capacity-building efforts. Pipeline programs that foster Navajo children's interest in science, technology, engineering, and mathematics, as well as their interest in public policy and community development, are helping create the next generation of leaders. Additionally, the University of Arizona won a $3 million National Science Foundation award to support undergraduate training on ways to improve food, energy, and water security in the Navajo Nation (University of Arizona 2017).

In addition to these types of capacity-building programs that can support meaningful stakeholder engagement, academic partners can also help facilitate engagement by bringing together stakeholders and conducting community-engaged research. Conferences are one way to bring stakeholders together to build both trust and technical capacity. For example, there have been three Tribal Environmental Health Summits held (Tribal Environmental Health Summit 2018, 2017). These have been useful opportunities for academic researchers, government representatives, other professionals dedicated to tribal environmental health goals, and members of various First Nations to come together to share cutting-edge research, build relationships, and generate solutions. At the 2016 conference hosted by Northern Arizona University, the meeting was opened with a blessing from a Navajo elder. Beyond conferences, academics can act as facilitators for stakeholders to generate shared understanding of complex issues. For example, the Water Resources Research Center at the University of Arizona recently brought together stakeholders for a social learning project designed to understand different perspectives and shared values of water interest groups. Through a series of focus groups, workshops, and interviews, the steering committee created a report, the *Roadmap for Considering Water for Arizona's Natural Areas* (Lacroix et al. 2016). One reason this project was successful was that it followed the core principles of community-engaged research and joint fact-finding.

Academics, as well as others acting in the role of facilitator, have a responsibility to consider meaningful engagement of affected community members in their work. In research with First Nations, part of this is ensuring that the sovereignty of the nations is respected and that appropriate tribal research offices and tribal elected officials approve all research-related activities (Chief et al. 2016). But another part is carefully considering the role of community members throughout the research process. Community members should have input into the research questions asked, methods used, interpretation of the results, and dissemination and use of the findings. This community-engaged research approach takes advantage of local insight and lived knowledge of community members, which improves the quality and relevance of the science (Wallerstein and Duran 2010). For example, it would be extremely challenging to identify unregulated sources of drinking water without community members pointing these sources out. Community members' insight is also critical in understanding narratives of water use in the region. In return, community members deserve to know the results of water quality testing and to have some say in how the results might be used to inform policy. Critically, research should only be done when communities can directly benefit from the research and when the research does not divert resources from investments or programs that would have addressed the problem.

Identifying creative and contingent solutions

As suggested by the Water Diplomacy Framework, moving beyond trust building and fact finding into solution generation requires an accounting of both the numbers and the narratives generated in the earlier phases of the process (Islam and Susskind 2013). Together, the numbers and narratives provide relevant constraints to the solution space. For example, while increasing population density may make water provisioning simpler in some contexts, it is not a reasonable strategy in this situation, because residents of rural, isolated areas are typically tied to their land by cultural, spiritual, and practical considerations. More creative solutions are necessary. Identifying mutually beneficial creative solutions and ensuring that these solutions are adaptable as the needs of stakeholders change over time are critical challenges.

The process of identifying mutually beneficial solutions must be inclusive. Top-down approaches to allocating funding for infrastructure projects in return for water rights specifications have not worked well in the past due to concerns about transparency and accountability. A recent example of this was in the failed Navajo-Hopi Little Colorado River Water Rights Settlement Act of 2012. Arizona Senator Jon Kyl introduced a bill to enable the Navajo and Hopi to end a lengthy adjudication process regarding water rights. While the goals were well intentioned, the process was problematic. Community members were largely excluded from the development and discussion of this proposal. For example, when Kyl and fellow Arizona Senator John McCain held meetings about the proposed settlement on the Navajo Nation, community members were excluded. This generated a strong backlash among both Hopi and Navajo grassroots activists. This backlash, along with a lack of trust that the US government would follow through on promised water infrastructure projects, resulted in the Navajo Tribal Council voting 15–6 against the settlement (Associated Press 2012; Krol 2017). Future efforts to resolve this issue, as well as efforts to address water provisioning more generally in rural, isolated parts of the Navajo Nation, will need to carefully consider the role of grassroots activists.

Different methods are possible to provision water to rural, isolated communities. One is to maintain and strengthen the status quo. Perhaps devoting more resources to water-hauling efforts – for example, increasing the number of individuals who routinely drive water trucks to isolated homes on the Navajo Nation – could work in the near term. Strategic well drilling could help make these efforts more efficient (Tory 2015). Other near-term measures might include better investments in water quality testing and point-of-use water filtration systems that are appropriate to the Navajo Nation (Dilks et al. 2014). While these and other similar near-term solutions may be necessary as part of the solution, they do not necessarily fulfill the equity and sustainability criteria. Generating and implementing creative, longer-term solutions is critical. Sustainable solutions will likely rely at least in part on legislative actions and other large-scale investments. This is the case, for example, with the Navajo-Gallup Water Supply Project.

The Navajo-Gallup Water Supply Project represents one of the most important recent advances in the efforts to ensure water security on the Navajo Nation. The negotiation and planning process for this agreement took four decades and resulted in a project designed to deliver nearly 38,000 acre-feet of water annually from the San Juan Basin to 250,000 people by 2040. The Navajo–Gallup Water Supply Project was authorized as part of the Navajo Nation San Juan River Basin Water Rights Settlement in New Mexico (US Bureau of Reclamation 2017). It was made possible through appropriations under the Omnibus Public Land Management Act of 2009 (P.L. 111–11) and the Claims Resolution Act of 2010 (P.L. 111–291; Stern 2017). The project is currently underway, and a nearly $62 million contract was awarded in September 2017 from the Bureau of Reclamation to Oscar Renda Contracting Inc. (Duke 2017). The fact that this settlement was finalized and that promised projects are being realized is critical to ongoing relationship-building efforts among stakeholders.

By choosing to pursue a negotiated settlement instead of litigation, stakeholders involved in the Navajo Nation San Juan River Basin Water Rights Settlement had more flexibility to find solutions with tangible benefits for everyone. One advantage of settlements is that they can involve the appropriation of funds and authorization of water infrastructure projects, whereas litigation cannot, because courts can only specify rights to water on paper. Settlements also have the advantage of allowing more nuance and adaptation planning. For example, rather than limiting the potential uses for water, some settlements allow for water leasing and marketing (Stern 2017). Water leasing and other similar mechanisms allow water rights holders to respond to changes in water demand over time (Crammond 1996). Non–zero–sum problem solving adds value to agreements because all stakeholders feel their interests are represented. In the Indian water rights settlements, some stakeholders found increased water security from the quantification of water rights or from additional investment in water infrastructure. Nevertheless, as settlements are negotiated, some stakeholders have expressed concerns about agreeing to promises of investments that they feel are unlikely to be realized, investing in water infrastructure projects that could have adverse ecological consequences, or limiting water rights that had not previously been limited, especially when water use is expected to increase due to climate change and population growth (Stern 2017).

Conclusion

Moving forward, stakeholders will have to jointly decide if settlements such as the one leading to the Navajo–Gallup Water Supply Project could generate mutually beneficial opportunities that ensure access to safe drinking water for rural, isolated residents of the Navajo Nation. The Water Diplomacy Framework could offer guidance for how to successfully reach an equitable and sustainable agreement. First, the Water Diplomacy Framework could help stakeholders actionably define the problem, accounting for legacy effects, transboundary interactions, varying sources of uncertainty, and interdependencies among human and natural systems. Second, the Water Diplomacy Framework suggests methods that could build trust and social capital and could help redistribute power among actors with a history of asymmetric, exploitative relationships. Finally, by reframing the problem-solving approach to be more inclusive and to focus on non–zero-sum negotiations, the Water Diplomacy Framework could facilitate the development of contingent solutions that meet the current needs of all stakeholders while offering opportunities to adapt to changing conditions over time.

Key question

Addressing water access in rural, isolated parts of the Navajo Nation requires a consideration of how the asymmetry of power among stakeholders influences negotiations and how the negative effects of this power asymmetry can be mitigated. Historically, the exploitation of power asymmetries led to a lack of trust among stakeholders that complicates current negotiations. Furthermore, stakeholders who have not traditionally held formal power have asserted their claims to be part of the process. When these stakeholders were excluded, the negotiations failed. This happened in the Navajo-Hopi Little Colorado River Water Rights Settlement Act of 2012. In contrast, negotiations succeeded in producing agreements when stakeholders were engaged meaningfully throughout the process. This happened with the Navajo Nation San Juan River Basin Water Rights Settlement. Engaging community members and others with local knowledge but without formal power in processes to identify and address complex water security problems increases the likelihood that the process will be equitable and will produce sustainable, mutually beneficial solutions.

Acknowledgments and disclaimer

The author is not Diné. The views expressed in this chapter do not necessarily reflect those held by the Navajo and are not meant to represent an official position. The chapter benefited from the feedback of Tommy Rock, Esther Erdei, Laura Read, and Tahira Syed. Funding was provided by the National Science Foundation (0966093), NHLBI (T32-HL-125232), a P.E.O. Esther Garrett Edgerton Endowed Scholar Award, and the Tufts University Office of the Vice Provost for Research.

References

Arnstein, S.R., 1969. A ladder of citizen participation. *Journal of the American Institute of Planners*, 35 (4), 216–224.

Associated Press, 2012. Navajo lawmakers reject water rights settlement. *The Denver Post*. Available from: www.denverpost.com/2012/07/05/navajo-lawmakers-reject-water-rights-settlement

Brougher, C., 2011. *Indian Reserved Water Rights Under the Winters Doctrine: An Overview*. Congressional Research Service, No. Congressional Research Service 7–5700.

Brugge, D., and Buchner, V., 2011. Health effects of uranium: New research findings. *Reviews on Environmental Health*, 26 (4), 231–249.

Brugge, D., and Goble, R., 2002. The history of uranium mining and the Navajo people. *American Journal of Public Health*, 92 (9), 1410–1419.

Chenoweth, W.L., 1985. *Historical Review Uranium-Vanadium Production in the Northern and Western Carrizo Mountains, Apache County, Arizona*. Arizona Geological Survey Open File Reports. Available from: http://repository.azgs.az.gov/uri_gin/azgs/dlio/457

Chief, K., Meadow, A., and Whyte, K., 2016. Engaging southwestern tribes in sustainable water resources topics and management. *Water*, 8 (8), 350.

Clay, R.F., 2014. Tribe at a crossroads: The Navajo nation purchases a coal mine. *Environmental Health Perspectives*, 122 (4), A104–A107.

Corlin, L., Rock, T., Cordova, J., Woodin, M., Durant, J.L., Gute, D.M., Ingram, J., and Brugge, D., 2016. Health effects and environmental justice concerns of exposure to uranium in drinking water. *Current Environmental Health Reports*, 3 (4), 434–442.

Crammond, J.D., 1996. Leasing water rights for instream flow uses: A survey of water transfer policy, practices, and problems in the Pacific Northwest. *Environmental Law*, 26, 225–264.

Dilks, C., Cummings, D., Weir, T., Sun, Y., and Gremillion, P., 2014. *Final Design Report: Water Filter for Uranium, Arsenic, and Bacteria Removal*. Sublime Engineering.

Duke, M., 2017. *Interior Announces $62 Million Construction Contract on Navajo-Gallup Water Supply Project*. Available from: www.doi.gov/pressreleases/interior-announces-62-million-construction-contract-navajo-gallup-water-supply-project

Electric Light and Power, 2017. *Navajo Generating Station Reaches Agreement to Keep Operating*. Available from: www.elp.com/articles/2017/10/navajo-generating-station-reaches-agreement-to-keep-operating.html

Environmental Justice Atlas, 2016. *Water Rights of the Dineh-Navajo Tribe, USA*. Environmental Justice Atlas. Available from: http://ejatlas.org/conflict/water-rights-of-the-dineh-navajo-tribe-usa

Fisher, P.A., and Ball, T.J., 2003. Tribal participatory research: Mechanisms of a collaborative model. *American Journal of Community Psychology*, 32 (3–4), 207–216.

Frisch, I., 2017. The end of coal will haunt the Navajo. *Bloomberg.com*. Available from: www.bloomberg.com/news/features/2017-10-13/the-end-of-coal-will-haunt-the-navajo

Hoover, J., Gonzales, M., Shuey, C., Barney, Y., and Lewis, J., 2017. Elevated arsenic and uranium concentrations in unregulated water sources on the Navajo Nation, USA. *Exposure and Health*, 9 (2), 113–124.

Hurlbert, M., and Gupta, J., 2015. The split ladder of participation: A diagnostic, strategic, and evaluation tool to assess when participation is necessary. *Environmental Science & Policy*, 50, 100–113.

Hurlbut, D., Haase, S., Barrows, C., Bird, L., Brinkman, G., Cook, J., Day, M., Diakov, V., Hale, E., Keyser, D., Lopez, A., Mai, T., McLaren, J., Reiter, E., Stoll, B., Tian, T., Cutler, H., Bain, D., and Acker, T., 2016. *Navajo Generating Station and Federal Resource Planning; Volume 1: Sectoral, Technical, and Economic Trends*. National Renewable Energy Lab. (NREL), Golden, CO (United States), No. NREL/TP-6A20–66506.

Innes, J.E., and Booher, D.E., 1999. Consensus building and complex adaptive systems. *Journal of the American Planning Association*, 65 (4), 412–423.

Islam, 2017. Is principled pragmatism a viable framework for addressing complex problems? *Water Diplomacy Network Blog*. Available from: http://blog.waterdiplomacy.org/2017/03/is-principled-pragmatism-a-viable-framework-for-addressing-complex-problems

Islam, S., and Susskind, L.E., 2013. *Water Diplomacy: A Negotiated Approach to Managing Complex Water Networks*. New York: RFF Press.

Jett, S.C., 1992. An introduction to Navajo sacred places. *Journal of Cultural Geography*, 13 (1), 29–39.

Kahn-John Diné, M., and Koithan, M., 2015. Living in health, harmony, and beauty: The Diné (Navajo) hózhó wellness philosophy. *Global Advances in Health and Medicine*, 4 (3), 24–30.

Kane, J., 2013. Uranium mining companies descend upon Navajo Nation. *Farmington Daily Times*. Available from: www.daily-times.com/ci_22702691/uranium-mining-companies-descend-upon-navajo-nation

Kriebel, D., Tickner, J., Epstein, P., Lemons, J., Levins, R., Loechler, E.L., Quinn, M., Rudel, R., Schettler, T., and Stoto, M., 2001. The precautionary principle in environmental science. *Environmental Health Perspectives*, 109 (9), 871–876.

Krol, D.U., 2017. Water settlement for Navajo and Hopi tribes inches forward. *Water Deeply*. Available from: www.newsdeeply.com/water/articles/2017/06/15/water-settlement-for-navajo-and-hopi-tribes-inches-forward

Lacroix, K.E.M., Xiu, B.C., and Megdal, S.B., 2016. Building common ground for environmental flows using traditional techniques and novel engagement approaches. *Environmental Management*, 57 (4), 912–928.

Leake, S.A., Macy, J.P., and Truini, M., 2016. *Hydrologic Analyses in Support of the Navajo Generating Station: Kayenta Mine Complex Environmental Impact Statement.* Reston: U.S. Geological Survey, USGS Numbered Series No. 2016–1088.

Lustgarten, A., 2015. Navajo generating station powers and paralyzes the Western U.S. *Scientific American*. Available from: www.scientificamerican.com/article/navajo-generating-station-powers-and-paralyzes-the-western-u-s

Markstrom, C.A., and Charley, P.H., 2003. Psychological effects of technological/human-caused environmental disasters: Examination of the Navajo and uranium. *American Indian and Alaska Native Mental Health Research*, 11 (1), 19–45.

Minard, A., 2012. Navajo Hopi water rights debate continues as Navajo council denies agreement; Hopi approves. *Indian Country Media Network*. Available from: https://newsmaven.io/indiancountrytoday/archive/navajo-hopi-water-rights-debate-continues-as-navajo-council-denies-agreement-koCcx32fA0uwVsOfieFnjw/

Minard, A., 2016. Let the water flow! Navajo in Utah closer to water rights settlement. *Indian Country Media Network*. Available from: https://indiancountrymedianetwork.com/news/politics/let-the-water-flow-navajo-in-utah-closer-to-water-rights-settlement

Mitchell, R.K., Agle, B.R., and Wood, D.J., 1997. Toward a theory of stakeholder identification and salience: Defining the principle of who and what really counts. *Academy of Management Review*, 22 (4), 853–886.

National Research Council, 1999. *Health Effects of Exposure to Radon: BEIR VI.* Washington, DC: National Academies Press (US).

Navajo Division of Health and Navajo Epidemiology Center, 2013. *Navajo Population Profile 2010 US Census.*

Navajo EPA, 2015. *Navajo Nation Safe Drinking Water Act (NNSDWA).* Available from: http://navajopublicwater.org/nnsdwa.html

Navajo EPA, 2017. *Public Water Systems Supervision Program.* Available from: http://navajo-publicwater.org/PWS.html

Navajo Nation Council, 2002. *Council Resolution CN-69-02.* Available from: www.navajo-courts.org/dine.htm

Navajo Nation Department of Water Resources, 2011. *Water Resource Development Strategy for the Navajo Nation.* Available from: https://data.globalchange.gov/report/navajonatdwr-draftwaterresource-2011

Navajo Tribal Utility Authority, 2015. *2015 Consumer Confidence Report: It's About Your Water.*

Navajo Tribal Utility Authority, 2017. *Navajo Tribal Utility Authority – About Us*. Available from: www.ntua.com/aboutus.html

Necefer, L., Wong-Parodi, G., Jaramillo, P., and Small, M.J., 2015. Energy development and native Americans: Values and beliefs about energy from the Navajo nation. *Energy Research & Social Science*, 7 (Supplement C), 1–11.

Panikkar, B., and Brugge, D., 2007. The ethical issues in uranium mining research in the Navajo nation. *Accountability in Research*, 14 (2), 121–153.

Pearson, J., 1980. Hazard visibility and occupational health problem solving the case of the uranium industry. *Journal of Community Health*, 6 (2), 136–147.

Price, M.E., and Weatherford, G.D., 1976. Indian water rights in theory and practice: Navajo experience in the Colorado river basin. *Law and Contemporary Problems*, 40 (1), 97–131.

Randazzo, R., and Lyn Smith, N., 2019. Navajo nation ends bid to buy Navajo generating station coal power plant. *AZcentral*. Available from: www.azcentral.com/story/money/business/energy/2019/03/22/navajo-nation-ends-bid-buy-navajo-generating-station-coal-power-plant/3246913002/

Raymond-Whish, S., Mayer, L.P., O'Neal, T., Martinez, A., Sellers, M.A., Christian, P.J., Marion, S.L., Begay, C., Propper, C.R., Hoyer, P.B., and Dyer, C.A., 2007. Drinking water with uranium below the U.S. EPA water standard causes estrogen receptor – Dependent responses in female mice. *Environmental Health Perspectives*, 115 (12), 1711–1716.

Richards, C., Blackstock, K., and Carter, C., 2007. *Practical Approaches to Participation*. Socio-economic Research Group, No. SERG Policy Brief 1.

Rothman, K.J., 1990. A sobering start for the Cluster Busters' conference. *American Journal of Epidemiology*, 132 (Supplement 1), 6–13.

Schwarz, M.T., 2013. *Navajo Lifeways: Contemporary Issues, Ancient Knowledge*. 1st ed. Norman: University of Oklahoma Press.

Smithsonian National Museum of the American Indian, 2016. *Nation to Nation*. Available from: http://nmai.si.edu/nationtonation/navajo-treaty.html

Stern, C., 2017. *Indian Water Rights Settlements*. Congressional Research Service, No. R44148.

Tory, 2015. The woman who brings drinking water to remote Navajo homes. *High Country News*. Available from: www.hcn.org/issues/47.5/the-woman-who-brings-drinking-water-to-remote-navajo-homes

Tribal Environmental Health Summit 2018, 2017. *Tribal Environmental Health Summit 2018*. Available from: http://blogs.oregonstate.edu/tehs2018

Tsosie, H., 2012. Navajo water settlement is an exercise of sovereignty. *Indian Country Media Network*.

UNDESA, 2014. *International Decade for Action 'Water for Life' 2005–2015. Focus Areas: The Human Right to Water and Sanitation*. Available from: www.un.org/waterforlifedecade/human_right_to_water.shtml

University of Arizona, 2017. *UA Leads STEM Traineeship to Address Needs of Navajo Nation*. Available from: https://research.arizona.edu/stories/ua-leads-stem-traineeship-address-needs-navajo-nation

US Bureau of Reclamation, 2017. *Navajo-Gallup Water Supply Project*. Available from: https://web.archive.org/web/20170928210557/www.usbr.gov/uc/rm/navajo/nav-gallup

US EPA, 2015. *Final Rulemaking for the Navajo Nation Class II Underground Injection Control Program*. Available from: www.epa.gov/uic/final-rulemaking-navajo-nation-class-ii-underground-injection-control-program

US EPA, 2016a. *Providing Safe Drinking Water in Areas with Abandoned Uranium Mines*. Available from: www.epa.gov/navajo-nation-uranium-cleanup/providing-safe-drinking-water-areas-abandoned-uranium-mines

US EPA, 2016b. *Cleaning Up Abandoned Uranium Mines*. Available from: www.epa.gov/navajo-nation-uranium-cleanup/cleaning-abandoned-uranium-mines

US EPA, 2016c. *Five-Year Plan to Address Impacts of Uranium Contamination*. Available from: www.epa.gov/navajo-nation-uranium-cleanup/five-year-plan-address-impacts-uranium-contamination

US EPA, 2017. *Northern Trust: Legal Document and Settlements*. Available from: www.epa.gov/navajo-nation-uranium-cleanup/northern-trust-legal-document-and-settlements

US EPA and NNEPA, 2014. *Federal Actions to Address Impacts of Uranium Contamination in the Navajo Nation*. Available from: www3.epa.gov/region09/superfund/navajo-nation/pdf/nn-five-year-plan-2014.pdf

US Government Accountability Office, 2014. *Uranium Contamination: Overall Scope, Time Frame, and Cost Information Is Needed for Contamination Cleanup on the Navajo Reservation*. No. GAO-14–323.

US Senate, 2017. *115th Congress. 1st Session. S.664, Navajo Utah Water Rights Settlement Act of 2017. As Introduced in Senate*. Available from: www.congress.gov/bill/115th-congress/senate-bill/664/text

US Supreme Court, 1978. *United States v. New Mexico, 438 U.S. 696.*

van Rees, C., and Reed, J.M., 2015. Water diplomacy from a Duck's perspective: Wildlife as stakeholders in water management. *Journal of Contemporary Water Research & Education*, 155 (1), 28–42.

Voyles, T.B., 2015. *Wastelanding: Legacies of Uranium Mining in Navajo Country*. Minneapolis: University of Minnesota Press.

Wallerstein, N., and Duran, B., 2010. Community-based participatory research contributions to intervention research: The intersection of science and practice to improve health equity. *American Journal of Public Health*, 100 (S1), S40–S46.

12 Coupling and complexity of natural and human systems

A case study from the southwest Bangladesh delta

Wahid Palash, Kevin M. Smith, and Shafiqul Islam

Introduction

The natural resource and disaster manager are increasingly recognizing the complexity of coupled natural and human (CNH) systems. When systems incorporate both human and natural elements, they often become complex as interconnectedness and interdependence between elements dominate the system dynamics (Maguire et al. 2006; Ramalingam et al. 2008; Jeffrey 2011; Allen 2014; Cooper and Islam 2015). The individual elements of uncoupled systems can be studied in isolation in order to predict system-level behavior. However, this is not possible for complex systems, whose high degree of interconnectedness and interactions leads to emergent system-level behaviors that cannot be predicted from their constituent parts (Holland 2002; Fromm 2005; Palash et al. 2018). Simultaneously, in a complex system, micro-level dynamics often influence the response at the macro level (Lorenz 1963; Li 2012; Binder et al. 2013). Yet it is almost impossible to establish a direct cause-and-effect relationship between those micro-level actions and macro-level responses. Currently, there is no generally accepted framework available to deal with complex systems in a coherent way (Lloyd 2001; Bonabeau 2008; Cilliers et al. 2013). As a result, a predictable understanding of a complex system is difficult to achieve.

Many attempt to understand and manage a complex system by employing a mechanistic model assuming that they know the cause–effect relationship of the system based on the behavior of its individual parts. The key insight of contemporary complexity science, however, suggests that cause–effect relationships are often not known or not known to a level that we can model and predict. How do we, then, manage such a complex system? Identifying requisite simplicity for such a system may help us (Stirzaker et al. 2010; Palash et al. 2018). In many cases, a given complex system – from a given problem perspective and context – will have a requisite level of simplicity captured by a subset of elements, connections, and processes. Those connections and processes are important to identify and understand. Therefore, although a generalized solution for a complex system is difficult to achieve, identifying simplified system descriptions may lead to actionable outcomes and effective management for a chosen problem (Holling 2001).

This chapter discusses several key attributes that distinguish complex systems from those that are merely complicated (for a detailed discussion about simple, complicated, and complex systems, please see Chapter 2 of this book). It then presents a distinction between system complexity (i.e., where these attributes are intrinsic to the system) and modeling (i.e., where these attributes are artifacts of trying to model a system). To provide an example of the challenges posed by CNH systems, this chapter provides an example by analyzing water problems in Bangladesh's southwest delta region through the lens of complexity science. Important interconnections and interactions between and among the elements and subsystems of hydrology, water resource policy and management, agricultural practices, fisheries management, and public health within the region are considered in the discussion. External drivers such as disputes over water sharing in transboundary rivers and climate change are also discussed. In each case, evidence of nonlinear relationships, feedback, and emergence in southwest Bangladesh's CNH system is presented with examples.

Complex system and complexity science

Simple, complicated, and complex systems

Systems can be categorized as either simple, complicated, or complex. While simple and complicated systems can be intricate, their behavior can be ascertained from the properties of their constituent parts. Such an assessment is not possible for systems which exhibit "system complexity." After Cilliers (1998) and Cilliers et al. (2013), we define system complexity by particular characteristics of a system that arise because of interconnections, interactions, and nonlinear relationships among system elements. System complexity is an emergent property of the entire system rather than the individual elements. These peculiar properties emerge at the system level because complex systems cross multiple domains (i.e., natural, societal, political) and scales (i.e., space, time, jurisdictional, institutional; Islam and Repella 2015). Therefore, it is nearly impossible to predict the behavior of a complex system by reducing it to its constituent parts and studying them in isolation. This inadequacy of reductionist approaches to describe system behavior is at the heart of our distinction between complex systems and those that are merely simple or complicated. Figure 12.1 presents a comparison between aspects of simple, complicated, and complex systems adapted from the Cynefin framework proposed by Snowden and Boone (2007).

Fundamental features of a complex system

We discuss fundamental features of a complex system by using three major groupings of attributes: interconnection, interaction and interdependency; nonlinearity, feedback, and causal uncertainty; and emergence. This general

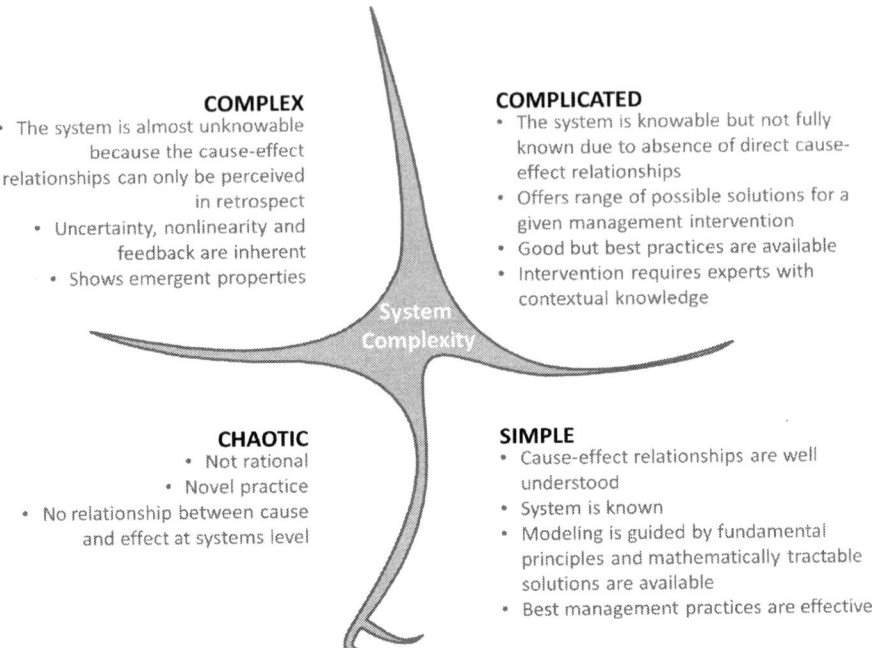

COMPLEX
- The system is almost unknowable because the cause-effect relationships can only be perceived in retrospect
- Uncertainty, nonlinearity and feedback are inherent
- Shows emergent properties

COMPLICATED
- The system is knowable but not fully known due to absence of direct cause-effect relationships
- Offers range of possible solutions for a given management intervention
- Good but best practices are available
- Intervention requires experts with contextual knowledge

CHAOTIC
- Not rational
- Novel practice
- No relationship between cause and effect at systems level

SIMPLE
- Cause-effect relationships are well understood
- System is known
- Modeling is guided by fundamental principles and mathematically tractable solutions are available
- Best management practices are effective

System Complexity

Figure 12.1 Cynefin framework showing distinction between simple, complicated, and complex systems (adapted from Snowden and Boone 2007; Islam and Repella 2015)

categorization of system properties can be used to identify characteristics of both nature–nature and nature–human coupled systems that exhibit system complexity.

Interconnection, interaction, and interdependency

Many coupled systems – be it nature–nature (e.g., hydrometeorology in a river basin and floods in an undeveloped floodplain; interactions between wetland or forest ecology, maritime ecology, etc.) or nature–human (e.g., when a flood reaches a community, village, or city; development activities in a wetland or mining in a forest; maritime fishing, etc.) – exhibit a high degree of interconnection, interaction, and interdependency among system elements. While these attributes may also be present in complicated systems, in complex systems, they are ill described by natural laws. For example, interactions between natural systems and human agents are seldom predictable in principle and contribute to the complexity observed in many CNH systems (Liu et al. 2007; Milner-Gulland 2012; Walsh and Mena 2016). The degree of interconnection, especially across hierarchical levels of the organization, often determines the impact of these interactions on system-level behavior. Likewise, the impact of interdependencies between elements and subsystems is a direct result of inter-connections. The degree and nature of these interdependencies define the system's response and adaptability to external changes, in particular whether these changes cascade throughout the entire system or if the impact is localized (Borge-Holthoefer et al. 2013; Islam and Susskind 2018). While complex systems can often exhibit a high degree of adaptability, their interdependencies also leave them vulnerable to sudden and catastrophic collapse.

Nonlinearity, feedback, and causal uncertainty

Nonlinearity, feedback, and causal uncertainty are the second set of properties we use to identify a complex system. The nonlinearity common to complex systems occurs when a change in one element of the system results in a non-proportional change in another (Snowden and Boone 2007). Coupled with the interconnections, interactions, and interdependences that exist in complex systems, the overall impact of a change in one system element can be difficult to understand, model, and predict (Palash et al. 2018). In other words, cause–effect relationships of a complex system are simply not known to a level that we want to represent in a model. As a result, nonlinear interactions between elements acts as a major source of uncertainty for understanding a complex system (Maguire et al. 2006; Ramalingam et al. 2008). These non-linear interactions may appear random or chaotic at first look. However, within that apparently random and chaotic space, there may be an underlying pattern that is important to identify and understand. However, it often remains elusive (Maguire et al. 2006).

Feedback occurs when interconnections create a looped response between input and output of system elements. Feedback loops, taken as a whole, can be either positive or negative. In a positive feedback loop, any change in one element of a system can lead to disproportionately large and continuous change in other parts of the system. On the other hand, a negative feedback loop counteracts and self-regulates changes. Feedback loops occur in simple and complicated systems as well, but they often exhibit predictable linear responses. In complex systems, the feedback loops are often nonlinear and can even appear random, leading to an overall lack of predictability.

Taken together, nonlinearity and feedback loops make a complex system extremely sensitive to initial conditions and very difficult to predict. This sensitivity to initial conditions was aptly described by Lorenz (1963) in his famous "butterfly effect" metaphor: a small change in the initial condition may result in a large-scale effect across the system. This sensitivity creates thresholds and tipping points (Hossain and Dearing 2016; Hossain et al. 2016) beyond which the system may move to an irrecoverable state. From an operational point of view, thresholds and tipping points along with "safe and just operating spaces" (beyond which the risk of unpredictability becomes very high) are critical in managing complex systems (Dearing et al. 2014). In the absence of causal cer-tainty, we may not be able to predict the responses of many CNH systems.

Emergence

When all of the attributes previously considered are active in a system, the result is often unpredictable system–level behaviors, which are commonly called emergence (Ramalingam et al. 2008). Not all system behaviors or responses are emergent in nature. Some of the behaviors may be intricate yet predictable; emergent behaviors, on the other hand, are fundamentally unpredictable on the basis of integration of well–understood system elements (Ramalingam et al. 2008). In other words, we cannot anticipate emergence based on local–level interactions and rules. As such, emergence is the property of the entire system, not the individual elements, making it an important attribute that distinguishes a complex system from a complicated or simple system. Reducing the system into its elements disrupts and obscures the system's emergent properties (Levin 1998, 1999; Snowden and Stanbridge 2004; Stirzaker et al. 2010; Roux and Foxcroft 2011; Cilliers et al. 2013; Islam and Repella 2015; Turner et al. 2016; Smith and Islam in Chapter 2 of this book).

Emergence and coupled natural and human systems

When natural systems become coupled with human systems, they often exhibit emergence. Let us explain this by making a distinction between the natural process of a "flood" and the CNH process of "flooding." Here, we use the word "flood" to describe a natural process involving a large number of variables and processes that span hydrometeorology, hydrology, river morphology, and hydraulics. We use the word "flooding," on the other hand, to describe a consequence of a CNH system involving human–social interventions, interconnections, and interactions that influence and are influenced by floods (i.e., the natural process).

Two important conditions dictate a flood's space–time scale and magnitude as a natural process: a catchment condition that controls the state of the catchment (e.g., soil, land use, and vegetation, etc.) and an atmospheric condition that provides the key forcing mechanism (e.g., precipitation; Palash et al. 2018). The social and human dimensions of flooding become important when streamflow exceeds the river's carrying capacity, overtops its bank, erodes dikes, spreads across the floodplain, and inundates communities, villages, or cities. Because of interactions with human settlements and involvement of human and social agents, a natural process of a flood is transformed into a CNH process, ultimately giving rise to a CNH system. As a result, coupled interactions between natural and human systems leads to an emergence of certain system behavior (flooding) that was otherwise absent in the natural process (flood).

Distinguishing between system complexity from modeling complexity

Developing a flood forecasting model for a natural process like a flood (e.g., modeling rainfall–runoff processes in a river basin) and is an example of modeling complexity. Developing a model for a CNH system, say to predict flooding, has both system complexity and modeling complexity. A key distinction between system complexity and modeling complexity is whether key observable complex attributes (e.g., emergence) are intrinsic to the system or are an artifact of the modeling process itself. Modeling systems using numerical tools may give rise to apparent emergent phenomena when observing results that are unpredictable, unanticipated, or not readily explainable. However, this "complexity," if not present in the underlying system, is merely an artifact arising from the limitations of available data, physical knowledge, modeling tools, and the precision of digital computers. Consider that for even a very well-known physical system – for instance, the rainfall–runoff in a river basin where we know nearly all the variables and processes – it is still very difficult to generate reliable flood forecasts beyond a three- to five-day lead-time. Now consider developing a distributed hydrological model with 1 km spatial resolution for a river basin having various types of soil, land use, or vegetation cover. Even if we know the infiltration equation perfectly for each combination of soil, land use, and vegetation types, the simulated infiltration rate for each of these soil and land types will be very different than the real value. This happens because we may not have soil or land use data at the scale of our model's spatial resolution; we may lack precipitation data at the scale of our model's temporal resolution; and we may not know the values of parameters that we need to fit for realistically simulating infiltration. In effect, we cannot develop a model for a river basin's rainfall–runoff process at the level and detail that we want to represent or reproduce. Our inability to develop such a detailed model, either due to lack of data or knowledge of the physical process, is what we refer to as modeling complexity.

Given the discussion above, it is important for us to make a distinction between modeling complexity and system complexity. We need to identify whether a problem is associated with a simple, complicated, or complex system. Islam and Repella (2015) and Smith and Islam (in Chapter 2 of this book) argued that the inability to make such a distinction could lead us to apply the wrong approach to the right problem or the right approach to the wrong problem. Once we can make these distinctions – attributes of modeling complexity and system complexity – clearer, we will be able to characterize and manage CNH systems with actionable and trackable outcomes.

The southwest delta of Bangladesh – an example of a coupled natural–human complex system

Here we present an example of a CNH system by describing the interconnectedness and interactions primarily between the water system and systems surrounding it including relevant major human activities in the southwest delta region of Bangladesh (Figure 12.2). We consider both local drivers (e.g., water and agriculture policies and practices, infrastructure development and urbanization, etc.) and external drivers (e.g., upstream water diversion, climate change, etc.) in this case example. The hydrology of the lower

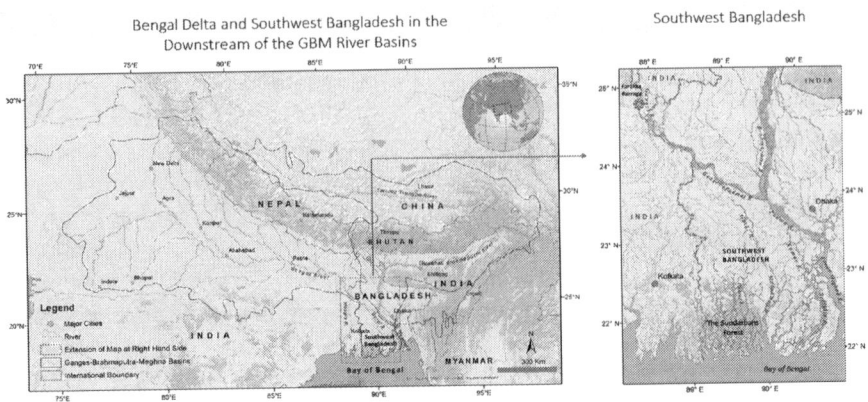

Figure 12.2 Southwest delta of Bangladesh.

Ganges, Brahmaputra, and Meghna Rivers' deltaic river system (i.e., the Bengal delta), along with the Bay of Bengal coast, and human settlement over the last several thousand years have led to the formation of complex CNH systems in this part of the world.

A conceptual and looped networked diagram of interconnections and interactions of various elements of water issues in southwest Bangladesh is shown in Figure 12.3. Such a conceptualization helps to identify key connections and feedback of southwest Bangladesh's water. Identifying key connections and nodes, nonlinearity, tipping points and thresholds, as well as feedback and emergence in a CNH system like this one is essential to understand how human and social systems will respond to changes and interventions (Liu et al. 2007; Milner-Gulland 2012). Past studies like Szabo et al. (2015), Hossain et al. (2016), and Borgomeo et al. (2017) introduced similar looped dynamic frameworks for household food security and social–ecological systems in southwest Bangladesh.

The distributaries of the Ganges River are the main source of freshwater inflow to the southwest delta region of Bangladesh. The southwest delta supports the Sundarbans, one of the largest mangrove forests in the world. The area is intersected by a dense network of tidal rivers and *khals* (small channels), mudflats, and small islands of salt-tolerant mangrove forests and provides habitat for a wide range of fauna, including the Bengal tiger and other threatened species like the estuarine crocodile and the Indian python (UNESCO 2017). Agriculture and fishing are the main production activities of the region. Since the 1960s, Bangladesh has adopted certain water and agriculture policies and new project planning and operation methodologies that have increased agriculture and fisheries production by a large margin and have helped the country to significantly improve its food security (Sala and Bocchi 2014; Yosef et al. 2015).

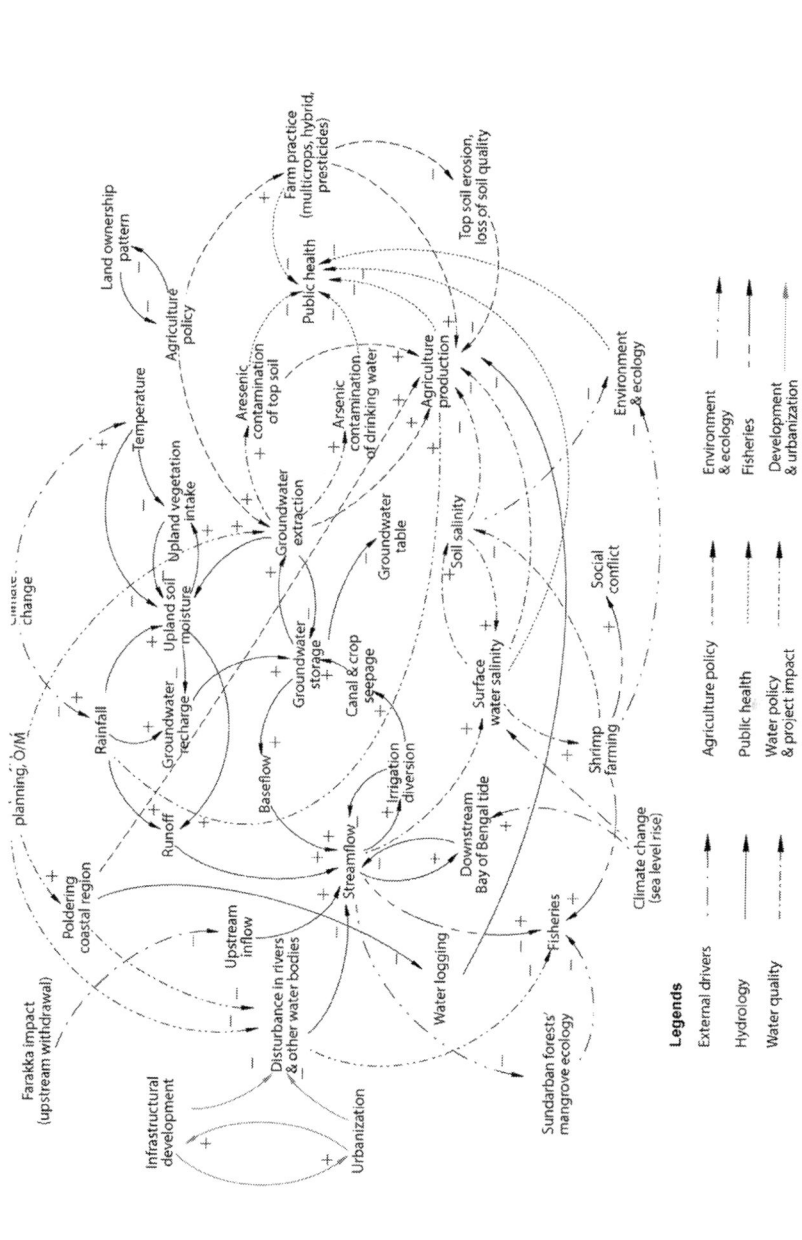

Figure 12.3 Interconnectedness and interactions primarily between the water system and systems surrounding it including relevant major human activities in the southwest delta region of Bangladesh. The positive and negative signs indicate positive and negative contribution respectively, not the positive and negative feedback.

One such policy shift – that helped achieve a doubling of rice production and significant reduction in infant mortality due to diarrheal diseases – was to promote use of groundwater for irrigation (i.e., to increase agriculture productivity) and drinking water supply (i.e., to fight against diarrheal disease). While both of these initiatives succeeded in fulfilling their respective objectives (i.e., increasing crop production and reducing diarrhea), the use of groundwater created a massive public health tragedy in the country that began to emerge in the mid-1980s. The increasingly excessive extraction of groundwater exposed a large number of people to arsenic-contaminated water and food, which emerged as a serious threat to public health in the country (Paul and De 2000; Edmunds et al. 2015; Naser et al. 2017). More than 20 million people in 59 out of 64 districts of the country are affected by arsenic contamination, and the southwest delta and southeast region are the two most affected areas in Bangladesh (Chakraborti et al. 2010; Flanagan et al. 2012). The story of arsenic poisoning in Bangladesh – and the events that both preceded and succeeded it – is an example of emergence in a complex system. The lesson is simple: it was nearly impossible to predict emergence of arsenic contamination when the use of groundwater was initiated with two attainable objectives.

India's diversion of a larger portion of dry season flow from the Ganges River by building Farakka Barrage in 1975 adds another level of complexity to southwest water problems (Mirza 2004; Gain 2014). Bangladesh and India signed the Ganges Water Treaty (GWT) in 1996. The GWT promised a minimum flow through the Padma River during the dry season (Dinar et al. 2007; Salehin et al. 2010). However, due to low flow availability at Farakka point in recent years, the stipulated minimum flow in the treaty could not be maintained on a regular basis. As a result, a sharp drop in historical average dry season flow due to the Farakka Barrage has not improved significantly during the posttreaty period. Figure 12.4 shows how a marginal improvement has been achieved since the Ganges treaty came into operation in 1997.

The Gorai River is one of the most important distributaries of the southwest delta region of Bangladesh. The Gorai serves as the primary freshwater inflow source for the western part of the region including the Sundarban forests. Low freshwater inflow from upstream results in high salinity intrusion from downstream and gradually converts rivers and canals, topsoil, and groundwater of the western part of the region from fresh to brackish (Palash et al. 2014; Kirby et al. 2015) and creates a series of adverse effects for agriculture, aquaculture, domestic, and industrial water use and public health (Palash et al. 2014; Dasgupta et al. 2015; Szabo et al. 2015; Johnson et al. 2016; Khanom 2016). On somewhat of a positive spin, this gradual conversion of a freshwater area to a brackish one has helped to expand shrimp farming over the last three decades in the western part of the southwest delta region. This growth in shrimp farming is often referred to as the "Blue Revolution" in the development sector due to its success in attracting foreign currency (EJF 2003). However, the "Blue Revolution" has also created some severe environmental

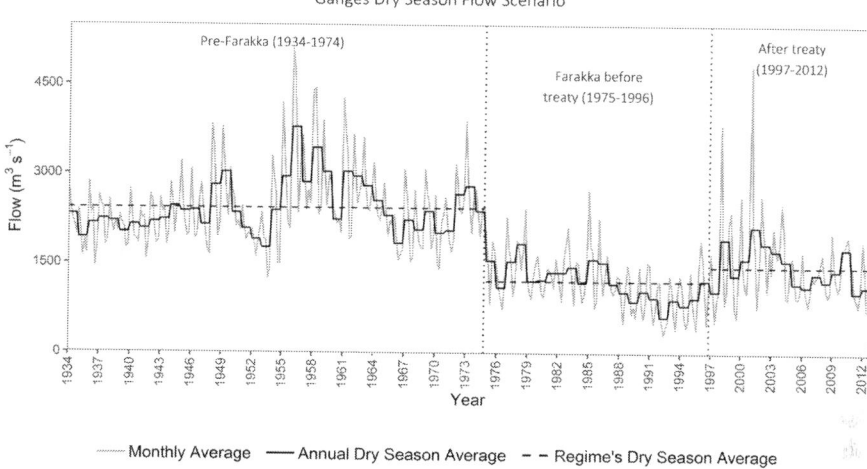

Figure 12.4 The Ganges dry season (January–May) flow from 1934–2012 illustrating three flow regimes: natural (1934–1974), post Farakka Barrage (1975–1996), post Ganges Treaty (1997–2012) period.

and ecological imbalances in the region and has triggered social conflicts (Hossain et al. 2013).

Building earthen embankments along the rivers and poldering (from the Dutch term *polder*) along the coastal areas is another intervention that led to emergent system response. The goal of these efforts – initiated after the major floods in 1958 – was to protect the coastal land from tidal and storm-surge inundation and increase agricultural production (Auerbach et al. 2015). These activities have indeed been very successful in protecting millions of lives from coastal flooding during cyclones and in boosting agriculture production over the last several decades (Rahman and Salehin 2013; Auerbach et al. 2015). However, they have also created a series of environmental and social impacts (Auerbach et al. 2015; Staveren et al. 2017). Notable impacts include embankment erosion, water logging, loss of soil quality, drops in agriculture production, destruction of natural fish habitat, deteriorating water quality, and increased public health risks (Mondal et al. 2013; Awal 2014; Borgomeo et al. 2017; Parvin et al. 2017).

Hossain et al. (2016b) examined the interconnectedness between natural variables (e.g., rainfall, streamflow, water, and soil salinity, etc.) and ecosystem services (e.g., rice, fisheries, and forestry production) for the southwest coastal region of Bangladesh. They summarized that rice production is significantly and positively correlated with temperature but negatively correlated to soil salinity. Fish production in natural habitats is negatively correlated with water salinity of the rivers and canals but demonstrates a positive relationship in closed water bodies, such as ponds or small lakes. Ecological production in forests is negatively related to water salinity, and the reason is directly linked to deteriorating mangrove forests, like the Sundarban, with the increase of both water and soil salinity. Sea level rise is positively correlated to groundwater depth (i.e., sea level rise raises the groundwater table) but negatively impacts water and soil salinity in the region (i.e., sea level rise increases water and soil salinity). In effect, sea level rise will exaggerate existing widespread salinity intrusion from the Bay of Bengal caused by low dry-season flow through the Gorai River due to the Farakka Barrage on the Ganges upstream.

In addition, one of the most noticeable examples of feedback in southwest Bangladesh exists in the subsystem related to shrimp farming (Hossain and Dearing 2016; Hossain et al. 2016). For example, high profits in shrimp farming "increase the possibility of conversion of rice farms into shrimp farms (Swapan and Gavin 2011) with further destruction of the mangrove forest (Azad et al. 2009), but the higher water salinity that conversion and destruction causes makes the shift from rice to shrimp farms even more likely" (Azad et al. 2009, p. 14).

Recall, to improve agricultural productivity and reduce diarrheal disease, the use of groundwater for agriculture and household use rapidly increased in the region starting in the 1960s. Almost three decades later, these two initiatives were identified as major contributors to the mobilization of groundwater to a depth where dissolved arsenic is extremely high, which led to widespread contamination of groundwater with arsenic (Harvey et al. 2002; Clarke 2003; Chakraborti et al. 2010). Observing the consequences, Clarke (2003, p. 255) pointed out that "the tragedy that has subsequently unfolded reveals in stark terms how 'solutions' to water–resource problems can go spectacularly awry if our knowledge of a region's hydrology and geology is incomplete." Such a lack of knowledge, particularly for complex systems, is seldom prospective and often only identified retrospectively.

Thus, we need to acknowledge that emergent phenomena like arsenic poisoning or the "Blue Revolution" will remain unanticipated at the beginning of the planning and implementation phase; consequently, we must be willing to consider and adopt interventions appropriate for particular contexts. Because it is nearly impossible to design and execute permanent solutions to complex problems like the ones in the southwest Bangladesh delta, we need to seek and find contingent approaches to resolve these water problems in light of changing circumstances.

Concluding remarks

System complexity arises when interactions and interdependencies exist, and they cannot be well described by natural laws; nonlinearity, feedback, and causal uncertainty are intrinsic; and the system exhibits emergence (Islam and Susskind 2018). Hydrological, climatic, ecological, social, policy, and governance – all these processes are involved and interact nonlinearly with feedback, tipping points, and thresholds in a complex system like the one we discussed in this paper for the southwest Bangladesh's coupled natural and human (CNH) system. A network view of such a complex water system (Figure 12.3) can be a helpful practical tool to understand and suggest how contingent decisions about water management can be made.

However, to make actionable and timely interventions in a complex system requires adopting a problem–driven approach. We argue that working to understand the interactions, interdependencies, and feedback loops relevant to a particular problem is a pragmatic approach to addressing those problems. Consider the water–related problems in a CNH system like southwest Bangladesh. Any chosen problem – for example, impact of salinity instruction on agriculture as well as the rise of the "Blue Revolution" – will require identifying and addressing interconnections and interdependencies relevant to the context of the problem. In reality, we cannot exhaustively numerate all variables, processes, actors, and institutions or their interactions for a CNH system into a model that attempts to replicate that system (Sterman 2000; Schlüter et al. 2014). Such an effort would essentially make the modeling complexity more intractable than the system itself and thereby would decrease the likelihood of creating actionable knowledge. We argue that not all interconnections and interdependencies of a complex system are equally important for a given problem. Identifying the subset of the important connections for a given problem is the key first step in diagnosing and resolving a problem within a CNH system.

A problem–driven approach leads to a requisitely simple description of a complex system. Requisite simplicity provides a framework that discards some details of a complex system but maintains conceptual clarity and scientific precision (Stirzaker et al. 2010). For example, going back to our southwest Bangladesh's CNH example, arsenic contamination in the food chain of the region is probably less critical than threats posed by drinking arsenic-contaminated water. Or we may observe that increasing Gorai inflow is perhaps more important than designing a mitigation plan against sea level rise to deal with the salinity intrusion in the region. In advancing the notion of requisite simplicity, we are not advocating for reductionism. Indeed, we maintain that classical reductionism is an inadequate response to an understanding and managing of complex systems. Rather, for complex problems, our challenge is to first identify the nature of complexity (system complexity and modeling complexity). Then we should decide what features are important relative to a particular problem while remaining aware about the numerous interconnections, interdependencies, feedback, and emergent behaviors. Such a problem–driven approach allows us to draw actionable insights even in the absence of a fully explicit or fully predictive theory. One such insight relative to the southwest Bangladesh delta is that activities like raising embankment height of coastal polders to fight against possibly more frequent cyclone and coastal flooding in the region due to global warming and sea level rise need to be complemented by other activities to facilitate sediment delivery and elevation recovery inside the polders. Specifically, simultaneous work needs to be done to systematically breach embankment sections (Auerbach et al. 2015) and promote the widespread application of tidal river management (Hossain et al. 2015; Gain et al. 2017). These particular actionable insights are easily lost in the high-level discourse of climate change impact mitigation in the region but become apparent by studying particular problems (i.e., polder resiliency).

Our challenge in resolving problems arising in complex CNH systems, such as the southwest Bangladesh delta is, therefore, not to look for universal solutions but to identify a requisitely simple set of interactions for a given problem that can be intervened upon, effectively impacted, and readily tracked with a measurable outcomes. While simplification can lead to opportunities for new perceptions and actions, the process and approach to simplification must be rooted in problem context and remain contingent and adaptive to changing circumstances. We hope that this case study will encourage readers to explore the interconnections, nonlinearity, feedback, and emergent behavior of complex systems and, moreover, to seek requisitely simple interpretations of system complexity to arrive at contingent yet actionable solutions.

References

Allen, P.M., 2014. Evolution: Complexity, uncertainty and innovation. *Journal of Evolutionary Economics*, 24 (2), 256–289.

Auerbach, L.W., Jr., Goodbred, S.L., Mondal, D.R., Wilson, C.A., Ahmed, K.R., Roy, K., Steckler, M.S., Small, C., Gilligan, J.M., and Ackerly, B.A., 2015. Flood risk of natural and embanked landscapes on the Ganges: Brahmaputra tidal delta plain. *Nature Climate Change*, 5, 153–157.

Awal, M.A., 2014. Water logging in southwestern coastal region of Bangladesh: Local adaptation and policy options. *Science Postprint*, 1 (1), 38.

Azad, A.K., Jensen, K.R., and Lin, C.W., 2009. Coastal aquaculture development in Bangladesh: Unsustainable and sustainable experiences. *Environmental Management*, 44, 800–809.

Binder, C.R., Hinkel, J., Bots, P.W.G., and Pahl-Wostl, C., 2013. Comparison of Frameworks for analyzing social-ecological systems. *Ecology and Society*, 18 (4).

Bonabeau, E., 2008. *Complexity: 5 Questions*. Copenhagen: Automatic Press, 21–26.

Borge-Holthoefer, J., Baños, R.A, González-Bailón, S., Moreno, Y., 2013. Cascading behaviour in complex socio-technical networks. *Journal of Complex Networks*, 1, 3–24.

Borgomeo, E., Hall, J.W., and Salehin, M., 2017. Avoiding the water-poverty trap: Insights from a conceptual human-water dynamical model for coastal Bangladesh. *International Journal of Water Resources Development*, 34 (6), 900–922.

Chakraborti, D., Rahman, M.M., Das, B., Murrill, M., Dey, S., Mukherjee, S.C., Dhar, R.K., Biswas, B.K., Chowdhury, U.K., Roy, S., Sorif, S., Selim, M., Rahman, M., Quamruzzaman, Q., 2010. Status of groundwater arsenic contamination in Bangladesh: A 14-year study report. *Water Resources*, 44, 5789–5802.

Cilliers, P., 1998. *Complexity and Postmodernism Understanding Complex Systems*. London: Routledge.

Cilliers, P., Biggs, H.C., Blignaut, S., Choles, A.G., Hofmeyr, J.S., Jewitt, G.P.W., and Roux, D.J., 2013. Complexity, modeling, and natural resource management. *Ecology and Society*, 18 (3), 1.

Clarke, T., 2003. Delta Blues. *Nature*, 422, 254–256.

Cooper, E., and Islam, S., 2015. *Exploring the Interconnections and Interdependencies in California's Water Problem*. Available from: http://blog.waterdiplomacy.org/2015/09/exploring-the-interconnections-and-interdependencies-at-play-in-californias-water-problem/

Dasgupta, S., Hossain, M., Huq, M., and Wheeler, D., 2015. Climate change and soil salinity: The case of coastal Bangladesh. *Ambio*, 44 (8), 815–826.

Dearing, J.A., Wang, R., Zhang, K., Dyke, J.G., Haberl, H., Hossain, M.S., Langdon, P.G., Lenton, T.M., Raworth, K., Brown, S., et al., 2014. Safe and just operating spaces for regional socialecological system. *Global Environmental Change*, 28, 227–238.

Dinar, A., Dinar, S., McCaffrey, S., and McKinney, D., 2007. *Bridges Over Water: Understanding Transboundary Water Conflict, Negotiation and Cooperation.* New Jersey: World Scientific.

Edmunds, W.M., Ahmed, K.M., and Whitehead, P.G., 2015. Environmental science processes & impacts a review of arsenic and its impacts in groundwater of the Ganges: Brahmaputra – Meghna delta. *Environmental Science: Processes & Impacts*, 17, 1032–1046.

EJF, 2003. *Smash & Grab: Conflict, Corruption and Human Rights Abuses in the Shrimp Farming Industry*. London: Environmental Justice Foundation.

Fromm, J., 2005. *Ten Questions About Emergence*. Technical Report. Available from: http://arxiv.org/abs/nlin.AO/0509049.

Flanagan, S.V, Johnston, B., and Zheng, Y., 2012. Arsenic in tube well water in Bangladesh: Health and economic impacts and implications for arsenic mitigation. *Bulletin of the World Health Organization*, 90 (11), 839–846.

Gain, A.K., Benson, D., Rahman, R., Datta, D.K., Rouillard, J.J., 2017. Tidal river management in the south west Ganges-Brahmaputra delta in Bangladesh: Moving towards a transdisciplinary approach? *Environmental Science & Policy*, 75, 111–120.

Gain, A.K., and Giuponni, C., 2014. Impact of the Farakka dam on thresholds of the hydrologic flow regime in the lower Ganges River basin (Bangladesh). *Water*, 6, 2501–2518.

Harvey, C.F., Swartz, C.H., Badruzzaman, A.B.M., Keon-Blute, N., Yu, W., Ali, M.A., Jay, J., Beckie, R., Niedan, V., Brabander, D., Oates, P.M., Ashfaque, K.N., Islam, S., Hemond, H.F., and Ahmed, M.F., 2002. Arsenic mobility and groundwater extraction in Bangladesh. *Science*, 298 (5598), 1602–1606.

Holland, J.H., 2002. Complex adaptive systems and spontaneous emergence. *In*: A.Q. Curzio, M. Fortis, eds. *Complexity and Industrial Clusters. Contributions to Economics*. Physica-Verlag HD, 25–34.

Holling, C.S., 2001. Understanding the complexity of economic, ecological, and social systems. *Ecosystems*, 4 (5), 390–405.

Hossain, F., Khan, Z.H., and Shum, C.K., 2015. Tidal river management in Bangladesh. *Nature Climate Change*, 5, 492.

Hossain, M.S., and Dearing, J.A., 2016. Recent changes in ecosystem services and human well-being in the Bangladesh coastal zone. *Regional Environmental Change*, 16 (2), 429–443.

Hossain, M.S., Eigenbrod, F., Johnson, F.A., and Dearing, J.A., 2016. Unravelling the interrelationships between ecosystem services and human wellbeing in the Bangladesh delta. *International Journal of Sustainable Development & World Ecology*, 24 (2), 120–134.

Hossain, M.S., Uddin, M.J., Fakhruddin, A.N.M., 2013. Impacts of shrimp farming on the coastal environment of Bangladesh and approach for management. *Environmental Science and Biotechnology*, 12, 313–332.

Islam, S., and Repella, A.C., 2015. Water diplomacy: A negotiated approach to manage complex water problems: University council on water resources. *Journal of Contemporary Water Research and Education*, 155, 1–10.

Islam, S., and Susskind, L., 2018. Using complexity science and negotiation theory to resolve boundary-crossing water issues. *Journal of Hydrology*, 562, 589–598.

Jeffrey, G., 2011. Emergence in complex systems. *In*: P. Allen, S. Maguire, and B. McKelvey, eds. *The SAGE Handbook of Complexity and Management*. London: SAGE Publications, 65–78.

Johnson, F.A., Hutton, C.W., Hornby, D., La, A.N., and Mukhopadhyay, A., 2016. Is shrimp farming a successful adaptation to salinity intrusion? A geospatial associative analysis of poverty in the populous Ganges – Brahmaputra – Meghna delta of Bangladesh. *Sustainable Science*, 11, 423–439.

Khanom, T., 2016. Ocean & coastal management effect of salinity on food security in the context of interior coast of. *Ocean and Coastal Management*, 130, 205–212.

Kirby, J.M., Ahmad, M.D., Mainuddin, M., Palash, W., Quadir, M.E., Shah-Newaz, S.M., and Hossain, M.M., 2015. The impact of irrigation development on regional groundwater resources in Bangladesh. *Agricultural Water Management*, 159, 264–276.

Levin, S.A., 1998. Ecosystems and the biosphere as complex adaptive systems. *Ecosystems*, 1, 431–436.

Levin, S.A., 1999. *Fragile Dominion: Complexity and the Commons*. Cambridge: Perseus.

Li, B., 2012. From a micro: Macro framework to a micro – Meso – Macro framework. *In*: S. Christensen, C. Mitcham, B. Li, and Y. An, eds. *Engineering, Development and Philosophy*. Dordrecht, The Netherlands: Springer.

Liu, J., Dietz, T., Carpenter, S.R., Folke, C., Alberti, M., Redman, C.L., and Provencher, W., 2007. Coupled human and natural systems. *A Journal of the Human Environment*, 36 (8), 639–649.

Lloyd, S., 2001. Measures of complexity: A nonexhaustive list. *IEEE Control Systems Magazine*, 21 (4), 7–8.

Lorenz, E.N., 1963. Deterministic nonperiodic flow. *Journal of the Atmospheric Sciences*, 20 (2), 130–141.

Maguire, S., McKelvey, B., Mirabeau, L., and Ötzas, N., 2006. Complexity science and organization studies. *In:* S.R. Clegg, C. Hardy, T.B. Lawrence, and W.R. Nord, eds. *The SAGE Handbook of Organization Studies*. 2nd ed. London: SAGE Publications, 165–214.

Milner-Gulland, E.J., 2012. Interactions between human behaviour and ecological systems. *Philosophical Transactions of the Royal Society B: Biological Sciences*, 367 (1586), 270–278.

Mirza, M.M.Q., 2004. The Ganges water diversion: Environmental effects and implications – An introduction. *In*: M.M.Q. Mirza, ed. *The Ganges Water Diversion: Environmental Effects and Implications*, 1–12. Dordrecht: Springer.

Mondal, M.S., Jalal, M.R., Khan, M.S.A., Kumar, U., Rahman, R., and Huq, H., 2013. Hydro-meteorological trends in southwest coastal Bangladesh: Perspectives of climate change and human interventions. *American Journal of Climate Change*, 2, 62–70.

Naser, A.M., Martorell, R., Narayan, K.M.V., and Clasen, T.F., 2017. First do no harm: The need to explore potential adverse health implications of drinking rainwater. *Environmental Science & Technology*, 51 (11), 5865–5866.

Palash, W., Jiang, Y., Akanda, A.S., Small, D.L., Nozari, A., and Islam, S., 2018. A streamflow and water level forecasting model for the Ganges, Brahmaputra and Meghna rivers with requisite simplicity. *Journal of Hydrometeorology*, 19, 201–225.

Palash, W., Quadir, M. E., Shah-Newaz, S. M., Kirby, M. D., Mainuddin, M., Khan, A. S., and Hossain, M. M., 2014. *Surface Water Assessment of Bangladesh and Impact of Climate Change*. Bangladesh Integrated Water Resources Assessment Report, 150 pp.

Parvin, G.A., Ali, M.H., Fujita, K., Abedin, M.A., Habiba, U., and Shaw, R., 2017. Land use change in southwestern coastal Bangladesh: Consequence to food and water supply. *In*: M. Banba and R. Shaw, eds. *Land Use Management in Disaster Risk Reduction: Disaster Risk Reduction (Methods, Approaches and Practices)*. Tokyo: Springer.

Paul, B.K., and De, S., 2000. Arsenic poisoning in Bangladesh: A geographic analysis. *Journal of the American Water Works Association*, 36, 799–809.

Rahman, R., and Salehin, M., 2013. Flood risks and reduction approaches in Bangladesh. *In:* R. Shaw, F. Mallick, and A. Islam, eds. *Disaster Risk Reduction: Approaches Bangladesh.* Tokyo: Springer.

Ramalingam, B., Jones, H., Reba, T., and Young, J., 2008. Exploring the science of complexity: Ideas and implications for development and humanitarian efforts. *Development,* 16, 89. Available from: www.odi.org.uk/rapid/publications/RAPID_WP_285.html

Roux, D.J., and Foxcroft, L.C., 2011: The development and application of strategic adaptive management within South African National Parks. *Koedoe*, 53 (2), 1049.

Sala, S., and Bocchi, S., 2014. Green revolution impacts in Bangladesh: Exploring adaptation pathways for enhancing national food security. *Climate and Development*, 6 (3), 238–255.

Salehin, M., Khan, M., Prakash, A., and Goodrich, C., 2010. Opportunities for transboundary water sharing in the Ganges, The Brahmaputra and The Meghna Basin. *In:* Infrastructure Development Finance Company, ed. *India's Water Future Is in Danger If Current Trends in Its Use Continue: India Infrastructure Report*. Oxford: Oxford University Press.

Schlüter, M., Hinkel, J., Bots, W.G., and Arlinghaus, R., 2014. Arlinghaus application of the SES framework for model-based analysis of the dynamics of social-ecological systems. *Ecology and Society*, 19 (1), 36.

Snowden, D.J., and Boone, M.E., 2007. A leader's framework for decision making. *Harvard Business Review*, 85 (11), 69–76.

Snowden, D.J., and Stanbridge, P., 2004. The landscape of management: Creating the context for understanding social complexity. *Emergence: Complexity and Organization*, 6 (1–2), 140–148.

Sterman, J.D., 2000. *Business Dynamics: Systems Thinking and Modeling for a Complex World.* New York: McGraw-Hill.

Stirzaker, R., Biggs, H., Roux, D., and Cilliers, P., 2010. Requisite simplicities to help negotiate complex problems. *Ambio*, 39 (8), 600–607.

Swapan, M.S.H., and Gavin, M., 2011. A desert in the delta participatory assessment of changing livelihoods induced by commercial shrimp farming in Southwest Bangladesh. *Ocean & Coastal Management*, 54 (1), 45–54.

Szabo, S., Hossain, M.S., Adger, W.N., and Matthews, Z., 2015. Soil salinity, household wealth and food insecurity in tropical deltas: Evidence from south-west coast of Bangladesh. *Sustainability Science*, 11 (3), 411–421.

Turner, B.L., Menendez III, H.M., Gates, R., Tedeschi, L.O., and Alberto, S.A., 2016. System dynamics modeling for agricultural and natural resource management issues: Review of some past cases and forecasting future roles. *Resources*, 5 (4), 40.

UNESCO, 2017. *The Sundarbans.* Available from: http://whc.unesco.org/en/list/798

Van Staveren, M.F., and Warner, J.F., 2017. Bringing in the tides: From closing down to opening up delta polders via tidal river management in the southwest delta of Bangladesh. *Water Policy*, 19 (1), 147–164.

Walsh, S.J., and Mena, C.F., 2016. Interactions of social, terrestrial, and marine sub-systems in the Galapagos Islands, Ecuador. *PNAS*, 113, 14536–14543.

Yosef, S., Jones, A.D., Chakraborty, B., and Gillespie, S., 2015. Agriculture and nutrition in Bangladesh: Mapping evidence to pathways. *Food and Nutrition Bulletin*, 36 (4), 387–404.

Part III

Looking back and looking forward

Reflections and lessons from the Tufts program on Water Diplomacy

13 Evaluation of an interdisciplinary graduate program

Lessons learned from the Tufts Water Diplomacy program

Glenn G. Page and Shafiqul Islam

Context and rationale

Nearly every grant-funded project will come with requirements regarding reporting and evaluation. For large organizations such as the National Science Foundation (NSF), the routineness of evaluation and the desire to compare performance across programs contributes to formulaic assessment strategies and a heavy reliance on output-oriented metrics. For an interdisciplinary research program, especially one in its early stages, these traditional approaches are unlikely to capture the developmental progress of the program, its impact on students and faculty, or its consequences outside of the academy. This mismatch can lead to frustration and lost opportunities.

While we seldom question the importance of regular reflection and self-evaluation in our personal and professional lives, our reaction to external evaluation can be much more averse. Rote reporting to a funding agency can become a task with little value if it is perceived as an unnecessary exercise that simply satisfies the funder. Such exercises can move from simply time consuming to adversarial when external evaluators render judgment that are not in alignment with views and perceptions of program leaders regarding performance and effectiveness.

Even *The 2010 User-Friendly Handbook for Project Evaluation*, produced under contract from the NSF, acknowledges those responsible for reporting often perceive evaluation as "threatening, disruptive, and not very helpful to project staff" (Westat et al. 2010). This frustration with evaluation often masks its capacity as a transformative practice that can help ensure innovative research programs adapt and thrive. In order to realize this potential, program evaluation must move beyond its number-centric conceptualization as a tool for cross-comparison. Program administrators need more flexibility to measure success on terms they feel reflect the program's goals, and participants must be allowed to have their narratives weighed alongside their numbers.

When program evaluation is thus aligned with program context, regular reporting need not to be seen as a Sisyphean task, but as a genuine opportunity for growth. Since every program is different, the challenge is to match the evaluation with appropriate methods and attention to process. In sharing our stories and experiences with alternative program evaluation, we hope to demonstrate what is possible and help readers avoid potential pitfalls.

How has our thinking, planning, and implementation of the program assessment evolved?

Our Water Diplomacy Integrative Graduate Education and Research Traineeship (IGERT) experience has significantly shaped our views and perspectives about assessment and evaluation: we have evolved from being assessment-agnostics to assessment–believers. In fact, our first interaction with Sustaina-Metrix (our external evaluator) in 2009 was quite revealing. Initially we saw assessment as a bookkeeping tool to track classes, publications, and conference participation rather than a practice that could improve the experience for participants and promote mission–aligned outcomes. Yet at the end of our IGERT, we were impressed by the transformative experience for students, faculty and administrators that was shepherded along via high quality external assessment. This jointly authored contribution by our external evaluator (Page) and principal investigator (Islam) is evidence of our adaptive learning.

We now have a better appreciation about how our formative assessment activities led to changes in our program and our intellectual culture; and how our external advisory committee helped shape the growth and on-the-go adjustment of initial program elements through adaptive feedback and careful guidance from our external evaluation team. The external assessment of the Water Diplomacy IGERT was innovative and served as a springboard for longer term institutional change as the SustainaMetrix Team was asked to work with Tufts Institute for the Environment to conduct transformative organizational scenario planning. This exploratory process generated new research directions and helped launch the new Sustainable Water Management M.S. program in 2018.

What is needed to make evaluation of interdisciplinary training effective?

Evaluation of an interdisciplinary training program is a challenging endeavor. Measuring genuine success involves not only evaluating whether the principal investigators and collaborating faculty have created a training program in which the student participants (trainees) have developed the expected core competencies but also the capacity to compose program elements in ways that generate new knowledge and insight for both theory and practice. From an evaluation perspective, it is not enough to put together a set of scholarly training experiences that fuse different disciplines, methods, and tools into academic activities, even if there are perceived needs for this fusion. The program needs to demonstrate these elements have been combined in a way that enables program graduates to apply their acquired competencies to creatively address novel problems. Trainees from any graduate program need to be sufficiently competent with their disciplinary knowledge and expertise. For participants in interdisciplinary programs, it is also essential for them to know what their disciplinary expertise can't provide and why they need complementary expertise to excel in their interdisciplinary niche – Water Diplomacy in our case.

Evaluation and assessment of interdisciplinary scholarship encounters difficulties when approaches based in single disciplines or traditional methods of program evaluation are applied without fully appreciating the nuances of interdisciplinarity. Too often, bibliometrics such as number of peer-reviewed publications in high-impact journals and subsequent citations are used as the "gold standard" to evaluate effective research and by extension graduate research trainee programs. We think assessment of interdisciplinary scholarship requires criteria and metrics that are specifically chosen to be aligned with the goals and objectives of the program. In an interdisciplinary context, this often means negotiating a balance between the recognizable aspects of traditional program evaluation (which are important to funding agencies as well as individual departments) with novel approaches that provide sufficient flexibility for the program to grow and adapt in alignment with its specific goals.

What was done at Tufts University and how?

An interdisciplinary Water Diplomacy Program was conceived and designed to educate a new generation of water professionals called Water Diplomats. A Water Diplomat is an interdisciplinary professional who contributes to the resolution of complex water problems that include: (i) multiple stakeholders with conflicting or divergent interests and (ii) sources of uncertainty in either the scientific or sociopolitical domain (or both). Water diplomats are trained to become competent in multiple water-related disciplines and applied fields. A key challenge was how to measure the development of these competencies as an interdisciplinary skill set. While there has been significant progress in the design and use of assessment protocols for undergraduate learning, there are few established models for assessment of student progress in interdisciplinary doctoral-level graduate education. This chapter explores the results of a novel approach to identify, track, and measure the development of key interdisciplinary competencies by each student to become a Water Diplomat. Problem-focused principled pragmatism drove the process – both the development of the Water Diplomacy Framework and the evaluation of the program. In this chapter, using recent findings from the developmental evaluation literature and our decade-long experience at Tufts, we will show by examples of lessons learned about what to do, what not to do, when to adapt, and how. These lessons and related insights are expected to have wider applications in designing and implementing other interdisciplinary graduate education programs.

Moving beyond traditional evaluation frameworks

Under conditions of interdisciplinary exploration and innovation, traditional evaluation methods and frameworks are not useful (Patton 2016). These approaches are often categorized as "formative" or "summative." Formative evaluations examine the design or ongoing progress of a program and suggest improvements to its model. Summative evaluations, on the other hand, are used to assess a program retrospectively and assess whether the program produced the desired outcomes and to what degree those outcomes can be attributed to the program model versus exogenous factors. Both types of assessment have their place and merits, but as Patton (2011) notes, "evaluation practitioners have become sloppy about what actually constitutes a formative or summative evaluation and the connection between the two." Designed to clarify the practice of evaluation and make it more effective, the developmental evaluation framework proposed by Patton (2012) has proven itself a useful evaluation approach in recent years.

Based on reviews of literature (Patton 2011; Preskill and Beer 2012; Dickson and Saunders 2014; Lam and Shulha 2015), we determined that a synthesis of traditional formative and summative assessment could be fused with recent findings from developmental evaluation approaches to create an appropriate evaluation framework to assess the evolution of our interdisciplinary Water Diplomacy Program. We recognized that there were aspects where formative evaluation approaches were appropriate, and suggestions garnered through formative evaluation led to the improvement of a team-taught course in the Water Diplomacy curriculum. We also identified key summative moments where rendering judgment on the merits and worth of the overall funded model was justified and created a summative report as a capstone for the end of the five-year funding cycle. We also felt that to assess the evolution of the Water Diplomacy graduate program and the progress of institutionalizing the Water Diplomacy Framework into the Tufts University academic model, a different set of criteria and metrics was needed. As such, traditional evaluation that advocates for clear, specific, and measurable outcomes that are to be achieved through logical and linear processes was not appropriate. Implementing an integrative training program requires ongoing innovation and adaptation. Traditional methods required a level of up-front specificity that we felt did not work under conditions of high innovation, exploration, uncertainty, and emergence that were signatures of this traineeship. On the other hand, one does not want to overeagerly dispense with certain early commitments, such as the goals and mission of the program. From this perspective, we found that our take on the developmental evaluation framework provided the necessary ingredients: evaluative rigor and timely feedback to inform adaptive development while maintaining fidelity to the initial ideas conceived by the program. This evaluation process allows one to: (i) ask evaluative questions; (ii) apply evaluation logic; and (iii) gather and report evaluative data to support and refine the goals and objectives of the program in an adaptive way.

How did we align evaluation with our program's pedagogical structure?

The Water Diplomacy Framework (WDF), serving as the overarching pedagogical structure, is based on a simple inquiry: who decides, who gets water, and how? The WDF defined a set of core competencies of the trainees so that they can understand how to better define, govern, and manage water problems with conflicting needs and priorities. The trainee is expected to recognize the interconnectedness of values, interests, and tools as they relate to framing of problems and formulation of policy within a politically contentious world. Competencies include recognizing competing values and interests of the political community and interdisciplinary perspectives of the knowledge community to develop scientific and technological solutions that are contextually relevant and actionable. Given the contingent nature of problem diagnosis and policy implementation, the WDF requires a principled and pragmatic approach to negotiation for more effective resolution of complex water problems. Developmental evaluation is based on a set of attributes that can be tracked as trainees build their competencies; for example, what specifically they were doing to develop interdisciplinary competencies, why they are choosing the specific activities such as internships and research directions, and how they are demonstrating their competencies. This exploration was identified as an alternative to the more traditional methods for evaluation of interdisciplinary programs that is largely reduced to bibliometrics that quantify the number and type of intellectual outputs that are collaborative. A new approach was needed and was developed by our partners at SustainaMetrix, who applied a mixed methods approach with numbers and narratives that are interpreted through a systems lens that examines scale, gives attention to inevitable and often arbitrary boundaries, accounts for a variety of perspectives, and examines the dynamic interrelationships that occur in an interdisciplinary graduate degree program. The hope is to move away from attribution analysis and the "tyranny of metrics" toward a more holistic contribution analysis. This provided a new way to assess interdisciplinary graduate education.

What are the key attributes of our proposed evaluation and assessment model?

The external evaluation was led by SustainaMetrix, a global consulting firm based in Portland, Maine, with previous experience in evaluation of interdisciplinary graduate research. The External Evaluation Team and Tufts Water Diplomacy group discussed the challenges of evaluating interdisciplinary education programs, the limitations of available methods, and the desire for the evaluation to contribute to the ongoing development of the Water Diplomacy graduate program. The SustainaMetrix team recognized that doctoral programs already have an established suite of assessment instruments to evaluate individual student development, and these are typically administered by the advisor, examining committees of graduate faculty, or academic departments and programs. Key assessment events include qualifying exams, development of dissertation proposal, and defense of dissertation research. Furthermore, doctoral programs have relied on a traditional suite of systemwide metrics to evaluate program efficacy (Golde and Dore 2001). Far too often, quantitative measures focus on retention/attrition rates and time to graduation. However, these traditional metrics do not speak to the heart of outcomes assessment of doctoral programs, that is, the ongoing development of a trainee and their evolution from "senior learner" to "junior colleague" to "disciplinary steward" (Walker et al. 2008). Based upon their experience with other IGERT programs, the SustainaMetrix team placed an emphasis on the core competencies of the student and the faculty in working in interdisciplinary settings with multiple perspectives. They recognized the common lenses of disciplinary bias, which complicates assessment and adaptation of curricula toward something that is useful for students and faculty from many disciplinary homes. A set of attributes on how developmental evaluation would be applied for the Tufts Water Diplomacy IGERT was developed and implemented as follows:

- **Developmental purpose:** a key goal of the evaluation was to illuminate, inform, and support the development of the Water Diplomacy Framework and training program by identifying the aspects of the training that are most effective.
- **Evaluation rigor:** by applying evaluation logic and appropriate methods, the evaluation was designed to gather, collaboratively interpret, and report data.
- **Utilization focus:** the intended use of the evaluation was to help the principle investigators (PIs) further develop an integrative framework for training graduate research in Water Diplomacy and for the external evaluation to be useful for participants.
- **Innovation niche:** Water Diplomacy was conceived as an innovative approach to define, govern, and manage intractable water problems, to influence systems change, and to provide rapid response under crisis conditions – the evaluation was envisioned to support the innovation and adaptation that was needed to shape the training program.
- **Complexity perspective:** the evaluation applied a set of complexity science concepts to make sense of the dynamic and emergent challenges of implementing the training program as part of a PhD experience.
- **Systems thinking:** both the evaluators and the PIs were deeply invested in systems thinking by considering interrelationships, multiple boundaries, and other dimensions of the interdisciplinary training program.
- **Co-creation:** the evaluators and PIs formed an interdependent and iterative process so that the results of the assessment were truly co-created.
- **Timely feedback:** the evaluators provided timely feedback during the external process to inform ongoing adaptation as needs, findings, and insights emerged rather than only at the end of the assessment period.

The remainder of this chapter will examine how this synthesis of evaluation methods was applied, how the team gathered data regarding what was unfolding and emerging, and how numbers and narratives of the program's evolution provided insights and rationale to the students and faculty. It will also show how to better adapt and learn to make a better program during the evolution of the program as well as what key lessons were learned that may be transferrable to other interdisciplinary settings.

Developmental purpose

An overarching objective of the traineeship was to co-develop a Water Diplomacy Framework (WDF) with the faculty and students. This was accomplished through an iterative process that began with a model initially proposed in Islam and Susskind (2013) and allowed for ongoing critique and feedback as to the effectiveness, appropriateness, and usefulness of the framework by student participants from this program and water professionals through the annual Water Diplomacy Workshop.

For example, one of the students remarked on the notion of principled pragmatism as a way to operationalize the WDF:

> "*I really like the definition of principled pragmatism that we have used, and really feel that this described my goals for my work well. I would refine 'WDF explicitly recognizes interconnectedness of values, interests, and tools as they relate to framing of problems and formulation of policy within a politically contentious world' to include a more explicit acknowledgement of societal systems, as I think we are subjects to not just our interests and tools as we approach problems but also to our embeddedness in social norms and values.*"

Another student described the developmental nature of the WDF:

> "*Who decides who gets water, how and what for? The WDF is a way to understand and manage water problems in regions with conflicting needs and priorities around this natural resource. WDF recognizes the competing values and interests of the political community and the interdisciplinary perspectives of the knowledge community to develop scientific and technological solutions that are contextually relevant ... Given the contingent nature of problem diagnosis and policy implementation, WDF requires a principled pragmatic approach to negotiation – for effective resolution of complex water problems – that is based on equity and sustainability as the guiding principles.*"

Evaluation rigor

The program featured several examples of applying evaluation logic and appropriate methods in order to gather, collaboratively interpret, and report data. In fact, a novel method was further developed and refined that allowed the trainees to define their interdisciplinary pathway. As the program unfolded, the key evaluation question that emerged was to what degree did the Tufts Water Diplomacy IGERT contribute to equipping student participants with the training and academic experience needed to build breadth in applying the Water Diplomacy Framework and depth in their chosen discipline to become interdisciplinary scholars and leaders in this emerging field? With this question in mind, the evaluation team proposed methods for tracking our students' development in terms of both breadth and depth. A conceptual framework was developed to define the explicit learning goals of students and track how they were related with disciplinary and interdisciplinary educational and research activities. The framework featured methods for the graduate students to define their developmental pathways to interdisciplinarity based on their aspirational goals for breadth and depth. Some of these methods are captured in the description of the "T" metaphor and exercise in the following section. The results have yielded a quantitative and qualitative measurement of a student's self-perception of his/her knowledge, interest, and development in interdisciplinary research across multiple disciplines over the duration of their graduate studies.

The "T" metaphor and exercise

A simplifying metaphor (Newell 2012) of a "T" was used as part of the external evaluation of the IGERT, which itself was used as a foundational concept to help develop different educational and research components of the Water Diplomacy Program. The T model provides a framework to articulate learning goals, define perceived current and aspirational future balance between disciplinary and interdisciplinary endeavors. The T model is also used as a quantitative and qualitative means to evaluate current and future learning goals and overall graduate program effectiveness. The T tool has considerable utility in measuring knowledge and interest in interdisciplinary research across a range of disciplines and graduate programs. Students report maturing of their learning by doing interdisciplinary scholarship and completing the program with a greater sense of their own breadth and depth dimensions of interdisciplinarity that also helped them to define their own educational path to create and navigate.

This exercise was developed by the SustainaMetrix Team in partnership with the Coastal Institute IGERT at the University of Rhode Island (August et al. 2010). The exercise involves asking the trainees to allocate 20 imaginary "blocks" on the vertical and horizontal axes of the T that represent their perception of the shape of their current T. They are also asked to reflect on their aspirations for the development of this T at the end of the IGERT experience and 5 to 10 years after attaining their doctorate. Each of the 20 blocks represents a unit of perceived interest, knowledge, and capability in disciplinary or interdisciplinary research. A T with 10 blocks vertically and 10 blocks horizontally represents an ideal balance of disciplinary and interdisciplinary interests and expertise. Tall and narrow Ts represent highly disciplinary emphasis, while short and broad Ts represent highly interdisciplinary emphasis.

This process yields a suite of simple, quantitative metrics that allow us to capture current and aspirational allocation of time and emphasis in disciplinary and multidisciplinary focus among our students. In addition, the T tool provides rich qualitative data as trainees engage in self-reflection and articulate the rationale behind their T construct. Aspect ratio (i.e., breadth divided by depth) simplifies the metaphor of the T into a single metric (Figure 13.1). When students build a T that increases in aspect ratio (broadens) when compared with previous Ts, this demonstrates their perception of an increased ability to perform interdisciplinary research. Conversely, students whose T aspect ratios decrease demonstrate their perception that they are focusing on enhancing depth in their chosen discipline. The results are not viewed as a value judgment; rather, they serve to provide students and faculty with a clear sense of individual goals and an ability to measure progress in meeting those goals. The T metaphor provides students with a vocabulary and nomenclature to identify current learning profiles and develop career goals as well as the opportunity for regular self-reflection. It also provides a shared vocabulary to communicate with peers coming from varied academic disciplines with different sets of academic and research challenges.

Within the first few weeks of starting the program each year, every incoming graduate student is asked to describe their perceived T expertise and interest with 20 blocks to allocate among three categories: vertical (disciplinary), horizontal (interdisciplinary), and "currently unallocated." This exercise is understood as a trade-off, as the trainees need to design and navigate between developing disciplinary depth and building interdisciplinary competencies with a focus on Water Diplomacy. An example from one trainee is presented

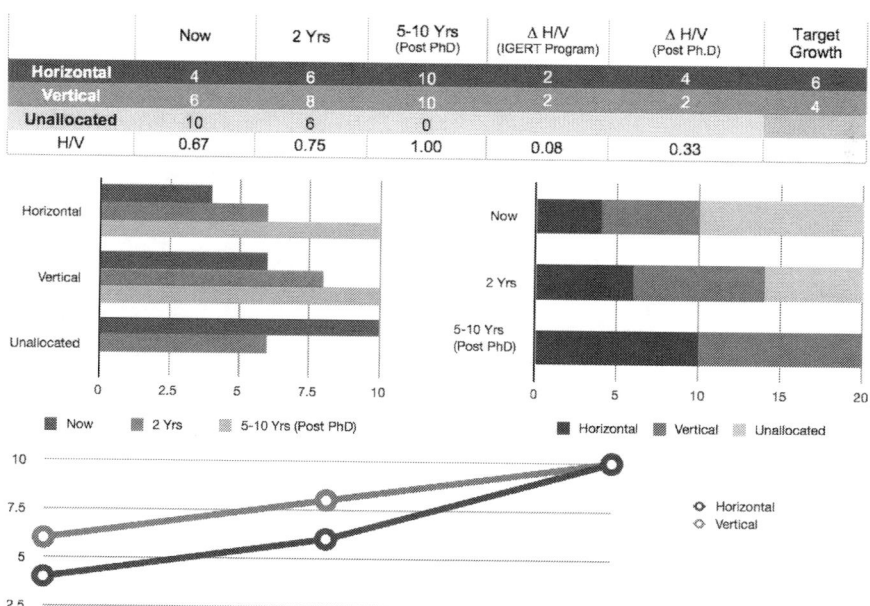

	Now	2 Yrs	5-10 Yrs (Post PhD)	Δ H/V (IGERT Program)	Δ H/V (Post Ph.D)	Target Growth
Horizontal	4	6	10	2	4	6
Vertical	6	8	10	2	2	4
Unallocated	10	6	0			
H/V	0.67	0.75	1.00	0.08	0.33	

Figure 13.1 Example of a T-Competency Summary that captures a student's progress benchmarked against their original conceptualization of their interdisciplinary journey.

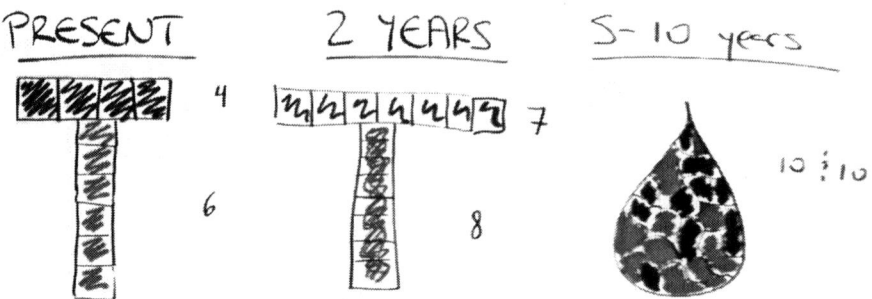

Figure 13.2 Example of a student's illustration of their interdisciplinary trajectory as conceptualized during the T exercise. An initial T with disciplinary emphasis morphs into a more balanced interdisciplinary T over time with an aspiration to become an "interdisciplinary water droplet" that brings both the breadth and depth dimension of an effective Water Diplomat.

below that illustrates trainee aspiration to become balanced between depth and breadth over time: the student started with (4, 6, 10) and had evolved into (7, 8, 5) in two years with an aspiration to become (10, 10, 0) in the future. An initial T with disciplinary emphasis has morphed into a more balanced interdisciplinary T over time with an aspiration to become an "interdisciplinary water droplet" that brings both the breadth and depth dimension of an effective Water Diplomat (Figure 13.2).

The following are comments, reflections, and insights from students regarding the T assessment exercises. These short commentaries, which include specific examples from our Water Diplomacy Program, highlight the utility of T expertise when conducting interdisciplinary scholarship:

- *"I think that it was helpful to sit down and think about which fields we are really digging into in depth."*
- *"The answers to the questions in the T assessment were not obvious and really required me to think deeply about what strengths I have developed and what areas I want to really dig deeply into."*
- *"I liked the goal-setting exercise at the beginning of my time in the program. I had some goals in mind, but I think the T-assessment helped me to clarify and make them more specific and measurable ... On one hand, it pushed me to think bigger than I otherwise would have and to be explicit about some larger dreams. At the same time, this framework took some of the pressure off by allowing me to be realistic and avoid seeing 100% of my goals as outcomes that I am absolutely expected to achieve."*
- *"The T-assessment process was a challenging, out-of-the-box 'exercise' that pushed me to define my professional life in a way I had not foreseen before. The vision of myself, my time, my career, in other words, my life, as a collection of 20 blocks was quite frightening at first. However, after a thinking period needed to overcome the unknown and the new, I saw the exercise as an opportunity for a reinvention."*

- "*The self-reflective aspects were by far the most valuable.*"
- "*The most valuable aspect was thinking about my present knowledge (both in and outside my discipline) and how I wanted that knowledge to develop over the course of my PhD. The T-assessment provided a road map for me to identify and explore topics outside of engineering that will help me become a successful engineer that can communicate and collaborate with experts from a variety of disciplines.*"
- "*The most valuable aspect of the T-Assessment was to think through, at the beginning of my PhD, what balance of disciplinary depth and breadth I was aiming to achieve. I think this helped me be more purposeful in my selection of classes, projects and other learning opportunities.*"
- "*Once I started thinking about my depth in terms of the type of analysis and research methods I want to have expertise in, rather than specific content knowledge, it became easier to think about my T and identify the steps I would take to gain expertise in my skill-based depth. Having honed in on this depth early on was really helpful in developing my dissertation research proposal because I knew which skills I wanted to focus on in my project and that helped me narrow down my ideas.*"
- "*I have a tendency to go broad, such that I am more 'inter' than 'disciplinary.' Foremost, the T exercise was useful in stressing the importance of building in depth. Laying out specific blocks helped me be more realistic about what could be done in the span of a few years, and through a life-long career. Lastly, having a visual check-in has been helpful to revert back to and to see how well I am meeting my initial goals, and if not, more intentionally examine why and if that's ok.*"
- "*The T assessment laid out a really realistic 'block' allotment, which helped discipline my big appetite. . . . The T Assessment helped create a structure within which I could imagine what areas I want to develop further.*"

Utilization focus

Adopting a utilization focus means the program is concerned with maximizing its utility to its participants. By adopting a concern for both breadth and depth, students engaged in interdisciplinary research must take care in investing their most limited resource: time. The limit of 20 blocks in the T exercise was one way we attempted to help students maximize the benefits they received in exchange for their investment of time into the program. In order to support students in coping with the competing demands of breadth and depth, we also asked them to routinely report on the core integration challenges they faced. In addition, we created other measures to help them plan, reflect, and adjust how they were investing their time. These included a set of "graduate progress markers" based on the Outcome Mapping exercise developed by the International Development Research Centre in Ottawa, Canada (Earl et al. 2001). We used these progress markers along with the following questions to guide students in this process:

- What do you EXPECT TO SEE as an outcome of your interdisciplinary Water Diplomacy training? Consider 2–3 outcomes that you see as the low-hanging fruit which would be relatively easy to accomplish.

- What would you LIKE TO SEE as an outcome of your interdisciplinary Water Diplomacy training? Consider 3–4 outcomes that may require more active learning or engagement, as well as the involvement of others but you would nevertheless like to accomplish.
- What would you LOVE TO SEE as an outcome of your interdisciplinary Water Diplomacy training? Consider 2–3 outcomes that would be truly transformative. These may be even be outcomes that are difficult to imagine but if they did happen would be evidence of profound change.

Students would track these developments in a journal and share the results with the external evaluator to discuss progress toward their interdisciplinary goals.

Finally, students were asked to identify and interview another member from the IGERT student cohort, someone with at least a year of experience in the program. Aside from a few priming questions, this conversation was left open-ended. Students found value in these focused conversations, which often clarified questions of what to expect and what has or has not worked with the program. All students were asked about how they valued these reflective exercises, and the results below show that students found use of the T metaphor and 20-block T exercise to be valuable for their interdisciplinary development (Figure 13.3). Defining the outcome challenge was also considered valuable by most, while the other exercises were considered valuable or extremely valuable.

Innovation niche

The Water Diplomacy Framework (WDF) was conceived as an innovative approach to define, govern and manage intractable water problems, to

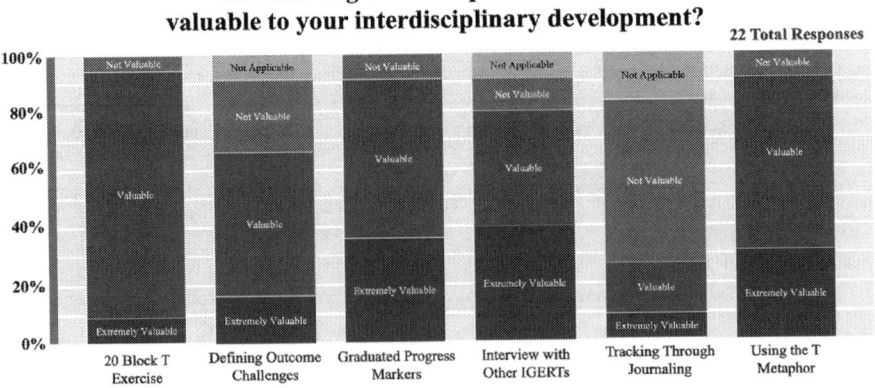

Figure 13.3 Example feedback demonstrating how students valued evaluation exercises in terms of positive impact on their interdisciplinary development.

influence systems change, and to provide rapid response under crisis conditions. It was critical that our evaluation framework would support rather than hinder the innovation and adaptation needed to shape the training program. The WDF aims to help resolve water-related conflicts. It acknowledges that traditional problem-solving frames are adequate to address simple and complicated problems where reasonable scientific certainty and consensus about intervention exists. The WDF hypothesizes the counter-productivity of these traditional frames when water challenges stem from complex – interconnected, uncertain, unpredictable, and boundary-crossing – system dynamics with feedback. A recurrent factor for such limitations is that traditional problem-solving frames often separate the observation-based technical (what is) from the value-based sociopolitical (what ought to be) dimensions of the problem. The WDF further hypothesizes that when dealing with complex problems, these dimensions cannot be de-coupled. The WDF acknowledges both the limits to knowledge – objectivity of observations and subjectivity of interpretation – and the contingent nature of our action.

The WDF proposes a principled pragmatic approach to negotiation that is based on credible scientific knowledge mediated through equity and sustainability as guiding principles for policy action. The WDF approach emphasizes that when addressing complex water problems, all parties have a legitimate right to have a voice about the evidence used and its interpretation, evaluate the future implications of an intervention, define measures of equity and sustainability, and be critical of the proposed package of actionable solutions. These parties include users and producers of water knowledge, managers, technical experts, policy makers, decision makers, and politicians. Furthermore, the WDF asserts that parties need to seek consensus and mutual benefits when negotiating a resolution. From an evaluation perspective, there is abundant evidence that this framework was innovative and has undergone adaptive enhancement over the duration of this program. At the end of the program, all program students were asked to define their degree of agreement with the WDF at the time of the survey and showed strong agreement but also some elements of disagreement, which led to further refinement of the WDF (Figure 13.4). The results are evidence of data being collected to measure consensus on the innovation.

Complexity perspective

The evaluation applied a set of ideas from complexity science to make sense of the dynamics present within a university context. These included incentives and reward structures that enabled or hindered collaboration across the disciplines as well as consideration of the unique challenges inherent in implementing the training program as part of a more disciplinary-focused PhD experience. This is illustrated in the methods used to encourage reflexivity as the trainees navigate their interdisciplinary path as well as methods that were used to identify patterns, explore and understand how the training program

Degree of Support for DRAFT Water Diplomacy Framework

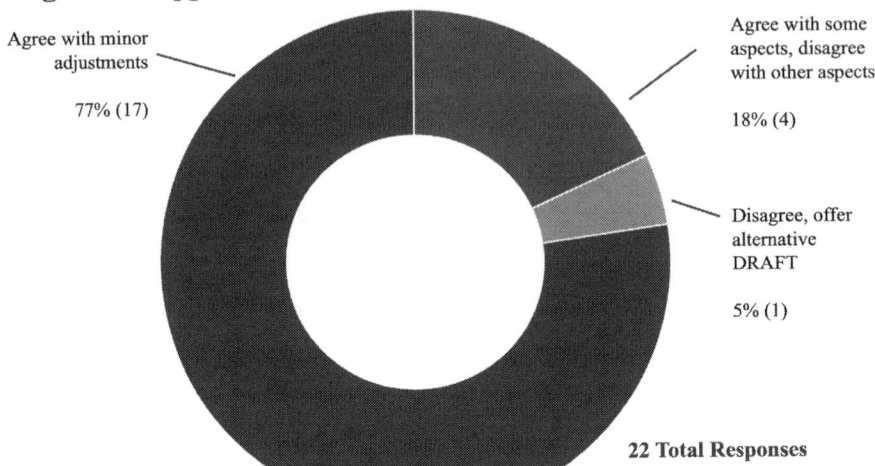

Agree with minor
adjustments

77% (17)

Agree with some
aspects, disagree
with other aspects

18% (4)

Disagree, offer
alternative
DRAFT

5% (1)

22 Total Responses

Figure 13.4 Feedback on the Water Diplomacy Framework, which co-evolved with student and faculty input since its formal introduction by Islam and Susskind (2013).

evolves, and observe how learning by doing is integrated into day-to-day activities. When evaluating our program and thinking about adaptations, we have viewed our program as a complex adaptive system. Our complexity lens colored the way we thought about adaptation and innovation within the structure of our program. It gave us a consistent framework for interpreting student feedback, contextualizing apparent shifts in collective thinking, testing changes in program structure, and accounting for exogenous and endogenous risks to program implementation.

For example, a major evaluation challenge was to explain and reaffirm the developmental purpose and appropriateness of the evaluation approach to new faculty, students, and other key stakeholders such as the provost and other senior administrative personnel. We were especially alert to changes in personnel that would require immediate relationship building to confirm the focus on a developmental perspective with a complexity mindset.

Systems thinking

Both the evaluators and the PIs were deeply invested in systems-thinking by considering interrelationships, multiple boundaries, and other dimensions of the interdisciplinary training program. This is best illustrated with the systems thinking contained within the concept of principled pragmatism that was developed through the course of the traineeship:

"Water Diplomacy is a principled pragmatic approach to negotiation that is based on credible scientific knowledge mediated through equity and sustainability as guiding principles for policy action. When addressing complex water problems, students are trained to appreciate that all parties have a legitimate right to have a voice about the evidence used and its interpretation, the past evidence and future implications of an intervention, metrics of equity and sustainability, and the package of actionable solutions. These parties include users and producers of water knowledge, managers, technical experts, policy makers, decision makers, and politicians. The training was designed to develop competencies to better engage with users and producers of water knowledge, managers, technical experts, policy makers, decision makers, and politicians. To seek consensus and mutual benefits when negotiating a resolution."

Co-creation

Both the evaluators and PIs formed an interdependent and iterative process so that the results of the assessment were truly co-created. The following illustrates what the students valued most about the collaborative creation process:

- *"I value the quality of scholarship and collaboration across the cohort it has engendered at Tufts."*
- *"The funding provided through this program has been incredibly valuable because it validates and provides space for the importance of developing your own interdisciplinary water approach . . . I really love the program's focus on our work as that of 'principled pragmatists'"*
- *"The retreats and other activities alike were very helpful to get to know fellows, professors and program managers in depth."*
- *"The variety of students that it attracted and the collaborations that were possible because of the program."*
- *"Involvement in the IGERT shaped the way I think about water challenges. I think three things led to that: (1) Shared office space with students from a range of departments, (2) Core IGERT courses and (3) Colloquium discussions on student presentations."*
- *"I value the introduction to 'diplomacy' from a non-technical perspective. I wish that we had more technical courses that had a water negotiation aspect."*
- *"I value the exposure to interdisciplinary work, including its value and the challenges and complexities involved."*
- *"The recognition that we are working on complex problems that exceed the capacity of traditional disciplines and techniques and that this work is messy and sometimes does not have a clear solution. The IGERT provides space for people working on these problems to share knowledge and burdens. It also lends a shield of legitimacy against people sceptical of the importance of the issues we study or our approach and stated problematics."*

Timely feedback

Feedback was provided to the PI and to the senior faculty team on a close-to-real-time basis as the external evaluator would brief the committee while on site to share immediate feedback that was gained in the evaluation process. While there were reports, the vast majority of information was either delivered verbally while on site or via email within 3 to 5 days of the field visit to inform ongoing adaptation as needs, findings, and insights emerged rather than only at the end of the assessment period.

The Water Diplomacy program has actively promoted adaptive learning to continuously improve how we teach, communicate, and collaborate in this interdisciplinary program. It has challenged existing thinking and practice by focusing on defining, refining, and implementing key features of the Water Diplomacy Framework in their dissertation research. We expected students from our Water Diplomacy Program to develop a broader understanding of key ideas, assumptions, and principles of the Water Diplomacy Framework through individualized curricula, the weekly colloquium, a PhD dissertation that fulfilled the discipline-specific requirement but contained a Water Diplomacy issue, joint advising on the dissertation by both a natural and societal domain faculty, and an internship experience with relevance to Water Diplomacy. Here, we highlight a few examples of learning by doing through continuous feedback from different stakeholders – students, faculty, administrators, and external advisory board of the program.

Following the 2013 External Advisory Board meeting, a set of curriculum suggestions emerged from a thoughtful and spirited interchange with Water Diplomacy (WD) students and subsequent discussion amongst the board. Several students suggested that the third course in the Water Diplomacy series of core courses (WD III: a synthesis course) could usefully be scheduled first, before WD I (natural science of water) and WDII (societal aspects of water). The students argued that WD III could provide an early synthesis of the key and distinctive ideas of the Program, motivating engagement by seeing the big picture at an earlier stage of their engagement. The board also suggested that the two domain courses, WD I (natural science) and WD II (social science), should no longer be optional for those students with skills in that domain. Engagement with both classes could facilitate greater collaborative learning and group cohesion for the cohort of Water Diplomacy students. If WD I and WD II are taught after WD III, it is likely that these classes can also move from a semiremedial status, gaining intellectual challenge and subject specificity from the prior learning of key Water Diplomacy ideas in WD III. These changes in the sequencing of WD courses were implemented from fall 2014.

The external board suggested that students be encouraged by the program to make full use of their second advisor. In addition, the board suggested that the core program faculty regularly review student contact with advisors, to ensure that faculty are actively engaged in student learning. This aspect of dual advising and follow-up was actively pursued and implemented throughout the program.

Similar to many interdisciplinary initiatives, the Water Diplomacy Program was challenged during the first few years by a lack of a shared language across disciplines. Based on feedback from students, faculty and external advisory board, we have agreed that a shared framework is needed to identify and understand a problem and discuss methods to meet one's research goals. As a way to achieve this goal, the format of the weekly colloquium was restructured. The initial plan for the colloquium was to discuss one case study per semester in detail. In the revised structure, weekly colloquium supported both topical explorations with invited guest speakers with relevant interdisciplinary expertise and presentation opportunities for students' own evolving research. Students were invited to bring rough presentations and half-baked ideas to clarify, refine, and improve with their peers and professors at the colloquium. Many students found the colloquium experiences to be uniquely beneficial in sharpening their research agenda. It also helped develop group projects and provided a community setting for collaboration and development of a shared language over time. The revised format allowed participants to learn from each other and to actively employ and engage with concepts and methodologies from disparate disciplines. It also allowed students to apply these interdisciplinary understandings to their own problem context and effectively communicate their research questions, analysis, and findings to a broader audience.

Concluding remarks: key lessons learned

Interdisciplinary research and practice is not easy. It must be supported by a wider range of stakeholders – students, faculty, university administrators, and employers – and funding mechanisms to address real-world challenges. Despite wide-scale acceptance of interdisciplinarity as an effective mode of addressing complex problems, we need to recognize the challenges and opportunities provided by the context. The following lessons summarize some of the key learnings from our decade-long journey within the context of the Tufts Water Diplomacy Program that we suggest are transferrable to other interdisciplinary settings.

Combine tools for navigation with flexibility in milestones

The program must provide students with navigational tools (such as the 20 block T-Assessment process) to define their interdisciplinary goals and track their evolution over time. This will help them manage the expectations and requirements of an interdisciplinary program and their disciplinary home department. Paying attention to the expected level of engagement in various interdisciplinary activities by students from different disciplines is important. Specifically, program structure and procedures should flexible to assist students who are pulled in different directions because of meeting their disciplinary requirements, expectations, and milestones. Program flexibility in terms of meeting requirements and the timing of doing so at different points in their PhD process is essential.

Create an open intellectual forum for candid interdisciplinary conversation

Attracting and recruiting students with a wide range of research interests but with an overarching programmatic framework is critical. In this case, the focus and the framework is Water Diplomacy, while the research topics covered by the students included a wide range of disciplines. Such a diversity is interesting and enlightening to students; yet we need to recognize that it also makes interdisciplinary activities difficult to coordinate and implement within and across cohorts. We have created an open intellectual forum through our weekly Water Diplomacy Colloquium. Every week, students present their research ideas and seek feedback from a diverse group of faculty and students from multiple schools and disciplines. Weekly colloquium is an effective way to operationalize this pairing of intellectual diversity and research topics with a shared framework.

Provide a shared physical space for "accidental interdisciplinary" encounters

Students unanimously agreed that the common Water Diplomacy office space created for them a unique environment for highly productive and often accidental interdisciplinary encounters that led to many new ideas, publications, and research proposals. Shared office space and informal conversations created room for the free exchange of ideas after an exciting seminar or invited talks. The formal activities (such as shared classes and the annual retreat) also helped to build a sense of belonging to an interdisciplinary Water Diplomacy community.

Make the courses relevant to the chosen framework

Ensure that required interdisciplinary classes are strong on practical applications of the chosen framework. Use the classes to emphasize the need to develop a shared vocabulary that is transferrable across disciplines and contexts. For example, the first required Water Diplomacy course draws on a number of ideas from complexity science and negotiation theory. It begins with the Water Diplomacy Framework; then it uses three key ideas from complexity science (interdependence and interconnectedness; uncertainty and feedback; emergence and adaptation) and three from negotiation theory (stakeholder identification and engagement; joint fact-finding; and value creation through option generation) to show how application of these ideas can help enhance the effectiveness of water management.

Forge strong program ties with university administrators

Since interdisciplinary conversations in academic settings often exist between departments, interdisciplinary programs need institutional support to encourage and incentivize cross-departmental collaboration from the university at the highest level. Support in the form of teaching opportunities, post-NSF funding, and reward structures that incentivize tenure–track faculty engagement are critical in allowing the program to thrive. This dimension can ebb and flow depending on the individuals who serve as department chairs, deans, and provosts. It is essential to develop and maintain a communications strategy to keep them informed of benefits and the added value that interdisciplinary programs offer as well as national trends regarding student demand as well as job opportunities for students with cross disciplinary training.

Think early about building institutional memory

Too often, students move on after their funding ends. It is crucial to empower students who have graduated to maintain and strengthen their presence on campus and host events to familiarize other students and faculty about the opportunities and importance of interdisciplinary work. Our attempt to publish this book is an example of operationalizing the notion of building institutional memory.

Recruit and support a strong program coordinator

The role of an effective program coordinator is essential and often underrecognized. Selecting the right candidate and providing the necessary support through training and administrative power will allow the program coordinator to excel and support the complex tasks that are needed for a fully functional interdisciplinary program.

Take time for an annual retreat to bond, share stories, and adapt

At the start of each academic year, we welcomed the new cohort by hosting a weekend retreat off site for all members of the Water Diplomacy Program. This allowed the new cohort to bond as a group as well as develop relationships with faculty and existing students in an informal setting. While there was time for group recreation at the retreats, we also used the time to openly discuss what was going well and what could be improved and to set expectations for the upcoming year. Coordinating this event was a challenge, but it was well worth the investment.

Don't underestimate the importance of small gestures

If you can't afford to take your team off site for a retreat, consider ways you can utilize limited funds to improve the experience for students in small ways. While we don't have any official data on the matter, we suspect that the fact that we had food and coffee at our weekly colloquium went a long way in making the experience enjoyable, relaxed, and inviting. Other than food, we also invested in small gifts – water bottles, T-shirts, etc. – that helped keep a sense of unity and team spirit as the program adapted and evolved.

References

August, P.V., Swift, J.M., Kellogg, D.Q., Page, G., Nelson, P., Opaluch, J., Cobb, J.S., Foster, C., and Gold, A.J., 2010. The T assessment Tool: A simple metric for assessing multidisciplinary graduate education. *Journal of Natural Resources & Life Sciences Education*, 39, 15–21.

Dickson, R., and Saunders, M., 2014. Developmental evaluation: Lessons for evaluative practice from the SEARCH Program. *Evaluation*, 20 (2), 176–194.

Earl, S., Carden, F., and Smutylo, T., 2001. *Outcome Mapping: Building Learning and Reflection into Development Programs*. Ottawa, ON: IDRC.

Golde, C.M., and Dore, T.M., 2001. *At Cross Purposes: What the Experiences of Today's Doctoral Students Reveal About Doctoral Education*. The Pew Charitable Trusts.

Islam, S., and Susskind, L.E., 2013. *Water Diplomacy: A Negotiated Approach to Managing Complex Water Networks*. New York: RFF Press.

Lam, C.Y., and Shulha, L.M., 2015. Insights on using developmental evaluation for innovating: A case study on the cocreation of an innovative program. *American Journal of Evaluation*, 36 (3), 358–374.

Newell, B., and Proust, K., 2012. *Introduction to Collaborative Conceptual Modelling*. Australian National University. Open Access Research. Available from: https://openresearch-repository.anu.edu.au/handle/1885/9386

Patton, M.Q., 2011. *Developmental Evaluation: Applying Complexity Concepts to Enhance Innovation and Use*. New York: Guilford Press.

Patton, M.Q., 2012. *Developmental Evaluation Workshop*. Washington, DC: The Evaluators Institute.

Patton, M.Q., 2016. What is essential in developmental evaluation? On integrity, fidelity, adultery, abstinence, impotence, long-term commitment, integrity, and sensitivity in implementing evaluation models. *American Journal of Evaluation*, 37 (2), 250–265.

Preskill, H., and Beer, T., 2012. *Evaluating Social Innovation*. Center for Evaluation Innovation.

Walker, D.H.T., Bourne, L.M., and Shelley, A., 2008. Influence, stakeholder mapping and visualization. *Construction Management and Economics*, 26 (6), 645–658.

Westat, J.F., Mark, M.M., Rog, D.J., Thomas, V., Frierson, H., Hood, S., Hughes, G., and Johnson, E., 2010. *The 2010 User-Friendly Handbook for Project Evaluation*. National Science Foundation. Available from: http://www.evalu-ate.org/resources/doc-2010-nsfhandbook/

14 Reflections on the Tufts experiment with interdisciplinary Water Diplomacy research

Kent E. Portney, J. Michael Reed, and Amanda C. Repella

Water Diplomacy from a public policy perspective

Kent E. Portney

In 2009, Tufts University received a multimillion-dollar grant from the National Science Foundation to fund an Integrative Graduate Education and Research Traineeship program on Water and Diplomacy. With nearly a decade of experience in the earnest effort to meld a number of disciplines into a unified "Water Diplomacy" doctoral program, the time seems ripe to reflect on the successes and challenges that this effort experienced. I write this reflection from the perspective of one of the "gang of five" original grant co-PIs and executive committee faculty members overseeing the program responsible for guiding, in design and practice, the public policy and social science dimensions of Water Diplomacy, broadly defined. After all, diplomacy itself is a human process requiring extensive exposure to policy and social science issues.

In broad brush, the intent underlying the Water Diplomacy Program was to jointly develop technical and scientific expertise around water and knowledge of the social, political, and economic issues in which that water expertise is inevitably embedded. By design, the program sought to engage doctoral students from policy and behavioral sciences and from civil and environmental engineering and to create opportunities for these students to interact with each other, share a curriculum, and ultimately to engage in the research endeavor in such a way as to achieve various types of syntheses.

Presumably, the engineering students would bring the depth of understanding of hydrology and hydrologic sciences to the table, while the social and behavioral science students would bring knowledge of policy and social processes to the table. The various specific mechanisms used to accomplish this are discussed elsewhere in this volume. My effort here is directed toward reflecting on how well the program worked and what kinds of challenges were incurred. Here I highlight a small set of specific substantive issues that seem nearly universal at this point in time rather than highlighting the idiosyncratic and institutional issues in this particular program.

Different perspectives and mindsets

Common wisdom suggests that the primary impediment to interdisciplinary research of the sort sought in the Water Diplomacy Program is the disciplinary commitment of faculty in particular academic departments. But this description is somewhat limited. A primary challenge encountered in the Water Diplomacy Program was rooted in the nature of the students entering the program – the mindsets they brought regardless of whether those mindsets could be said to be disciplinary-based or not.

Students entering the Water Diplomacy Program were met with significant explicit challenges. Applicants to Water Diplomacy through the civil and environmental engineering program were evaluated along with all other applicants to the School of Engineering. The result was that the accepted students already had extensive education background in engineering, and most had significant work or field experience as practicing engineers. Most of these students came into the program with outstanding math skills and the ability to apply those skills to various kinds of natural system data analyses. A small number of science students were admitted through non-engineering departments or programs, and these students tended to share these characteristics with the engineering students. On the other hand, students coming into the program from the social and behavioral sciences were far more diverse.

Most were stellar students as undergraduates and had extensive experience primarily in the nonprofit and NGO world. They did not necessarily bring with them analytical skills comparable to those of the engineering and science students. Some had never been explicitly exposed to research on policy processes, to policy evaluation research, or to quantitative methods in general. Very few had ever been involved in conducting original empirical research. Moreover, few of the students from any prior background had exposure to economic analysis of water policy. All of this implies that each student came to the program with a set of priors – particular mindsets about the most appropriate and effective ways to address significant transboundary water problems. The hope of the program was that whatever appeared to be gaps in students' backgrounds would be filled by the common core Water Diplomacy curriculum, and the result would be the development of a new mindset – a synthesis of their priors and the Water Diplomacy curriculum.

From mindsets to synthesis

In the best of all worlds, the program perhaps would have liked to take engineering students and teach them how to think like social scientists and economists and to take social science students and teach them how to apply hydrologic sciences to their research. But the program was more realistic than that in the sense that it sought to ensure that students understood important aspects of each even if they would never become experts in the concepts and methods from fields they theretofore had not been exposed to. It did hope to ultimately stimulate students' creativity in the search for synthetic approaches to the dissertation research in ways that no participating faculty member would be willing or able to tackle. Converting the various mindsets into practice was the challenge.

Tensions around "systems thinking"

One of the inevitable lessons embedded in the engineering-based part of the curriculum was the importance of understanding systems analysis and systems modeling. Indeed, the students with backgrounds in engineering understood this lesson. The challenge in this is that many of these students then sought to apply systems approaches to public policy and policy-making analysis, an idea that has been present in the policy sciences literature for many decades. With some notable exceptions, however, such applications of systems modeling have not turned out to be fruitful approaches to understanding public policy processes or results. This was an explicit topic of conversation and reading in the Water Policy and Economics course, one of the core courses in the program. Yet the commitment to systems analysis and to the idea that such approaches to research might serve as the foundation for the kind of interdisciplinary syntheses persisted despite the fact that none of the participating faculty from the social sciences and policy was an advocate for systems analysis.

A consequence of the dominance of systems approaches to Water Diplomacy often led students to seek to create "tools" designed to provide information and guidance to water policy makers and managers. And inevitably this was done without any specific regard to any sort of theory of water policy making or theory of water governance and decision making. It was based more on the assumption that if better information is provided to policy makers and managers, they will make better decisions. Of course, this is a common subject for research on science communication more broadly, and existing research on this issue defies simple conclusions. However, it was not beyond the scope of the program to imagine that a student might study the efficacy of a tool or set of alternative tools designed to provide better information. But defining a piece of research of this sort required that such students think and act like social scientists, and that was often a difficult path for students to justify.

Alternatives to systems thinking?

While the engineering elements of the curriculum were heavily imbued with systems analysis, the nonengineering curriculum did not contain any dominant set of theory or approaches to research. Because the students who came into the program from nonscience or nonengineering backgrounds had such diverse priors, developing a small number of common themes was somewhat challenging. Indeed, simply conveying the importance of conducting systematic empirical analysis of water-related policies and programs, building the capacities to engage in such research, and providing the levels and types of support to enable original policy-related research were never adequately accomplished.

Developing clear-cut alternatives to the systems thinking approaches became a weak link in the Water Diplomacy Program. Many students sought to delve into Water Diplomacy from the perspective of behavioral sciences, sometimes in the U.S. and sometimes in other nations. Such issues included, for example, understanding how to change individual people's or households' water consumptive behaviors or under what conditions individuals would be willing to support particular policies or programs to resolve water disputes. This required exposing students to an array of research issues and often involved developing expertise on the cultures and history of the relevant places. Other students took a decidedly "institutional" approach with the intent of understanding the policy-making institutions, organizations, and governance processes of particular places. The Water Policy and Economics course sought to address these issues by exposing students to the works of the eminent political scientist Elinor Ostrom. (See Ostrom 1990, 2007, 2009, 2011, for example.) The intent was to use her "institutional analysis and design" (IAD) approach as the foundation for channeling interest in melding understandings of behavior and institutions as applied to specific water policies, initiatives, and conflicts. Although IAD seems well-tailored to appeal to students who understand systems analysis, it proved to be too generic to serve as a guide to students looking for alternatives. In retrospect, relying on IAD approaches did not take students far enough and did not provide a clear pathway to defining their own original research.

Another alternative that evolved in its importance was embedded in negotiation and mediation techniques with special emphasis on mutual gains and other non–zero-sum game approaches (Islam and Susskind 2013). While this provided a highly useful way for students to understand how their knowledge of hydrology and policy-making institutions and processes could be applied in many real-life situations, it never produced a clear-cut agenda for students' research, although one Water Diplomacy student (Laura Kuhl, now Assistant Professor of International Affairs at Northeastern University) and colleagues successfully made use of this approach in defining ways of expanding its use beyond issues of water, per se, into "sustainable development diplomacy" (Moomaw et al. 2017). Certainly not intended, gravitation toward negotiation and mediation and away from public policy analysis may well have contributed to a sense that applied Water Diplomacy is sufficiently disparate from needed empirical research that the two need not be or cannot be merged.

Successes

With all of the disciplinary and institutional challenges described above and elsewhere in this volume, it is easy to overlook the ways in which the program was successful. Indeed, some of the chapters in this volume highlight the kinds of synthetic analysis that the program can take credit for producing.

One such success, in particular, deserves mention here. The works of Margaret Garcia, a Water Diplomacy student now an Assistant Professor of Sustainable Engineering at Arizona State University, serves as an outstanding example of how the concepts from multiple disciplines can be brought together in service to research that is truly interdisciplinary. Garcia's doctorate was granted in civil and environmental engineering, and indeed her research is very well grounded in systems modeling. Yet she has found creative ways of incorporating important policy-making aspects into her analyses.

A prime example of successful synthesis is found in the research Garcia conducted for her dissertation and that yielded a number of refereed-journal publications. Professor Garcia's dissertation, is based on extensive analysis of information from water policies and decision making in Nevada with special reference to metropolitan Las Vegas (Garcia 2017) and led to the development of a hypothetical "Sunshine City" example in which hydrologic modeling is used to understand coupled human–natural systems of water in a major hypothetical metropolitan area (Garcia et al. 2016). The analysis of Sunshine City is designed to provide insight into the dynamic processes associated with different decisions as they affect a number of water system characteristics over time. The modeling takes into account a wide array of variables measured at different scales not found in standard hydrologic models.

Should water system policy makers and managers try to meet demand as long as sufficient supplies exist, or should they adopt a policy that serves as a hedge in anticipation of an expected water deficit? What is the long-term impact of the decisions made at one point in time on the sociohydrologic system? Pursuing one approach rather than another seems to produce different consequences for water demand and pursing the hedging policy dampens the tendency for water systems to experience oscillations in the need to conserve or restrict water usage. These effects are not apparent when using traditional hydrologic models. This sociohydrologic model adds important but generally unconsidered dimensions to assist in the development and evaluation of water policy.

Conclusions and lessons

No one who sets out to create a truly interdisciplinary program does so with an expectation that it will be easy. Even when armed with familiarity of experiences from other such endeavors and with a deep commitment to their prospective virtues, interdisciplinary curricular projects face numerous challenges. Many of these challenges become apparent in the implementation of the program, and this is certainly the experience with Water Diplomacy. This program inevitably had to tackle appropriate and effective ways of integrating key (and sometimes disparate) concepts and research approaches embedded in the designated disciplines. With all the challenges faced by the Water Diplomacy Program, in retrospect, its successes are significant and set the stage for new research agendas that are not beholden to existing academic disciplines.

Reflections from an ecologist on an interdisciplinary IGERT program on Water Diplomacy

J. Michael Reed

I was one of the original "Gang of Five" in our Water Diplomacy Program – that is, one of the five PIs originally on this grant. As an ecologist, I was the only person in our leadership whose primary interest was (still is) in wildlife rather than in people. Our goal was to train a suite of graduate students in both breadth and depth in Water Diplomacy (Chapters 1 and 2). That is, we wanted students to have professional interests outside a single discipline and to have sufficient knowledge in other disciplines (ecology, engineering, social sciences, economics, etc.) so they could successfully speak the others' languages and engage in creative solutions to problems. We know that language can be an uncommon but regular barrier to meaningful action – every discipline has its professional words, and when the same words mean different things across disciplines, it can cause confusion. Even more basic were philosophical differences in our disciplinary world views; we often did not even agree on such fundamental concepts as differences between societal and natural domains (Chapter 1). As an ecologist, I view agricultural areas as part of the societal domain, while some of the original Gang viewed them as part of the natural domain – the remaining members carefully stayed neutral on the issue. Despite these disciplinary barriers, I think we were successful in most of our goals.

I had two students for which I was the primary advisor go through the Program, and they benefited greatly from it. Along with the funding, training, and internships, a great program benefit is that they started projects (some ongoing) that they would not have done were it not for the extensive interactions with each other and the other students in the program. This work resulted in producing material that will be published, but outside of their dissertations. One potential concern in a program like ours is that being deeply involved in a program outside of their discipline (ecology), the students had to work hard to budget their time and to not start too many side projects, which could derail one's formal degree. There was pressure to be committed to both the IGERT program and to the student's home department, which could be difficult for some students, particularly if they do not have a supportive graduate committee.

The complexity of many water-based issues, particularly if you pay attention to wildlife needs within the natural domain, as well as to human issues, is such that for the most part a single person cannot leave graduate school with all the tools to solve them. That was not our goal. We all knew professionals who had successfully bridged disciplines and did Water Diplomacy, but they all took their own different pathways to get there, largely due to decades of experiences. Rather, our goal was to draw on these people and their experiences, as well as on specialists in particular disciplines, to create a training program that would give students the capacity and appreciation to know when and how to collaborate on the solution of a problem. For the most part, I think we achieved this goal. What follows is a brief overview of what I think worked in our program and what did not work, in the hopes that it will inform future interdisciplinary programs. I limited myself to what I view as the three or four most important examples of each.

What worked

Getting students from different disciplines to talk and work with each other

This program could not have succeeded without extensive and intensive inter-action among the students (as well as with faculty). As anyone in any institution knows, being physically separated rapidly results in diverging cultures, not knowing your colleagues, and not exchanging ideas and information. This cannot be overcome without a pointed effort to create face-to-face time, and from my perspective, this is the biggest barrier to collaborative research of any kind. Our program succeed because of personal interactions – graduate students shared one large office space; they took classes together with their incoming cohort; we had weekly meetings to share research and program ideas with students and faculty; annual weekend retreats; after the first year, students were involved in interviewing graduate candidates for subsequent cohorts; we had peer-advising programs across years; and there was planned and unplanned socializing. Students talked to each other – a lot. This generated familiarity with one another's disciplines, projects, values, and ideas. Disappointingly, some students made the deliberate choice to not integrate into the program (see below).

The T exercise

One of the requirements of our grant was to engage an outside evaluator with whom we would meet regularly to update on the program, work on planning, and receive feedback. Get a good person for this – specifically, one whose goal is to help you succeed. Our evaluator was Glenn Page, who runs SustainaMetrix (www.sustainametrix.com). One of the exercises he made students do is the classic T exercise, whereby one identifies the discipline in which one wishes to develop depth and the topics in which one wishes to develop breadth, which can be drawn in a T shape (Figure 14.1a). This concept of depicting breadth and depth had its origin in depicting sharing knowledge across an organization while primarily supporting one's own work unit (Hansen von Oetinger 2001). It seemed corny to me at first, but it really engaged the students, and over time I came to appreciate its effectiveness. Taking this concept one step further, Morello (2005) introduced the concept of a "Versatilist," which is someone who is a specialist in a discipline but who can also work across disciplines because of their breadth of

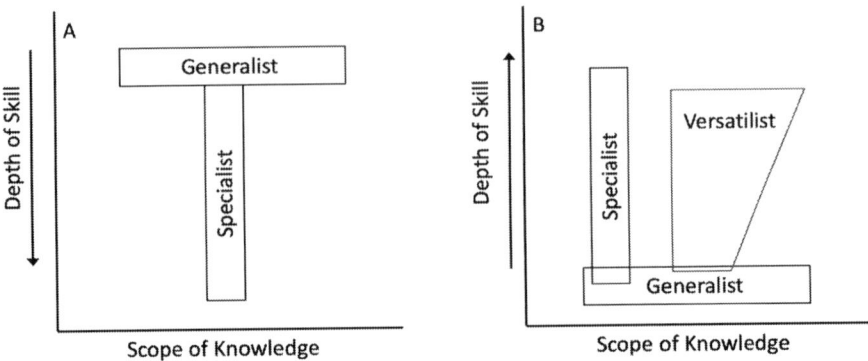

Figure 14.1 (A) The T-shaped view of combining specialist and interdisciplinary knowledge and the (B) view of a Versatilist, who combines generalist and specialist knowledge – this is how I view our Water Diplomats. The latter is modified from Morello (2005).

knowledge and experiences (Figure 14.1b) – this is how I view our Water Diplomats and, I think, how most of them view themselves.

Internships

Each of our graduate students was given a significant amount of funding to support an internship. This was to support interdisciplinary and disciplinary research related to the Water Diplomacy Program. I think it was wildly successful. Using these internships, our students engaged in real-world problems, created and expanded professional networks, and got excellent immersion in a wide variety of experiences. For all of the students, results of the internships became part of their dissertations, and for many, they led to job or postdoc opportunities. These internships allowed the students to put into practice what they learned in the classroom.

Responsive leadership

Water Diplomacy is an emerging field. Although the greatest contributions to date appear to be from civil engineers and policy wonks, our program showed that significant contributions could be made from a variety of additional fields. From my perspective, the professional field is in flux, it is in many ways quixotic (reflecting the specific personalities running the program), and it is being forged by professionals with traditional training – typically with strong personalities. Consequently, one can easily imagine that the leadership would have a strong opinion about how students should be trained. And this was true for us. What was unexpected (at least by me, at first), was the degree to which the program's PI (Islam) sought out student feedback about the various courses and training and modified them for training subsequent cohorts of students.

As a specific example, the order of two of the courses was reversed because students thought the material and content would work better in reverse order. It would have been much easier to stick dogmatically with the original plan if, for no other reason, the folks in charge "knew better" how to train students on what the emerging field was all about. However, Islam actively and formally set out to get feedback on various parts of the training, and students saw that the program changed in response to their feedback. Along with improving the program, this generated a feeling of ownership, at least in the students who really engaged in the program.

What did not work

Not all students engaged in the program

Despite being given two years of salary, tuition waiver, funding for an internship, support to attend conferences, excellent cross-disciplinary training, and other opportunities for professional development, some students were in the program only for themselves. Once their financial support ran out, they largely disengaged. They were interested strictly in what they could get from others in the program and to bolster their CVs, but they were not interested in volunteering their time, participating in group activities or projects, or helping to maintain the cohesiveness of the program. It was obvious to students in the program – possibly even more so than it was to the faculty – and it put a drag on a small program (5–7 new students per year). It meant that the heart of the program, from the student realm, was driven by only a subset of students. Is there blame to be apportioned? Perhaps a little, but the goal is to learn for future programs. I would say that some people were allowed in the program despite warning signs that they were not team players. Some students took their cues from disengaged faculty (see next section). But for the most part, I am not sure we could tell ahead of time. One wants a variety of student types – not just obviously outgoing students, because some of the quiet students bloom later – so it is probably inevitable that some students will not engage. In that case, I suspect that the key is to have a large enough program to absorb the drag caused by these students.

Engaging a broader portion of faculty

For a program that lasted only seven years (given no-cost extensions on the grant), I was surprised at the faculty turnover and the general lack of faculty interest outside the program. In retrospect, the latter surprises me less now than it did then. Of the original co-PIs, only two of the five stayed in the program longer than 4 years. Departures were due to a philosophical conflict, a new job, and retirement. These faculty were replaced by others that became mentors to students they had supported in the program. However, we almost never had a meeting at which all the faculty mentors were present. I can see faculty outside the program might not be strongly interested in our graduate students – after all, faculty are very busy – but there were opportunities to become part of the program and have a graduate student or two supported for 2 years. I thought this would be sufficient incentive, but few faculty took advantage, and not all who took advantage by having a student supported actually joined in with any of the supporting activities. Money must be available to initialize engagement, but it is not always enough to secure more than superficial interest. I don't know how to solve this problem, which may be widespread in academic institutions.

Getting more than initial buy-in from the university administration

I guess that like many shiny toys in a world of short attention spans, investing money in a new, cross-cutting program is exciting, but investing to keep it going and expand is less exciting. Initially we received interest and support from the university for our program, being promised space and resources for development. After all, an NSF IGERT grant is prestigious. Once the money showed up, however, there were continual battles to keep our space and to get other resources to improve our program. As I mentioned above, space for graduate student offices – all together in one area – was one of the key features that allowed this program to thrive. Space is a perennial problem on college campuses, but having a long-term commitment to a program is necessary for it to succeed – even if this just means we stop putting up barriers. One example of a barrier that surprised me is that our grant supported students from both the College of Engineering and the College of Arts and Sciences, and it was never easy sharing resources – even though we are the same university! Something as apparently simple as getting an expense reimbursed could sometimes take months, part of which was arguing who was in charge of the paperwork. The senior university administration could have reduced the barriers between its own colleges.

Acknowledgments

I thank K. Reed and J. Rozek for comments on this essay.

It is always about the people: human effort and the Water Diplomacy program

Amanda C. Repella

While innovation in interdisciplinary PhD education may be the greater legacy of the Water Diplomacy experiment, the Tufts Water Diplomacy Program had multiple arms and aims. The NSF IGERT funded interdisciplinary research program was the keystone to its formation and progression. The NSF funded Water Diplomacy Research Coordination Network (2012–2016) worked to catalyze and strengthen new collaborative research relevant to Water Diplomacy. The Water Diplomacy Workshop (2011–2016) brought a cohort of international water practitioners to Boston for a week each summer to explore Water Diplomacy concepts. The Water Diplomacy Aquapedia Database captured case studies of boundary-crossing water problems written by more than 50 volunteer contributors.

These activities, though distinct from the PhD program, blended well with opportunities for students to assist with trainings, write case studies, and participate in the Boston Water Group, a unique water research and practice focused network connecting researchers and practitioners working in "water" across the greater Boston metro area.

There are a range of outputs and outcomes discussed elsewhere in this book. However, under all of these reportable and measurable items is a truly unmeasurable amount of human effort contributed by students, faculty, workshop alumni, Water Diplomacy Network and Boston Water Group Members, and staff. Human effort is not just the intellectual capacity that was converted to course syllabi, research, papers, presentations, and dissertations but also the community building, communication across and between Water Diplomacy activities, and the support and assistance that helped move projects forward and encouraged individuals to continue to volunteer their time to our causes. Here, I share my two takeaways from my experience with Water Diplomacy at Tufts.

Technological tools continue to require human buy-in and human capacity

The Aquapedia Case Study Database was envisioned as a new tool for sharing case studies about complex water problems and their resolutions (or lack thereof). This wasn't just a repository of cases but an organizational system that would be more operationally useful for sharing knowledge and learning and collaboratively building a better understanding of complex boundary-crossing water problems.

A general problem with case studies is that they capture a history up to a point in time, and then they are published and abandoned. Cases are limited in terms of the perspectives or framing for the case problem. A traditional case study represents what might only be a small piece of a larger puzzle that requires additional context, new perspectives, and updates over time to truly understand. Aquapedia was seen as an opportunity to make case studies of boundary-crossing water problems more like living documents, with the ability to add and update. The goal was to create a place where cases could represent multiple viewpoints and be edited over time to add greater context. Rather than attempting to only include consensus views or those informed by our own research and biases, we wanted to make space for unusual or opposing takes on the same problem. By using software that allowed for semantic classification and search, the hope was to connect cases by themes that aren't typically used as organization tools in hierarchal databases of historical events (e.g., geography, time period, type of water resource) but instead by the underlying issues of the case (e.g., types of constraints, types of systems impacted by the resource or conflict, types of stakeholders, types of governance issues, etc.). The most important aspect was to make every case freely available online and editable by any real person with something to contribute to a case's story.

The vast majority of cases contributed to Aquapedia came from Tufts or Massachusetts Institute of Technology (MIT) graduate students who wrote them as part of a class requirement. Some came from Water Diplomacy Workshop participants, and a few were from undergraduates. Writing a case study is a time-consuming thing to do. That research and writing task is made much more difficult by asking our students to consider their written work to be a living document that can be altered, edited, and challenged in the future. We asked students to take a problem and communicate their take on it but to additionally consider the countless other perspectives that could also be taken and how the knowledge of that problem may change over time before writing their piece of that potentially endless story. We also asked them to consider how their case could relate to other water conflicts, whether those were in our database, or they were unknown, future problems. What could their case do toward informing individuals with interest or stakes in these other cases? What key issues or questions would forge a meaningful linkage between their case and other complex water problems?

Within the Water Diplomacy Program, in which we incorporated key ideas from complexity science into our basic assumptions, this was a reasonable request. However, in practice, asking students to publish their work (1) in an unfinished or incomplete state where anyone could read it; (2) in a format fully co-optable by someone else in the future; (3) in a process that must take into consideration a multitude of perspectives, some that may not yet be clearly articulated; and (4) use that process to try to connect that work to other works that are currently unknown and perhaps not fully knowable – is a big request. This request flies in the face of the writing instruction and practical report writing experiences students build from grade school onward.

In presenting the work requirement, we used stories. We borrowed examples from Wikipedia, showing the evolution of an article from a single sentence then edited by multiple contributors over multiple years to thousands of words. We discussed the importance of freely accessible, cited, and well-written information to individuals struggling to understand a problem without access to the wealth of library materials and experts available across our campuses. We used analogies and told stories to obtain buy-in to a writing process that was foreign to their student experiences thus far in their extensive educations.

Some students really embraced this process. Some students did not. This is where professors, the Aquapedia student assistant, or I stepped in to help shape the final work product in a way that met both the student's current need and the project's needs.

The database was built with Semantic MediaWiki, which is a powerful open-source extension to MediaWiki, the software that powers Wikipedia. Semantic MediaWiki turns a wiki website into a knowledge management system in which information within pages can be assigned to a semantic classification system. A challenge here, is much like in Wikipedia, there becomes an additional technical component to writing, as a specific markup language is used for formatting text and images. Aquapedia never broke free of the need for a dedicated editor to assist in transferring content written by case contributors to the system. Only a handful of case writers did not seek any technical assistance in adding their cases. The common thread of those writers: they had previously written content for Wikipedia or were particularly skilled in a different markup language (such as HTML) while also being generally curious about how to use new-to-them technologies.

There was a general hope that a critical mass of case studies in the database would be beneficial for supporting or refuting the assertions made in the Water Diplomacy Framework (Islam and Susskind 2013). I'm not sure if the database met that criteria. I also don't know how widely it is truly used as was originally intended. I believe it is still a strong example of a novel way to organize case-based knowledge as a network of functional themes rather than by a hierarchical organization of geographic or resource attributes. We used the best available open-source tool for creating and building Aquapedia, but it's not a sustainable platform to maintain these cases as living documents because of the required content gardening, necessary technical updates, and the hurdle of a learning curve for contributing content. Until there is a cost effective and largely automated process for both extracting semantic connections from case study documents and still formatting those cases for human reading, this type of approach may not be able to grow to the case study critical mass needed to use it as a mineable research tool.

It is still always about the people

Human systems require humans. Of course, by definition this is true. However, competing priorities for limited funding force grant seekers to attempt to find the greatest efficiency in their allocation of dollars. Grant makers often minimize the ability to fund staff roles in RFPs to maximize the dollars that go directly to the individuals and systems they hope to impact.

Through the combined grants that enabled this program, there were multiple goals that required the voluntary participation of researchers, practitioners, and students who made substantial contributions to the research and scholarship of Water Diplomacy. Some PhD students received two years of tuition, stipend, and internship support. However, the majority of individuals involved with Water Diplomacy across all of our activities were not directly compensated for their time. Some travel or conference funding was available, primarily for students. A few Aquapedia case study contributors did so in return for partial funding support for their Workshop participation. Some faculty participation was entirely in support of their students in the program.

Keeping so many people connected, engaged, and productively contributing time, effort, and intellect to the various Water Diplomacy activities required significant relationship management. Water Diplomacy represented a human system of students and professionals from academia and practice, both those at Tufts and network members in Boston as well as around the globe. Maintaining these relationships required consistent gardening to nurture, strengthen, and maintain the human connections that facilitated collaboration and productive work. There was plenty of this work to go around, and these actions and activities were performed by students, faculty, associates (particularly dedicated Boston Water Group participants) and program staff. This program benefited from its ability to fund two full-time supportive staff (Global Network Coordinator, IGERT Program Coordinator).

The annual Water Diplomacy Workshop supported two contract administrative roles: an outreach/enrollment/logistics manager and an on-site event manager. Aquapedia had a paid student assistant role for a significant portion of its active development. Each of these individuals were trained and experienced in fields that were relevant or complementary to Water Diplomacy. These were not calendar keepers and milestone watchers but integrated members of the team who were able to build strong relationships with program leadership and participants to help support continued forward momentum and make meaningful contributions to the collective body of work produced during this program.

The process of building Aquapedia, a freely editable open-access database of 74 case studies on transboundary water conflicts, required countless volunteer hours, primarily from graduate students at Tufts and MIT. However, each of those cases required technical assistance, editing and interactions with the volunteers along the way.

The Boston Water Group was fueled by curious members of our community who continued to voluntarily turn out to events to talk about water problems that might not have had any relevance to their professional work. Invested faculty and ongoing staff support helped to sustain the effort, but the value of those exchanges required enthusiastic participation from individuals who freely volunteered their time and intellect to the community. While our various communities across Water Diplomacy activities saw some attrition, our efforts around relationship management and communication helped to keep group members engaged.

There remains a great need for supportive roles filled by actual dedicated humans in these ambitious programs. Integrating intelligent, appropriately credentialed, skilled individuals who are granted the time and freedom to focus on communication, messaging, and relationship management is particularly valuable to the success of ambitious research and scholarship communities.

References

Garcia, M., 2017. *Infrastructure, Hydrology, and Policy: Socio-Hydrological Modeling of Urban Water Consumption Dynamics.* Tufts University.

Garcia, M., Portney, K., and Islam, S., 2016. A question driven socio-hydrological modeling process. *Hydrology and Earth System Sciences*, 20 (1), 73–92.

Hansen, M.T., and von Oetinger, B., 2001. Introducing T-shaped managers: Knowledge management's next generation. *Harvard Business Review*, 79 (3), 106–116.

Islam, S., and Susskind, L.E., 2013. *Water Diplomacy: A Negotiated Approach to Managing Complex Water Networks*. New York: RFF Press.

Moomaw, W.R., Bhandary, R.R., Kuhl, L., and Verkooijen, P., 2017. Sustainable development diplomacy: Diagnostics for the negotiation and implementation of sustainable development. *Global Policy*, 8 (1), 73–81.

Morello, D. 2005. *The IT Professional Outlook: Where Will We Go From Here*. Available from: www.ee.iitb.ac.in/~hpc/.old_studs/hrishi_page/outlook/report.pdf

Ostrom, E., 1990. *The Evolution of Institutions for Collective Action*. New York: Cambridge University Press.

Ostrom, E., 2007. A diagnostic approach for going beyond panaceas. *Proceedings of the National Academy of Sciences of the United States of America*, 104 (39), 15181–15187.

Ostrom, E., 2009. A general framework for analyzing sustainability of social-ecological systems. *Science*, 325 (5939), 419–422.

Ostrom, E., 2011. Background on the institutional analysis and development framework. *Policy Studies Journal: The Journal of the Policy Studies Organization*, 39 (1), 7–27.

15 Perspectives on Water Diplomacy

Key findings, remaining challenges, and future directions

Lawrence Susskind, Enamul Choudhury, and Greg Koch

Interdisciplinary approaches to Water Diplomacy: reflections on six sticky ideas

Lawrence Susskind

People involved in managing water resources, especially boundary-crossing water problems, must be able to get others to take their ideas and proposals seriously. Since individual users or managers are rarely in a position to impose their views on other stakeholders, even if they have the legal authority to make final decisions, they need to be able to work collaboratively or, at least, to convince others to let them do what they want. Under the best of circumstances, water users and managers would engage in joint problem solving leading to detailed water sharing and water management agreements. Each party's ability to acquire or control shared waters (along with related land and infrastructure) depends on their capacity to negotiate with other riparians, decision makers, and stakeholders, including some from very different backgrounds.

At Tufts University, with support from the U.S. National Science Foundation, our faculty team has spent more than 7 years training doctoral students with an interest in Water Diplomacy. Successful "Water Diplomats" must know how to interpret complex technical information, often drawing on their disciplinary skills, and integrate them with tacit local knowledge. They must also be able to facilitate or participate in water sharing and water management negotiations at different levels of governance.

In the Water Diplomacy field, we have been trying to deepen our understanding of how to manage conflicts over water allocation, especially at the food-energy-water nexus. For example, we have built Aquapedia, an online collection of detailed case studies of water conflicts around the world, to generate case studies describing the various ways in which boundary-crossing water conflicts tend to unfold. There is a great deal of science involved in water disputes; indeed, Water Diplomacy usually begins with an assessment of water resource availability and the complex systems involved.

Competing demands on a natural resource must also be viewed in the context of the interactions between these natural systems and the sociopolitical systems by which competing claims are typically resolved. Our goal has been to think about ways of empowering stakeholders to pursue their interests in the face of competing and apparently irreconcilable demands on limited water supply. The doctoral students who have been part of our Integrative Graduate Education and Research Traineeship (IGERT) program have learned how to generate useful prescriptive advice – **what to try and why and how to respond when what you try initially doesn't work** – for a wide range of situations that won't be defined primarily by the formal authority they might have.

To ensure that Water Diplomats will remember the most important things they need to know, we have identified six key ideas – drawn from the interdisciplinary field of negotiation, particularly water negotiation, and framed them in the most vivid way possible. We want them to be easy to recall in fast-breaking and sometimes uncomfortable circumstances. Framing them in the right way makes them "sticky," or memorable. Stickiness, in our experience, is most readily achieved when a prescriptive idea ("how to do something") is clearly linked to a broader objective ("what needs to be done and why"). Thus, our goal is to connect the "how" and the "why" of water management negotiation in the most memorable ways possible.

Six sticky ideas

Here are six sticky ideas intended for stakeholders, decision makers, and water managers to keep in mind while addressing complex water issues: (1) focus on the interests of others, not just your own; (2) invest heavily in the right kind of preparation; (3) be ready to play the game of "what if?"; (4) seek to create value (i.e., most water negotiations do not need to be framed as zero-sum battles, although they often are); (5) accept uncertainty as a given; don't try to eliminate it by allowing experts to dominate; and (6) create the most robust agreements possible (i.e., agreements should be easily modifiable in light of new information or changing conditions). These ideas are discussed in more detail in Islam and Susskind (2013). Now I will review each in a bit more detail (see Table 15.1 for a summary).

Focus on the interests of others, not just your own

All too often, water users and the stakeholders with whom they interact are preoccupied with their own goals and objectives. They tend to focus single-mindedly on the marching orders they were given by someone to whom they report (i.e., their "backtable"). Our research shows, though, that the best way to ensure your interests are met with regard to water allocation, especially in a situation in which others also have a say, is to make sure that the interests of other stakeholders are being met simultaneously. This requires trying to learn as much as possible about the concerns of others. Put yourself in their shoes. Once you understand their priorities, you'll know what you need to offer them, at a minimum, to win their support. The goal of Water

Table 15.1 The What and Why of the Six Sticky Ideas

What	Why
Focus on their interests and ideas, not just your own (even if they are contrary to your initial assumptions).	Listen carefully and make space for their views. Your initial assumptions about what the other side believes could be wrong. The other side may not be monolithic; you can capitalize on internal disagreements on their side. You might hear something you didn't expect that can lead to an outcome that is even better for you than what you intended to request.
Preparation is key for you and your side – learn as much as you can beforehand.	You want to be able to make arguments on the merits. You will be more convincing if you can appeal to "what any independent and qualified observer would agree on." The more prepared you are, the more confident (and less defensive) you will be.
Be ready to play the game of "what if?"	Brainstorming can open up new and better possibilities. To do this effectively, you need to agree to a period of time that can be spent "inventing without committing." When there is no implied commitment, negotiators will be more open to considering unfamiliar packages or deals. You are unlikely to stumble across mutually advantageous trades. They usually emerge when one side asks, "what if I offered A, would you give me B?"
Seek to create value – most negotiations over water management or water sharing do not need to be framed as zero-sum battles.	You are unlikely to create value unless you set out to do so. Add more issues to the discussion. Search for mutually advantageous trades. Link issues in creative ways. Exploit differences in the time value of money. The key to creating value is packaging.
Accept rather than try to eliminate uncertainty.	By spelling out contingencies, it is often possible to fashion agreements that include conflicting forecasts. "If my forecast is correct and X happens, we agree to proceed as follows. If your forecast is correct, we will shift to all the alternative provisions spelled out in Appendix A." Thus, the parties do not have to agree on the forecast, just the contingent conditions. Since each side assumes their forecast is correct, they should not object to including contingent alternatives as an appendix to the agreement.
Create robust agreements that won't collapse if something unexpected occurs or	Include dispute-resolution clauses in all agreements. The worst time to talk about the

What	Why
one side thinks the other has failed to comply.	rules for handling a disagreement is in the middle of the disagreement. Commit to maintaining relationships. Agree to monitoring implementation of all agreements and work to build trust through continued commitment to adaptive management.

Diplomacy is to meet your interests very well while meeting the interests of others reasonably well. The techniques for doing this are spelled out in Susskind (2014). Remember to make space for the ideas and interests of others, even if they are contrary to your own.

Preparation is key for you and your side

There is ample evidence to suggest that the best ideas, as has often been stated, "occur to prepared minds." This is especially true with regard to water management and allocation decisions. The more you understand the history and the context of the situation you are in, the more likely you are to come up with a way of proceeding that will meet interests on all sides and take account of the best scientific advice available. Preparation means understanding how all the relevant systems work, what alliances with other stakeholders might be relatively easy to build, and what the institutional obstacles are that must be overcome to get any agreement implemented.

Be ready to play the game of "what if?"

The best negotiators spend time at the beginning of every negotiation talking with the others about how they see the situation and how they think negotiations ought to proceed; that is, they negotiate how they are going to negotiate. This includes setting an agenda, a timetable and ground rules. For inexperienced negotiators this may seem counter-intuitive, since it puts an emphasis on cooperation at the outset, at a time when inexperienced negotiators often think they should be posturing to gain some advantage.

When experienced negotiators agree on a way to proceed, they can commit to a period of brainstorming, i.e., what Fisher and Ury (1981) call "inventing without committing." If you can brainstorm a range of possible elements of agreement before anyone locks into a firm position, negotiations are much more likely to lead to mutually beneficial outcomes. So remember to negotiate how you are going to negotiate.

Seek to create value

Too many water negotiators believe that water allocations are zero–sum games (i.e., any gain to one must reflect losses to the other party or parties). In real life, though, experienced water negotiators know this is not the case! The same water can be used in different ways, multiple times, as long as the parties involved cooperate. In a sense, more water can be created through efficiencies, recycling, reuse, and investment in new technologies and practices. These will produce more water for everyone, as long as the parties agree on the rules and adhere to them.

Water saved through collective efforts means there is more water for everyone. The creation of value in water negotiations – expanding the proverbial "pie" – should always precede efforts to allocate available supply. Along similar lines, adjacent countries that share boundary-crossing waters are more likely to enhance their own water security if they try to enhance their neighbors' water security as well. Unilateral efforts by one nation to enhance their own water security by attempting to control water along their border actually reduces their neighbors' security. This, then, causes their neighbors to act in ways that reduce the unilateral actor's water security. Value creation is possible but only if parties cooperate, so look for mutually advantageous trades.

Accept uncertainty rather than trying to eliminate it

Uncertainty about the future is inevitable. Parties rely on different experts (who often start with contrary assumptions) to generate conflicting estimates of future water supplies or different estimates of demand. They are more or less inclined to subscribe to assumptions that a new technology or practice will work as planned. Natural systems create unexpected impacts, like climate change. Given there is no way to eliminate uncertainty, and parties have different appetites for low-probability events, water-sharing agreements or water-sharing plans must accommodate these differences. Use contingent agreements to bridge differences in the assessment of risk.

Create robust agreements

Agreements that make no accommodation for unexpected changes or surprises are very brittle. Good water-sharing agreements include "dispute resolution" clauses that spell out how suspected violations or unexpected changes will be investigated and handled. They also include provisions that encourage parties to maintain communication (even if everything is going well). If underlying relationships are strong, disagreements about how to adjust to changing circumstances are much easier to work out. So work to build relationships and trust through continued interaction.

Keeping context in mind

Learning to merge the "what" and the "why" of Water Diplomacy is not easy. You can't master everything at once, no matter how many books you read or lectures you listen to. The only way to build a successful personal theory of practice is through trial and error and, of course, careful reflection and a commitment to learn from your own mistakes. A neophyte learns rules and tries to apply them exactly as they are prescribed; a more advanced learner focuses on the context in which they are operating and knows the basic rules well enough to know which ones to follow and which to ignore. Context (and culture) are absolutely essential to formulating and deploying appropriate Water Diplomacy strategies. The six sticky ideas I have presented offer a starting point, but knowing when and how to apply them requires careful attention to context.

One key element of context is water law. Water law prescribes a set of principles that have evolved over a long period of time (both domestically and internationally). Water law provides a starting point, but general propositions rarely decide concrete cases. Water law proclaims (1) the principle of equitable and reasonable use of international watercourses; (2) the obligation not to cause significant harm; (3) the obligation of notification, consultation and negotiation of planned measures; (4) obligations related to environmental protection (i.e., environmental minimum flow); and (5) the obligation to peacefully settle water disputes. The Water Diplomats and students we have trained know that these principles must be interpreted and applied in different constitutional and political settings. They know these principles, but they also understand enough about the importance of context to know that they have to be ready to make an argument for how they should be applied in particular situations.

Interdisciplinarity and Water Diplomacy

It is highly unlikely that a Water Diplomat (whatever the context) will know as much as all the other stakeholders involved about every aspect of a decision that has to be made. So being able to draw on the knowledge of others is important to Water Diplomacy. Indeed, regardless of how much technical information or background a stakeholder might have, they can always use more input. However, to take advantage of additional information, Water Diplomats need what might be called "metaknowledge" – an understanding of how to integrate what others have to offer with what they already know.

Having knowledge about one aspect of a water system is one thing; mastery of the metaknowledge required to structure and participate effectively in actual decision making is another. Being able to formulate or assess a forecast of water supply and demand is valuable, but it doesn't guarantee that the relevant legal, political, and institutional considerations are being taken into account effectively.

The metaknowledge we have shared with the Water Diplomats in the Tufts training program focuses on how to ask questions about the interests of others, how to listen carefully when information is shared, and how to control one's unconscious biases that cause us to see only what we expect to see and hear only what we want to hear. Those who are trained as interdisciplinarians are usually better at these metatasks because they have had to learn about cross-disciplinary communication. Those trained in water negotiation theory and practice ought to be better still because they have acquired the metaknowledge they need.

(Meta-) knowledge and skill in working with others, even former adversaries, requires pragmatism. Ideologues who think they can "win" water allocation debates by stressing the importance of arguments most meaningful to them are likely to fail. Unless and until the interests of others are met, they are likely to oppose any water-sharing agreement advocated by others on ideological grounds. The Water Diplomats we have trained have not only acquired interdisciplinary and metaknowledge; they know better than to let the "best (in a philosophical sense) become the enemy of the good."

Knowing when and how to present actionable knowledge to others is best learned experientially. That is, wherever possible, we have tried to teach Water Diplomats by referencing real cases, introducing them to experienced stakeholders, arranging for them to intern in water agencies or with water advocacy groups, and creating hypothetical (but realistic) opportunities for them to participate in role-play simulations. In our view, experiential, not just interdisciplinary learning, is crucial to the success of Water Diplomacy.

Coda: interdisciplinarity and Water Diplomacy

My colleagues at Tufts used National Science Foundation (NSF) funds to design, test and revise a Water Diplomacy curriculum. It is now available to others, particularly the role play simulation called Indopotamia and the case studies we have included in Aquapedia (http://aquapedia.waterdiplomacy. org). Most importantly, they learned it is possible to augment the capabilities of engineers and scientists by introducing them to interdisciplinary and metaknowledge about negotiation and the best ways of working in situations where power is distributed asymmetrically.

They also learned that students with law and applied social science backgrounds can master sufficient engineering and science understanding to be effective participants in water-management debates. These applied social science students are quicker to grasp the metaknowledge we offered to both scientists and social scientists. Both sets of students have learned the overarching importance of the unique context in which each water-management or water-sharing conflict occurs. I think the graduates of the Tufts IGERT training program and others who have access to the training materials I have described will be in a good position to avoid or resolve the water conflicts likely to emerge in the years ahead.

Reflections on Water Diplomacy – from theory to practice

Enamul Choudhury

For these reflections I was asked to consider the following premise:

> The Water Diplomacy Framework (WDF) aims to help resolve water related conflicts. It acknowledges that traditional problem-solving frames are

adequate to address simple and complicated problems where reasonable scientific certainty and/or consensus about intervention in natural processes exist. The WDF hypothesizes that when water policy challenges stem from complex – interconnected, uncertain, unpredictable, and boundary crossing – system dynamics with feedback, traditional frames for problem solving can be limiting or counterproductive. A recurrent factor for such limitations is that traditional problem-solving frames often separate the observation-based technical (what is) from the value-based sociopolitical (what ought to be) dimensions of the problem. The WDF further hypothesizes that when dealing with complex problems, these dimensions cannot be de-coupled. The WDF acknowledges both the limits to knowledge – objectivity of observations and subjectivity of interpretation – and the contingent nature of collective action. The WDF proposes a principled pragmatic approach to negotiation that is based on *credible scientific knowledge* mediated through equity and sustainability as *guiding principles* for policy action. The WDF approach emphasizes that, when addressing complex water problems, all parties have legitimacy to decide on the evidence used and its interpretation, the past evidence and future implications of an intervention, metrics of equity and sustainability, and the package of actionable solutions. These parties include users and producers of water knowledge, managers, technical experts, policy makers, decision makers, and politicians. Furthermore, the WDF asserts that parties need to seek consensus and mutual benefits when negotiating a resolution.

I was asked to provide an assessment of the premise including the conventional wisdom supporting or refuting it. I was also asked to look to the future of Water Diplomacy and reflect on what I see as the path forward for negotiating complex water problems. First, I will provide my assessment of the premise. The premise as above asks big questions and is quite broad in scope and scale. For my purpose here, I have decided to focus on three interconnected arguments:

1   WDF addresses water conflicts that are *complex*, for which traditional frames of problem solving are *limiting* and can be *counterproductive*.
2   WDF acknowledges the *limits to knowledge* in terms of the objectivity of observations, the subjectivity of interpretations, and the *contingent nature of action*.
3   Thus, under the *conditions* of complexity, limits of knowledge, and contingency of action, WDF *proposes a principled pragmatic approach to negotiation*. Here pragmatic policy action rests on negotiation that draws upon *credible scientific and technical understanding* of possibilities, being *guided* by the values of *equity and sustainability*.

There are two points of critique of the premise. The first is regarding the meaning of credible scientific understanding, and the second is on the meaning of guiding principles. The first critique is that the meaning of credibility beyond prediction or consensus has not been made clear. This relates to the second critique, where for policy action, it is unclear how the guiding principles of equity and sustainability relate to scientific credibility. My observation is that clarity will not come by considering scientific credibility as a factual issue and guiding principles as a value issue and then seeking to link the two. Rather, the policy action becomes enabling when facts and values are considered jointly. For instance, what is required is scientific credibility on sustainable practices that are arrived at equitably. This is the premise of principled pragmatism in negotiation.

A principled pragmatic approach to conflict resolution creates a space for all affected parties to have their voice in (a) collecting and interpreting the evidence; (b) considering past evidence and future implications of an intervention; (c) determining and agreeing on the meaning and measure of equity and sustainability that creates mutual benefits; and (d) developing the package of actionable solutions through negotiation. Since mediated negotiation allows for these actions to become not only operationalized but contextually specified, mediated negotiation in turn becomes the cornerstone of principled pragmatism.

Complexity here stands for the knowledge of reality resulting from interconnectedness of variables, processes, actors, and institutions. Complexity as a science focuses on the properties of interconnectedness, uncertainty, unpredictability, and emergence that characterizes the observable features of a system. Simple problems are understood as those that are fully specifiable under the knowledge of the parts. Here, the nature of the problem can be effectively represented, and solutions can be developed with sufficient technical capacity and organizational arrangement. Simple problems are those in which we can separate fact-based technical issues from value based sociopolitical issues.

The premise is an exploration to understand, characterize, and manage complex problems. This is more a change in thought style than a paradigm shift. It is rather a gestalt switch that foregrounds complexity and a pragmatic approach to problem solving. Thinking differently here involves crafting an interdisciplinary inquiry of the interaction between natural and societal processes in relation to water issues. This is not a novel approach, but it is an emerging feature of contemporary environmental thinking in general. Thus, the premise is *supported* by conventional wisdom.

However, the premise in its implications also goes beyond the conventional wisdom of addressing complex problems. Conventional approaches take a design approach to address complex problems using well-established research methods and tools. It is usually based on expert input and remains project based. In contrast, complexity thinking – as adopted in the Water Diplomacy Framework (WDF) – allows a process-based engagement of experts and stakeholders, where both learn from each other and act jointly to address the problem in a given context.

In doing so, the WDF faces the theoretical and methodological challenges of how to build bridges across domains: be it natural–societal, fact–value, science–society, politics–administration, planning–implementation, etc. In other words, there are significant challenges regarding how to operationalize WDF in practice. For example, how are we to bridge understandings of "complexity science" that may work well in physics into the social and policy domains?

The absence of bridges continues to challenge theory-practice synthesis for a given problem. The solution so far has been mainly in terms of designating systems as either hard or soft, and designating power as either hard or soft (both being a continuum rather than polar choices). This approach shows co-existence but not bridging or linkages based in interaction and learning based on feedback. Currently, metaphors and analogies are the conceptual devices used to make intuitive or even conceptual sense of the linkages. However, for operational use, we need to translate analogies and metaphors into contextual measures and meaning (such measures need not be wholly quantifiable). It is the absence of such measures that forms the ground for critique and the *refutation* of the premise.

One way to address this theoretical challenge is to focus inquiry on interaction as a **unit of both representation** and **analysis**. The meaning of interaction as the unit of representation and analysis will vary across scales and contexts. In this interactional frame of thinking, actions of natural variables and processes (for example, water flows downhill due to gravity) and intentions of societal actors and institution (for example, water may flow uphill to money) do not hold any autonomous or independent standing. Rather, both find their meaning and efficacy in and through the interaction process. This mode of interaction of values, interests, and tools is the centerpiece of complexity thinking. This type of thinking and reframing is quite different than the conventional frame of thinking.

In the conventional frame, interaction is understood as a byproduct, an outcome of the activities of otherwise autonomous causal processes and agents. The agents choose or create instruments and structures that constitute the interaction, which the agents then use to address their independent needs (think of market as a structure, incentive as an instrument, and agents' needs as utility satisfaction). In such a conventional framing, a system cannot recognize the emergence of goals as a property of interaction. That is, goals, incentives, and accountability have to be specified prior to interaction; consequently, such a framing can't address emergent patterns.

A possible path forward for resolving complex water problems

This reframing of the interaction effect for complex problems suggests it is important to explore possible solutions through negotiation. In the conventional frame, negotiation is a tool/instrument of conflict management/resolution in which disputing parties and interests constitute the starting point of negotiation, and the attempt is to bring the parties to interact constructively for de-escalation, conciliation, or peace. With the lens of complexity thinking, negotiation becomes a process of interaction and enactment, where the rules of interaction are the starting points of inquiry and interaction. Even the agreement reached is interactional, requiring the disputants to act jointly to define values, interests, and facts.

In order for such rules of interaction to account for and address the problems that result from the interaction and to make the interaction enduring requires the use of symbolic values as guiding principles. Thus, finding solutions to complex problems requires addressing not only the asymmetry of information and resources among the disputants but also the hard constraints on resources, information, and politics under which the disputing parties operate. Therefore, the initiation of constructive interactions among distrusting and deficient parties calls for enabling conditions that creates convergent interests from the initial position of strong differences.

Here symbols function not as a substitute of real policy, which is one reason why symbolic politics is viewed as deceptive, but as an enabling condition from which the policy emerges as contingent agreements through interaction and negotiation. As enablers, symbols function as aspirations or ideals, which guide action, but do not determine or define it. If symbols are allowed or relied on to define meaning, the policy becomes vague, subject to opportunistic manipulation and operationally ineffective. Equity and sustainability are two such guiding symbols of collective action that is mediated through negotiation.

Symbols in the sense used here are not used as universal or generalizable. Rather, each context and the participants involved therein define what can be considered as symbolic to gain the initial agreement to interact, and then the rules of interaction set the process through which operational definition of symbols emerges. In this sense, all actions emerge as negotiated outcomes. This view of the process clearly contrasts with the conventional approach of considering policy as a blueprint and action as the implementation of the plan. In this conventional approach, negotiation is conceived of as coming to agreement on the blueprint and a process for implementing it in a cost-effective way.

Symbols are often open-ended as to their meaning. However, the meanings of symbols need to be interactively constructed. Hence, not any value or meaning of a symbol can serve as a guiding principle. Only those meanings of a symbol that emerges from or are at issue or the product of interaction can be considered as guiding. This also will vary by context or domain of inquiry when addressing the interactions that I have previously noted as fact–value, nature–society dichotomy. In my judgment, the symbolic principle of equity and sustainability may satisfy the condition of being the rules of interaction. The WDF conceives of such symbolic values as guiding principles for negotiated agreements. Therein lies the path to move forward.

The Water Diplomacy IGERT at Tufts pursued a laudable initiative to cultivate the emergent approaches resulting from the interaction of natural and human systems in transboundary water management. Beyond the initiative, its effects have already borne some quick results – publications from the students, employment of graduates, and expert visitors to the program. However, more enduring results will occur when its graduates apply and continue to cultivate their learning in new ways and in varied contexts.

Reflections on Water Diplomacy from a veteran of corporate water stewardship

Greg Koch

The soft stuff is the hard stuff

This is the single most important lesson I learned from 30 years in working on water challenges. What you typically think of as the "hard stuff," the technologies, data, models, infrastructure, watersheds, pumps, piping, in other words, physical things, is not typically the crux of solving a water challenge.

No doubt, complicated, even difficult, calculations, research, and advanced manufacturing, as well as many smart people are needed in water. There have been many great advances in recent years, with more needed in the future.

However, what I have learned is that the "soft stuff" is the most difficult: getting alignment across multiple water users, the right policies, training, communication, developing and maintaining supply chains, and securing and dedicating a revenue stream to operate water systems over the long term. Listening, empathy, cultures, religions, debate, give and take, choices, rights, and people; all come to mind when considering the "soft stuff."

I've most often stated that "the soft stuff is the hard stuff" in community water access projects. For example, many a well-intentioned NGO or inventor would approach me with a preselected water treatment technology promising the wonders of how comprehensive ("cleans even raw sewage to pure water") and effective the technology was and even how inexpensive it was to install and maintain. I would grant them the benefit of the doubt in all these matters and then ask them questions along these lines:

1   Will the community understand and accept this advanced technology (maybe the most advanced technology they would have)?
2   How will the community raise, dedicate, and secure enough money to maintain the technology?
3   Where will they get the energy to run it, spare parts, etc.?
4   How will they communicate with that supply chain? How will they transfer funds to that often-remote parts supplier?
5   If they do save the money needed for future repairs, how are community dynamics and hierarchies changed when the group managing the funds now has the largest cash reserves in the community?

Address these challenges, at least in theory, and always in consultation with the community, and only select a technology at the end that best fits the situation.

The actors

Most water issues have a historic and well-defined set of actors: civil engineers, hydrologists, regulators, academics, civil servants, policy makers, NGOs, and politicians. All of these and more roles are required to solve water challenges. And yes, diplomats are also needed, but this role has historically been in the context of transboundary waters. The Nile, Mekong, and Danube are notable examples, and most of these agreements or treaties seek to resolve or avoid conflict over shared water use.

While these fall under the umbrella of "Water Diplomacy," we should be careful not to restrict Water Diplomacy to negotiations between foreign governments, since contentious water use agreements are debated intrastate, between communities, between industries, and even between individual water users.

The definition of diplomacy hints at this duality:

di·plo·ma·cy
noun

1   the profession, activity, or skill of managing international relations, typically by a country's representatives abroad.

synonyms: statesmanship, statecraft

2   the art of dealing with people in a sensitive and effective way.

synonyms: tact, tactfulness, sensitivity, discretion, subtlety, finesse, delicacy

The word originates from the French: *diplomatie*, from *diplomatique* (diplomatic). I like the Spanish translation better, especially "connection": "*diplomacia, comunicatividad, conexión.*" Both the international and local levels of Water Diplomacy can and should learn from each other to apply successful techniques of clear communication, transparency, data sharing and alignment, productive debate, and how to make fair but tough choices; all critical for effective diplomacy.

People

Unique to water and Water Diplomacy is the prevalence of emotions. I do not mean emotions in a pejorative sense. We all have emotions, and strong emotions will significantly affect the approach and outcome of our interactions with others.

Indeed, people do have emotions related to other environmental and social challenges, such as climate change, plastic waste, air pollution, and the like. However, unlike water, we do not have a long history with these topics, nor for the older generations were these topics studied in school. Though this is unfortunately changing, we do not yet have a widespread and engrained sense of spirituality or religion with such issues.

We do have that history and spiritual/religious connection to water. We also have a daily, visceral experience with water but not, as directly, with other issues. In addition, water can be a pro and a con when you think of having too much (a flood), having too little (a drought), it being too inexpensive (not enough revenue raised to maintain infrastructure), or it being too expensive (the poor pay the most), and the like. We recreate in water, move on water, clean ourselves in it, and hydrate our bodies with it.

These dualities and realties of water bear and strengthen the emotions we bring to a discussion or conflict on water and, therefore, should be anticipated, understood, and accepted among all who seek true solutions. My experience in Water Diplomacy, at the national and community levels, has taught me that I can have a chance at success only when I do my best to remember and apply this knowledge.

Future scholarship

Interdisciplinary studies and research on Water Diplomacy have a strong start in academia. Though I have been asked for recommendations and how to enhance such vital programs, I instead suggest that we must take what we know now into real-world water challenges. Water is in crisis or near crises in almost every country.

Climate change, population growth, urbanization, and economic development are quickly increasing water stress. If we only focus on financing technologies and infrastructure without engaging people and their emotions on water, we may well solve short-term crises while creating a much greater challenge for the future.

References

Fisher, R., and Ury, W., 1981. *Getting to Yes: How to Negotiate Without Giving In*. London: Arrow.

Islam, S., and Susskind, L.E., 2013. *Water Diplomacy: A Negotiated Approach to Managing Complex Water Networks*. New York: RFF Press.

Susskind, L., 2014. *Good for You, Great for Me: Finding the Trading Zone and Winning at Win-Win Negotiation*. New York: PublicAffairs.

16 Quo Vadis?

Shafiqul Islam and Kevin M. Smith

Water – it's what we think and do

The Academy – founded by Plato and transformed into the modern University – is an ideal institution to cultivate, create, and disseminate universal knowledge. However, many contemporary water problems arise from an intrinsic complexity that does not yield generalizable solutions or universal knowledge. Water Diplomacy is an idea – conceived by a group of reflective water scholars and professionals in Boston in 2006 – with an ambitious goal to explore new ways to "think and do" water scholarship on complex water issues and improve societal outcomes. The Tufts Water Diplomacy Program is an experiment to implement this idea through interdisciplinary pursuits, from theory to practice.

This concluding essay is a synthesis of our conversations, connections, and combinations of ideals, ideas, and implementation of interdisciplinary Water Diplomacy pursuits. In this book, we have briefly traced how ideals and ideas have changed over time to cultivate and create interdisciplinary scholarship for societal impact. Our Water Diplomacy experiment is an example test case to show and tell what we have done, what mistakes we have made, and what we have learned about interdisciplinary scholarship to translate ideas into actionable outcomes. We view our journey as an experiment in adaptive learning to address the complex problems of our time, using water as a motivating example.

It's not "either–or"; it's "and" as well as "mind the gaps"

From more than a decade of conversations and interactions, one key finding that emerged is that we need to focus on making our ideas clear for actionable outcomes and avoid getting mired in abstract debates over the primacy of one representational form over another. In such cases, we are faced with a duality of apparently competing representations, such as: numbers or narratives; natural or societal; particles or persons; explicit or tacit; deterministic or random; quantitative or qualitative; objective or subjective; facts or values; large-N or small-N; and fox or hedgehog. In reality, there is both the *space* for and a *need* for a spectrum of representations when working on complex water challenges.

So we need to replace *OR* with *AND*. This is where interdisciplinary conversations, connections, and combinations can play significant roles in achieving desirable outcomes. In a simple world – where cause–effect relationships are well understood – the *either–or* representation may work. For complex problems characterized by interconnections, uncertainty, and unpredictability, *either–or* frames for problem solving can be limiting or even counterproductive. We need to "mind the gap" between these two representations, because for many real-world problems, this gap is very large and requires a different way of thinking, framing, and addressing the problem. In Chapter 2 we explored how the identification of a system as simple, complicated, or complex reveals important information about whether the problem can be addressed by experts through an *either–or* representation (simple or complicated) or if it requires a synthesis of representations (complex) by an inclusive group of stakeholders engaged in fact–value deliberation, joint fact-finding, collective decision making, and adaptive management.

Stickiness of ideas for actionable outcomes

Our approach is problem-focused. We are concerned with developing a framework (i.e., the Water Diplomacy Framework) for effective practice rather than a general theory of the problem. We are not intrigued by the "beauty of the problem"; instead, we embrace its messiness by recognizing the disconnect among *values*, *interests*, and *tools* as well as *problems*, *policies*, and *politics*. We appreciate the power of scientific and technological solutions for efficiency and reliability; we are also cognizant of the power of political feasibility for a solution to be actionable. We don't dwell on an either–or paradigm; instead, we try to focus on a problem within a given system and identify its contextual capacity and constraints by asking: *who decides water for whom, at what cost, and what scale?* Asking these questions allows our ideas to "stick" and leads to options that are actionable and have measurable outcomes. In Chapter 3, we made the case for problem-driven approaches to interdisciplinary collaboration on complex water issues and the value of shared frameworks in facilitating these collaborations.

Embrace principled pragmatism in the search for actionable solutions

Many of our water problems are complex because they are interconnected and interdependent. These boundary-crossing water problems are dynamic and nonlinear and are often interconnected with other problems including food, energy, health, environment, and ecosystems. Decisions related to these problems – involving variables, processes, actors, and institutions – are likewise complex, making a range of interventions possible. But not all possible policy choices and decisions are actionable. When neither the certainty of a scientific solution nor the consensus of what intervention to implement exists, what we need is Water Diplomacy. Our approach to principled pragmatism – rooted in equity and sustainability as guiding principles for water governance and management – attempts to synthesize symbolic aspirations with a realistic assessment of the constraints and capacities of the problem context. Water Diplomats are water professionals who work with and help guide broad groups of stakeholders from the world of seemingly infinite possibilities to a subset of implementable options with trackable outcomes. In Chapter 4, we argued that in order to achieve this vision, Water Diplomats will need to employ a principled and pragmatic synthesis of four domains of knowledge: formal knowledge (episteme), practical wisdom (phronesis), practiced technique (techne), and thoughtful practice (praxis).

Quo Vadis? (where y'all goin'?)

Today, our world is globalized. Science, policy, and politics are interdependent. Change is inevitable. The pace and nature of change have accelerated at a rate that we have never experienced before. A few years ago, YouTube and Facebook did not exist; today, these technological innovations are shaping policies, changing the nature of politics, and redefining the dynamics of social networking. The world is continuously and dramatically being reshaped by the pace and intensity of scientific and technological innovations. *Has the evolution of our ideal Academy, with its emphasis on the production of generalizable and universal knowledge, kept pace with the changing nature of our recognition of the need for context-dependent and contingent approaches to address the complex problems of our time?*

In 2006, at the beginning of our journey, we argued that the nature of water as a resource is changing. A changing world requires a changing education in its ideals, ideas, and implementation. Science alone is not sufficient. Nor is policy making that doesn't take science and context into account. Sustainable solutions can emerge from interdisciplinary pursuits that consider science, policy, and politics within evolving water networks consisting of variables, processes, actors, and institutions.

Now in 2019, we recognize that many of our current and emerging water problems are complex because they are interconnected and interdependent. Our Water Diplomacy Program was designed to educate and train Water Diplomats how to frame, formulate, design, and implement research projects on complex water issues from beginning to end, with the goal of arriving at actionable outcomes that are grounded in the principles of equity and sustainability. What have we learned about interdisciplinary scholarship and practice? Where are we going from here?

Interdisciplinary research and practice are not easy. They must be supported by a wider range of stakeholders – students, faculty, university administrators, and employers – and funding mechanisms to address real-world challenges. Despite wide-scale acceptance of interdisciplinarity as an effective mode of addressing complex problems, we need to recognize the challenges and opportunities provided by the context. In Box 16.1, we summarize seven propositions for understanding and exploring resolutions of complex problems using water as a motivating example.

Box 16.1 Seven propositions

1 The complexity of water problems arises from the coupling of natural and human systems. For these problems, observation-based technical facts (what is) cannot be easily separated from the value-based socio-political (what ought to be) plurality of the problem description.

2 The solution space for these complex problems – with interdependent variables, processes, actors, and institutions – can't be prestated. The use of dualistic representations (numbers or narratives; facts or values; objective or subjective) for these problems is inadequate.

3 Complexity is neither generalizable nor specifiable.

4 Differentiate complexity from deterministic certainty and statistical uncertainty; instead, identify the conditions (rather than the cause) for emergent patterns.

5 Explore and adapt shared vocabulary and methods from appropriate disciplines and domains that are rooted in and bounded by scientific rationality and interpretive plurality.

6 Use the rigor of scientific methods as the principle to derive facts with an adherence to a negotiated application of sustainability and equity as guiding values to design and implement pragmatic interventions.

7 Focus on identifying and implementing societally relevant technological solutions given the context, constraints, and capacity of a given system.

In many ways, our experiment is just beginning. However, working on this book has offered us an opportunity to look back and reflect on a decade of "thinking" and "doing." At this stage in our experiment, 27 of our students have completed (or nearly completed) their PhDs, and we look forward to watching their careers unfold with great anticipation. Meanwhile, many of our Water Diplomacy fellows have written about their research in this book (see Chapters 6–12). While the problems addressed in these chapters included a wide range of disciplines and topics, they all share some aspects of the Water Diplomacy Framework. Our weekly Water Diplomacy colloquium has been an effective way to operationalize this pairing of intellectual diversity and research topics with a shared framework. We find the "T" metaphor (discussed in detail in Chapter 13) of interdisciplinary scholarship to be a useful tool for characterizing and tracking the evolution of interdisciplinarity for a Water Diplomat. The "T" metaphor of interdisciplinary scholarship attempts to reconcile disciplinary requirements (depth) with interdisciplinary expectations (width). Several lessons and learnings summarized in Chapters 1 and 13 – from our decade long journey within the context of Tufts Water Diplomacy Program – are transferrable to other interdisciplinary settings. We hope that the story and outcomes of our experiment will inspire other pursuits to create space for collaborative interdisciplinary scholarship on emerging complex problems at the interface of natural and human systems.

Where do we want to go, and what are the hurdles?

Our choice to use a complexity lens to understand, characterize, and manage problems arising from coupled natural and human systems is a significant departure from conventional "thinking." Thinking differently here involves crafting an interdisciplinary inquiry into the interactions between natural and societal processes that goes beyond conventional causality-based reasoning. Conventional methods take a *specify-design-implement* approach to address complex problems using currently established paradigms for research methods and tools. Such approaches are usually based on expert input and remain project-based; an approach that can work well for simple and complicated problems. In contrast, complexity thinking entails the active engagement of experts and stakeholders in collaborative processes, of which fact–value deliberation is an integral part. This is easier said than done, because to do so, we face the theoretical and methodological challenges of how to build bridges across dualities of representations: be it fact–value, natural–societal, objective–subjective, numbers–narratives, particle–people, etc. In other words, the challenge of how to operationalize the Water Diplomacy Framework in practice. For example, how do we translate an understanding of "complexity science" that may work well in physics (for particles) into the social and policy domains (for people)? Bridging this duality requires us to abandon an "either–or" mode of thinking and adopt an "and" mode of thinking. Our current calculus and toolbox are not yet mature enough for this adoption. More importantly, how do we translate our new way of "thinking" into "doing"? We have suggested principled pragmatism as a way to translate our ideas into action. Currently, metaphors and analogies are used to make bridges between facts and values, numbers and narratives, quantitative and qualitative, etc. However, for operational use, we need to translate analogies and metaphors into measures and metrics. It is the absence of such measures and metrics that will continue to be the hurdle that we need to address. Meanwhile, instead of looking for optimal solutions, we need to search for optimal spaces where certain conditions exist for implementing actionable solutions given the constraints the context imposes. We invite you to join us in this search.

Index